CLINICAL ENGINEERING

CLINICAL ENGINEERING

A Handbook for Clinical and Biomedical Engineers

Azzam Taktak

Paul Ganney

Dave Long

Paul White

AMSTERDAM • BOSTON • HEIDELBERG • LONDON
NEW YORK • OXFORD • PARIS • SAN DIEGO
SAN FRANCISCO • SINGAPORE • SYDNEY • TOKYO

Academic Press is an imprint of Elsevier

Academic Press is an imprint of Elsevier
The Boulevard, Langford Lane, Kidlington, Oxford, OX5 1GB, UK
225 Wyman Street, Waltham, MA 02451, USA
525 B Street, Suite 1800, San Diego, CA 92101-4495, USA

First edition 2014

Notice
No responsibility is assumed by the publisher for any injury and/or damage to persons or
property as a matter of products liability, negligence or otherwise, or from any use or
operation of any methods, products, instructions or ideas contained in the material herein.
Because of rapid advances in the medical sciences, in particular, independent
verification of diagnoses and drug dosages should be made

British Library Cataloguing-in-Publication Data
A catalogue record for this book is available from the British Library

Library of Congress Cataloging-in-Publication Data
A catalog record for this book is available from the Library of Congress

ISBN: 978-0-12-396961-3

For information on all Academic Press publications
visit our website at elsevierdirect.com

Printed and bound in the United Kingdom

14 15 16 17 18   10 9 8 7 6 5 4 3 2 1

Working together
to grow libraries in
developing countries

www.elsevier.com • www.bookaid.org

Dedication

In loving memory of my father Fouad. You will always be in our hearts.
Azzam

Contents

IV
REHABILITATION ENGINEERING AND ASSISTIVE TECHNOLOGY

Acknowledgements

Azzam Taktak would like to thank his wife Diane and his children Chris and Sarah for their continued love and support. He would also like to thank the following people who have had a major influence in shaping his career: Peter Rolfe, Paul Record, Iain Chambers, Alicia El-Haj, Justin McCarthy, Malcolm Brown, Tony Fisher, Steve Lake, Antonio Eleuteri, Paulo Lisboa and Bertil Damato.

Paul Ganney would like to thank his wife, Rachel, for her continued encouragement and his colleagues for many helpful discussions: especially Paul Ostro, Patrick Maw, Khalil Itani, Justin McCarthy and Bill Webster.

David Long would like to thank his wife, Fran, for her patience and encouragement, and to acknowledge the following people with whom he has had the privilege to work and who, in different ways, have influenced his thinking: Dave Calder, Paul Dryer, Barend ter Haar, Margaret Hannan, Rick Houghton, Henry Lumley, Linda Marks, David Mitchell, Wendy Murphy, Roy Nelham, Pauline Pope, David Porter, Pat Postill, Paul Richardson, Nigel Shapcott, Phil Swann, Linda Walker and Jon Ward.

Paul White would like to thank Tracey for her love, patience, encouragement and support for all the time associated with advancing Clinical Engineering and to our son Harry who makes it all worthwhile. He would also like to thank those that have inspired and mentored him throughout his career and the department of Medical Physics and Clinical Engineering at Cambridge University Hospitals Foundation Trust and to the Postgraduate Institute at Anglia Ruskin University who allow him to push forward boundaries in clinical research on an international stage.

Preface

This book is aimed at professionals, students, researchers, or anyone who is interested in clinical engineering. It provides a broad reference to the core elements of the subject for the reader to gain knowledge on how to successfully deploy medical technologies. The book is written and reviewed by professionals who have been working in the field of clinical engineering for decades. Many of the authors are clinical and biomedical engineers working in healthcare and academia and have also acted as trainers and as examiners on the subject.

As well as possessing engineering skills, clinical engineers must be able to work with patients and a range of professional staff. They need to keep up to date with fast-moving scientific and medical research in the field, and to develop their own laboratory, design, analytical, management, and leadership skills. This book is designed to assist the clinical engineer in this process.

The book is organized into four main sections. The first section covers generic aspects of the core skills needed to work in this area. It gives the reader a flavor of how to engage with research and development, data analysis and study design, and management and leadership. It also discusses in detail the important role engineers play in the healthcare environment.

The second section covers legislation relevant to information technology based medical devices and standards concerned with security, encryption, and data exchange. There is also material on software development/management and web development, which will be of interest to those working with these technologies across the entire field of clinical science and medical engineering.

The third section deals with clinical measurements and instrumentation. It starts with a quick overview of medical electronics theory before moving on to clinical measurements. It explains in detail the physics and engineering aspects involved in making useful and reliable measurements in the clinical situation. Examples of clinical measurements covered include cardiology, hematology, neurophysiology, and respiratory.

The forth section provides a comprehensive summary of the subject of rehabilitation engineering and assistive technologies. Topics covered include gait analysis, posture management, wheelchair and seating, and assistive technology. It is the first comprehensive and practical guide for engineers working in a clinical environment.

I would like to express my sincere gratitude to my coeditors who spent a considerable amount of time and energy recruiting authors and pulling together the material for their own sections, while working in such a demanding environment. I would also like to thank the authors and the reviewers for the fantastic effort they have put in.

I hope you enjoy reading this book and find it illuminating.

Foreword

Clinical Engineering is a broad arena and practitioners in this area need to understand a wide range of subjects, some in great detail, and others with just a working knowledge. In my experience although there are many separate books covering the complete subject area, there is no complete book that professes to cover the entire range of subjects, which can be a useful reference for the professional working in this field. Clinical engineers must have a working knowledge of the human body, both in how it functions and its anatomy. They must be able to work with patients, clinical staff and other health professionals. They need to be experts in their engineering areas, but keep up to date in the relevant research and innovations in this field. Finally they must be able to lead and manage, both themselves and their teams. This book seems unique in that the wide range of subjects mentioned is included, some in great detail, others necessary less so, but most chapters are referenced widely, with useful extra reading material presented for further study. There are some innovative parts of the book. For example, a section on leadership is not often included in text books such as this, but this particular chapter is very well presented, in a very personal style, with thought provoking exercises and sections. The excellent chapters making up the section on rehabilitation engineering are unusual to be included in a book such as this, but they make the book seem very complete. The web and computer sections give the book a very up-to-date feel.

Professor Azzam Taktak has edited the book and chosen with care some excellent co-editors and authors to contribute. His concept of the book came out of his vast experience in teaching the subject at his hospital and university, both in the classroom and using electronic learning. He has contributed to the new Modernising Scientific Careers (MSC) NHS programme and this experience has enhanced the book. It is interesting that the MSC course also includes leadership and professional issues as a key component, and it reassuring that this is included in this complete course on clinical engineering.

The contents of the book follow a logical sequence, that take the reader from a brief look at the anatomy and physiology of humans, to statistics, good clinical practice, the role of clinical engineers in hospitals, and information and computer systems. These subjects make up the first two sections of the book, which are about presenting the background 'core' areas and the legal processes involved. The final two sections of the book cover all the main areas of clinical measurement and rehabilitation.

I have had the pleasure of knowing Azzam for many years. It is hard to think of anyone with more knowledge and experience of clinical engineering in its widest form, and he has an extensive network of colleagues he can draw upon to contribute to this work. I have also had the pleasure of knowing most of the excellent authors in the book. Some chapters have been written by single authors, others by multiple ones. The variety in authorship gives a refreshing combination of styles, which keeps the writing alive and accessible.

The book will be a valuable resource for many engineers and clinicians working in this area, and also to refresh the many experts involved in the field of clinical engineering

Professor Mark Tooley
PhD FIET FIPEM FinstP FRCP
Consultant Clinical Scientist,
Royal United Hospital, Bath

List of Contributors

Tim Adlam Bath Institute of Medical Engineering

John Amoore Department of Medical Physics, NHS Ayrshire and Arran, Scotland, U.K.

Richard G. Axell Clinical Scientist, Medical Physics and Clinical Engineering, Cambridge University Hospitals NHS Foundation Trust, Cambridge, U.K. and Honorary Visiting Research Fellow, Postgraduate Medical Institute, Anglia Ruskin University, Chelmsford, U.K.

Dan Bader University of Southampton

Paul Blackett Lancashire Teaching Hospitals NHS Foundation Trust, Lancashire, U.K.

Tom Collins Queen Mary's Hospital

Donna Cowan Chailey Heritage Clinical Services

David Ewins Queen Mary's Hospital and University of Surrey

Paul S. Ganney University College London Hospitals NHS Trust, London, U.K.

Vicky Gardiner Opcare

Fran J. Hegarty Medical Physics & Bioengineering Department, St. James's Hospital, Dublin, Ireland

Mike Hillman University of Bath

Tim Holsgrove University of Bath

Paul Horwood Oxford University Hospitals NHS Trust

Robert Lievesley Kent Communication and Assistive Technology Service (Kent CAT)

David Long Oxford University Hospitals NHS Trust

Tori Mayhew Oxford University Hospitals NHS Trust

Justin P. McCarthy Clin Eng Consulting Ltd, Cardiff, U.K.

Ed McDonagh Diagnostic Radiology/PACS Royal Marsden Hospital NHS Trust, London, U.K.

Chris Morris University of Exeter

Ladan Najafi East Kent Adult Communication and Assistive Technology (ACAT) Service

Fiona Panthi East Kent Adult Communication and Assistive Technology (ACAT) Service

Sandhya Pisharody Varian Medical Systems, U.K.

Nicholas P. Rhodes Department of Musculoskeletal Biology, Institute of Ageing and Chronic Diseases, University of Liverpool, Liverpool, U.K.

Jodie Rogers East Kent Adult Communication and Assistive Technology (ACAT) Service

Anthony Scott Brown Royal Cornwall Hospitals NHS Trust, Truro, U.K.

Richard Scott Sherwood Forest Hospitals NHS Foundation Trust, Nottinghamshire, U.K.

Martin Smith Oxford University Hospitals NHS Trust

Ian Swain Salisbury NHS Foundation Trust

Azzam Taktak Royal Liverpool University Hospital, Liverpool, U.K.

Elizabeth M. Tunnicliffe University of Oxford Centre for Clinical Magnetic Resonance Research, John Radcliffe Hospital, Oxford, U.K.

Will Wade ACE Centre North

Merlin Walberg Phoenix Consultancy USA, Inc.

Paul A. White Cambridge University Hospitals NHS Foundation Trust, Cambridge, U.K. and Anglia Ruskin University, Chelmsford, U.K.

Duncan Wood Salisbury NHS Foundation Trust

GENERAL

Azzam Taktak, Anthony Scott Brown, Merlin Walberg, Justin P. McCarthy, Richard Scott, Paul Blackett, John Amoore, and Fran J. Hegarty

Overview

Over the past century, healthcare has become increasingly reliant on medical technology. Engineers play a pivotal role in the deployment and use of technology. To do this successfully they require solid knowledge of underpinning sciences and skills such as mathematics, physics, design, fabrication, and so on. In addition, clinical engineers require knowledge of some generic aspects related specifically to healthcare. This section gives an overview of such aspects with chapters on anatomy and physiology, research methodology, Good Clinical Practice, risk management, and healthcare technology management. More recently, there has been much emphasis on developing leadership skills of engineers working in the healthcare environment and this section includes a chapter on leadership, quoting many examples on how it can be a powerful tool in the workplace. The final chapter in this section brings all these topics together to highlight the important role clinical engineers play in applying their skills and knowledge in healthcare provision through appropriate deployment of the technology whilst containing cost and increasing access.

1

Anatomy and Physiology

Nicholas P. Rhodes

Department of Musculoskeletal Biology, Institute of Ageing and Chronic Disease,
University of Liverpool, Liverpool, U.K.

INTRODUCTION

This chapter summarizes the most basic and important principles of anatomy and physiology. It is intended to be just the starting point for your understanding of the subject area, rather than representing the full details of the biology of human beings.

CELL PHYSIOLOGY

The human body can be thought in terms of physiological systems, for example:

- Nervous system
- Endocrine system
- Myoskeletal system
- Cardiovascular system
- Lymphatic system
- Respiratory system
- Digestive system
- Urinary system
- Reproductive system

- Hematopoietic system (blood)
- Immune system or reticuloendothelial system (RES)
- Special senses (vision, hearing, etc.)

Each of these systems has unique and special properties that allow them to function in what seems an almost self-contained fashion, having positive and negative feedback loops, external sensing, and multiple action steps. However, each is constructed from many millions of specialized cells. The interesting feature about these cells is that almost all cells have very similar biology, with internal chemistry that could be difficult to differentiate. Study of a "typical" cell allows us to understand the processes occurring in many other cell types, and therefore tissues and physiological systems (Figure 1.1).

The cell can be thought of as an individual factory, having its own computer code and power station. Most cells contain the following:

- Cell membrane: Separates cell internals from external environment, provides

support for sensing receptors, and allows active uptake and output of chemicals (Figure 1.2)
- Nucleus: Houses copy of host master blueprints (DNA), handles copying of DNA to allow protein synthesis, performs cell replication
- Endoplasmic reticulum (ER): Fluid filled membrane system that synthesizes lipid (smooth ER) and proteins (rough ER)

- Golgi complex: Organizes trafficking of proteins and lipids to the external environment
- Mitochondria (plural of mitochondrion): Energy center for cells, derived from bacteria (evolutionarily)
- Lysosomes
- Microfilaments and microtubules
- Vesicles

For cells to undertake their primary function, they require energy. This is principally achieved by conversion of glucose in food to adenosine triphosphate (ATP), which cells use as an energy source, and CO_2.

The primary function of a cell generally requires it to do one or more of the following:

- Sense the environment, using surface receptors
- Synthesize proteins
  - Building blocks, e.g., collagen
  - Action molecules, i.e., enzymes
- Create and use energy
- Output an action
  - Create a force, e.g., muscle
  - Build new tissue
  - Dispose of unwanted cells or molecules

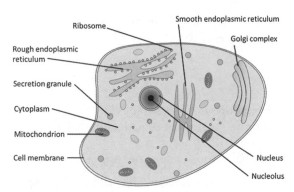

FIGURE 1.1　Generalized cell structure. *Source: Pixabay. com, http://pixabay.com/en/school-cell-help-information-48542.*

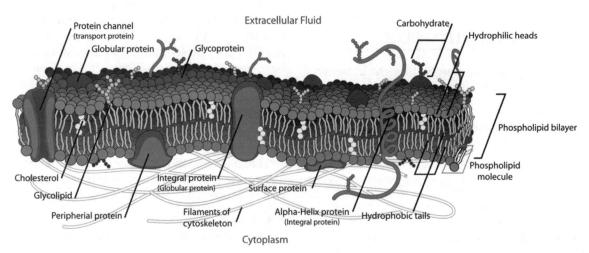

FIGURE 1.2　Structure of cell membrane. *Source: Pixabay.com, http://pixabay.com/en/science-diagram-cell-illustration-41522.*

Glycoproteins sense the environment external to the cell, using "lock and key" receptor-ligand fitting. "Activation" of such a receptor leads to a cascade of intracellular reactions to occur, resulting in upregulation of particular genes, transcribing of specific proteins, and an action (see previous list).

PRINCIPLES OF CELL REPLICATION

Organisms are organized in terms of their biology, from their simplest component parts to the more complex, as follows:

FIGURE 1.3 DNA, the code for life. *Source: Pixabay.com, http://pixabay.com/en/science-cartoon-double-helix-lie-24559.*

- DNA
- Proteins and peptides
- Cells
- Tissues
- Organs
- Whole organism (e.g., animal)

DNA is responsible for maintaining the organism in its current state and replicating, as it contains the code for life (Figure 1.3).

DNA has the following characteristics:

- Contains *all* information to build an organism
- Identical copy in *every* cell
- In humans, there are approximately 2 meters of DNA in each nucleus
- DNA is composed of only 4 types of nucleotide base
- Normally unraveled, but wrapped up into chromosomes during cell division
- DNA codes for proteins only
- Proteins are generally structural (e.g., collagen) or catalytic (enzymes, they do things)

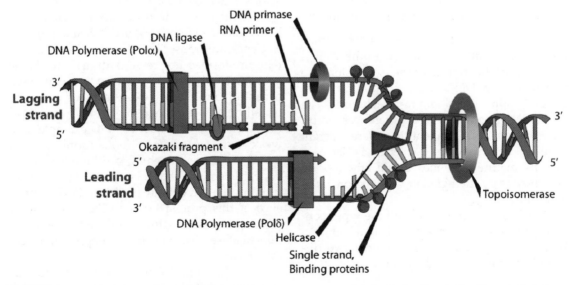

FIGURE 1.4 DNA replication by formation of complementary strands after helix dissociation. *Source: Pixabay.com, http://pixabay.com/en/diagram-illustration-dna-biology-41531.*

TABLE 1.1 Genetic Code—How Combinations of Bases Are Coded in DNA

First codon base letter		Second codon base letter				Third codon base letter
		U	**C**	**A**	**G**	
U		(Phe/F) Phenylalanine	(Ser/S) Serine	(Tyr/Y) Tyrosine	(Cys/C) Cysteine	U
						C
				Stop (Ochre)	Stop	A
		(Leu/L) Leucine		Stop	(Trp/W) Tryptophan	G
C		(Leu/L) Leucine	(Pro/P) Proline	(His/H) Histidine	(Arg/R) Arginine	U
						C
				(Gln/Q) Glutamine		A
						G
A		(Ile/I) Isoleucine	(Thr/T) Threonine	(Asn/N) Asparagine	(Ser/S) Serine	U
						C
				(Lys/K) Lysine	(Arg/R) Arginine	A
		(Met/M) Methionine				G
G		(Val/V) Valine	(Ala/A) Alanine	(Asp/D) Aspartic acid	(Gly/G) Glycine	U
						C
				(Glu/E) Glutamic acid		A
						G

- Each cell has DNA with approximately 3 billion base pairs
- Less than 1% is coding information (genes)
- Humans have approximately 25,000 genes
- Almost all genes in all people are identical

The question most people ask is "How then can people be different from each other?" It is all to do with the timing of the expression of a particular gene. DNA is a genetic library that encodes sophisticated timing machinery. The fourth dimension is where differences occur. The 99% of DNA content is where current scientific knowledge of genetics is lacking. As cells mature, enzymes chemically modify the DNA (e.g., methylation, acetylation, telomere shortening).

DNA has only four different base types connected together in a chain and attached to a complementary chain. There are only two different base pair (bp) combinations:

- Adenine—thymine
- Guanine—cytosine

Proteins are composed of amino acids (in the order of 100 in a typical protein). There are only 20 different amino acid types. Each amino acid is coded by a 3 bp sequence (Table 1.1).

Proteins are made up of amino acids covalently joined together (Figures 1.5 and 1.6).

Genes are transcribed continuously, so proteins are formed (expressed) all the time. Some proteins (and therefore genes) are expressed constitutively. Higher expression of one gene usually leads to higher expression of another (like a cascade or sequence). For example, in a weak bone osteoblasts (bone-forming cells) detect excessive stretching continuously (weak bone is bendy), and this turns on the bone-forming master gene (Cbfa/Runx2). The presence of this protein leads to the activation of other genes over time (collagen I, alkaline phosphatase, osteonectin, osteopontin, osteocalcin, etc.).

For a protein to be transcribed, double-stranded DNA in the nucleus is unraveled then a single-stranded copy of messenger

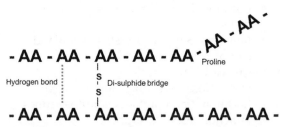

FIGURE 1.5 Amino acid structure. *Source: Nicholas P. Rhodes.*

FIGURE 1.6 Amino acids joined together in a typical dipeptide structure. *Source: Nicholas P. Rhodes.*

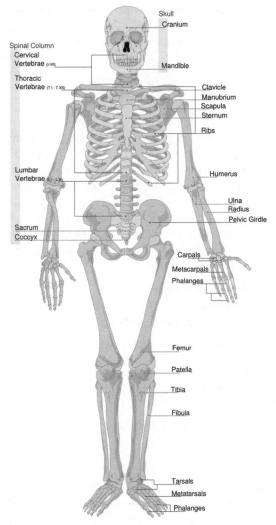

FIGURE 1.7 Major human bones. *Source: Pixabay.com, http://pixabay.com/en/back-model-science-diagram-kids-40500.*

RNA (mRNA) is created. This travels out of the nucleus where ribosomes attach, and attracts the correct transfer RNA (tRNA) molecule for each 3-base mRNA code. Each tRNA molecule has a different amino acid attached to it. In this way, a peptide is built by the ribosome as it travels along the mRNA decoding the base sequence.

Each time a gene is accessed, histones unravel the correct bit of DNA. Specific bits of DNA can be modified (e.g., acetylation, methylation). These can make transcription harder or easier over time (DNA binding around histones). Eventually, this can lead to the cell being killed off (apoptosis).

BONE AND SKELETAL PHYSIOLOGY

The skeletal system (Figure 1.7) can be described simply as comprising four different parts:

- Bones: Rigid support
- Cartilage: Flexible support
- Tendons: Bone-muscle attachment
- Ligaments: Bone-bone attachment

Bones are a structural support of the body, a connective tissue that has the potential to repair and regenerate. Bone is composed of a rigid matrix of calcium salts deposited around protein fibers. The minerals provide rigidity and the proteins provide elasticity and strength.

There are four main bone types:

- Long bones (e.g., femur)
- Short bones (e.g., finger bones)
- Flat bones (e.g., skull)
- Irregular bones (e.g., spine)

Long bones are hollow to save weight and are the engine of blood cell manufacture. They are a reservoir for the body's mineral content and are constantly remodeled. Long bones have a dense and rigid exterior of cortical compact bone surrounding a flexible, protein-rich interior of cancellous or trabecular or spongy bone (Table 1.2).

Bone consists of extracellular matrix and three main bone types:

- Osteoblasts (bone making)
- Osteocytes
- Osteoclasts (bone resorbing)

There are three main types of joint:

- Cartilaginous joints
- Fibrous joints
- Synovial joints

And three type of cartilage:

- Hyaline cartilage
  - Structure: Collagen and abundant proteoglycans
  - Location: Articular surfaces of bones; growing long bones
  - Function: support: Rigid yet flexible
- Fibrocartilage
  - Structure: More collagen (thick bundles) than proteoglycans
  - Location: Pubic symphysis, articular disks
  - Function: Support, resists tension and compression
- Elastic cartilage
  - Structure: Elastic fibers and collagen
  - Location: Ear, epiglottis, auditory tubes
  - Function: Flexible support

Bones can fracture in many different ways: complete, incomplete, comminuted, transverse, impacted, spiral, and oblique. The bone repairs itself by forming a hematoma around the break, the periosteum providing stem cells into the cavity, then callus formation, a substance rich in collagen fibers and cartilage. This callus then becomes ossified and ultimately remodeled into the same structure that existed before the facture occurred.

Under normal circumstances the structural integrity of bone is continually maintained by

TABLE 1.2   Description of the Properties of Different Bone Types in Long Bones

Bone Type	Physical Description	Location	% of Skeletal Mass	Strength	Direction of Strength	Stiffness	Fracture Point
Cortical	Dense protective shell	Around all bones, beneath periosteum; primarily in the shafts of long bones	80%	Withstand greater stress	Bending and torsion, e.g., in the middle of long bones	Higher	Strain > 2%
Cancellous	Rigid lattice designed for strength; interstices are filled with marrow	In vertebrae, flat bones (e.g., pelvis) and the ends of long bones	20%	Withstand greater strain	Compression; Young's modulus is much greater in the longitudinal direction	Lower	Strain > 75%

remodeling. Osteoclasts and osteoblasts assemble into basic multicellular units (BMUs). Bone is completely remodeled in approximately three years. Under normal conditions, the quantity of old bone removed equals new bone formed. When too much is removed, you get osteoporosis. The major factors involved in remodeling are hormones (estrogen or testosterone) and cytokines (growth factors, interleukins [1, 6, and 11], tumor necrosis factor-α, and transforming growth factor-β).

NERVE AND MUSCLE PHYSIOLOGY

The collections of nerve cells and supporting structures that are distributed throughout the body represent the nervous system (Figure 1.8). The central nervous system is encased in bone and comprises the brain and spinal cord. The peripheral nervous system is not encased in bone and has peripheral nerves and ganglia. The four types of nervous systems are characterized as follows:

- Autonomic nervous system: The afferent and efferent nerves that innervate the body organs to coordinate the internal environment (homeostasis)
- Somatic nervous system: The afferent and efferent nerves that innervate the musculoskeletal and integumentary systems for the purposes of motor function and sensation
- Enteric nervous system: The network of nerves that innervate the gut and coordinate gut function
- Vascular nervous system: The network of nerves that innervate the blood vessels

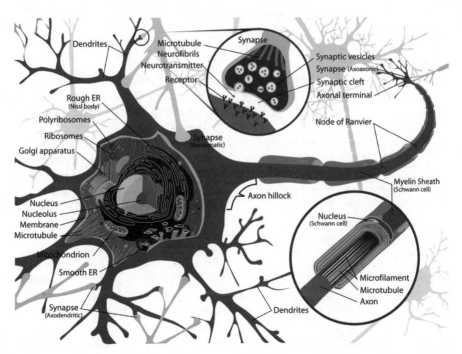

FIGURE 1.8 Structure of a nerve cell. *Source: Pixabay.com, http://pixabay.com/en/red-science-diagram-cell-41524/.*

and coordinate vascular smooth muscle function

The autonomic nervous system is divided into sympathetic and parasympathetic nerves. The sympathetic system is responsible for hyperarousal at a time of danger, whereas the parasympathic system activation results in promoting rest (Table 1.3).

The resting membrane voltage of a neuron is -70 mV. A nerve impulse is an electrochemical event that occurs in nerve cells following proper stimulation. It is an all-or-nothing process that is fast acting and quick to recover. The event is described by a voltage curve called an action potential. The nerve impulse can conduct itself along the entire length of a nerve cell without diminishment ("the domino effect").

There are three main muscle types: cardiac, skeletal, and smooth. Muscle fibers are innervated directly by axons deriving from the spinal cord. Each muscle fiber is composed of many myofibrils.

In skeletal muscle, thick filaments comprise mostly myosin, with thin filaments closely associated. The thin filaments are made up of G-actin, tropomyosin, and troponin complex. Force is generated in the muscle where myosin interacts with actin and undergoes a change in myosin head geometry under the action of ATP.

CARDIAC PHYSIOLOGY

The main functions of the circulation are: oxygenation, waste disposal, hormonal/signaling, and nutrition. The heart is the major organ within the circulation. Cardiac muscle cells are cylindrical in shape, shorter than skeletal muscle, and rich in mitochondria (up to 40% of cell volume). Cell fibers are branched. No nerves are involved in the spread of contraction through the muscle. Adjacent cells are interconnected end-to-end by intercalated discs.

In atrial systole (contraction) blood is forced through into ventricles due to the presence of valves. Ventricles contract as the atria relax (diastole) and blood is forced from the ventricles to the tissues (aorta) or lungs (pulmonary artery). Relaxation allows blood to flow into the different chambers. The following are mechanical characteristics of the circulation:

- Preload: Volume of blood returned to the heart from veins. An increase in blood volume stretches the cardiac muscles, increasing stroke volume.
- Afterload: Blood pressure in the circulation downstream of the aorta. An increase in blood pressure reduces volume of blood pumped.
- Starling's law: The strength of the heart's systolic contraction is directly proportional to its diastolic expansion.

The heart beats at a rate such that CO_2 is effectively replaced in the tissues by O_2. As CO_2 builds up, the heart beats faster. An action potential builds up in the sinoatrial node. This is transmitted via the cardiac muscle around the atria. The action potential reaches the atrioventricular node 40 ms later.

TABLE 1.3 Comparison between Sympathetic and Parasympathetic Nervous Systems

Sympathetic	Parasympathetic
• Fight or flight	• Rest and digest
• Thoracolumbar	• Craniosacral
• Short preganglionic nerves going to sympathetic chain	• Long preganglionic nerves going to organ associated ganglia
• Long postganglionic nerves going to organs	• Short postganglionic nerves going to organs
• Postganglionic nerves use norepinephrine (mostly)	• Postganglionic nerves use acetylcholine on muscarinic receptors (mostly)

VASCULAR PHYSIOLOGY

A system of blood vessels carries blood around the body principally to oxygenate tissues (Figure 1.9).

Arteries deliver blood that has been oxygenated by the lungs to the tissues, and veins carry carbon dioxide-rich blood back again to the lungs. Waste materials and toxins are removed from tissues and mostly processed in the liver. The circulation also acts as an efficient systemic signaling system, where small concentrations of hormones can bring about profound physiological changes.

Arteries are similar in structure to veins, except that the muscle layer (tunica media) is much thicker in arteries, to withstand the extra blood pressure that they are exposed to, being so much closer to the heart (see Table 1.4 and Figure 1.10). In addition, veins in the legs have valves to prevent back flow during cardiac diastole.

Blood is pumped through the arteries to the venous system. Blood perfuses tissue by way of muscular control of capillaries. Access to the tissues is opened in response to increases in:

- CO_2
- Lactic acid
- AMP
- K^+
- H^+

Although the vascular system is a leak-free system, hydrostatic pressure within the circulation means that a significant quantity of the water content travels through the tissues, returning on the venous side due to osmotic pressure (Figure 1.11).

The main regulators of blood pressure are the sympathetic nervous system, which causes vasoconstriction, and the kidneys, which modulate fluid removal and therefore blood viscosity.

Adaption within the juxtaposition of the leg muscles and the venous circulation allows a boost in venous blood pressure during walking and running. This allows blood to return to the heart more easily, and increases preload, a larger heart stroke volume.

The main morbidies of the circulatory system are:

- Heart valve disease, where valves are torn (allowing blood flow reversal), they don't close properly (causing regurgitant jets), or are stiff (requiring greater effort to pump blood normally).
- Atherosclerosis, where the vessel wall becomes less elastic, connective tissue builds up in the plaque, and calcium is laid down, leading to platelet activation, thrombosis, and embolism.
- Aneurysm, where the inner, muscular lining becomes breached, causing swelling of the artery prior to its catastrophic rupture.
- Stroke/seizure, where loss of blood flow to all or part of the brain is caused by hemorrhage, thrombosis, or embolism.
- Hypertension, possibly caused by poor diet. It can be the cause of heart failure, damage to kidneys, and an increase in atherosclerosis.

PULMONARY PHYSIOLOGY

The primary role of the lungs is:

- Exchange of oxygen and carbon dioxide
- Neurohormonal function
- Cellular processing
- Filtration of gases

The lungs are intimately associated, anatomically, with the heart, and are generally transplanted together if a recipient's heart function is poor (Figure 1.12).

The larger tubes leading into the lungs are known as bronchi, and the smaller tubes bronchioles, the inside of which are coated

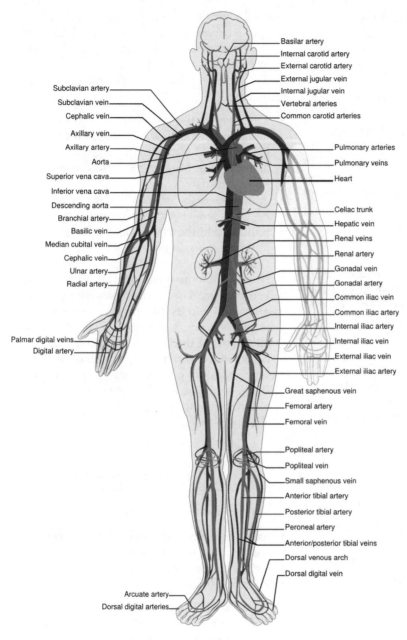

FIGURE 1.9 Major arteries and veins. *Source: Pixabay.com, http://pixabay.com/en/science-diagram-simple-kids-human-41523/.*

with many tiny cilia. These are responsible for mechanical filtration. The bronchioles branch off to millions of alveoli (Figure 1.13).

There is a highly dense network of capillaries on the surface of the alveoli, where oxygen is exchanged for carbon dioxide. The inside of

TABLE 1.4 Description of Diameters of Different Blood Vessels

Vessel	Diameter
Capillary	7–9 mm
Arteriole or venule	10–40 mm
Artery or vein	0.04–10 mm

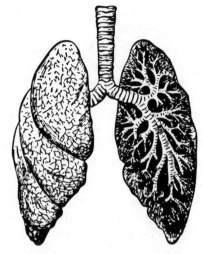

FIGURE 1.12 Human lungs. *Source: Pixabay.com, http://pixabay.com/en/diagram-outline-human-cartoon-37824/.*

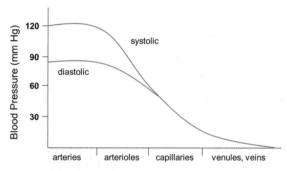

FIGURE 1.10 Typical blood pressures in different parts of the circulation. *Source: Nicholas P. Rhodes.*

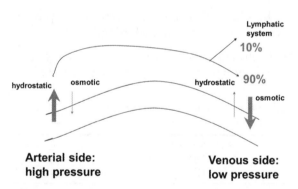

FIGURE 1.11 Fluid flow in the circulation. *Source: Nicholas P. Rhodes.*

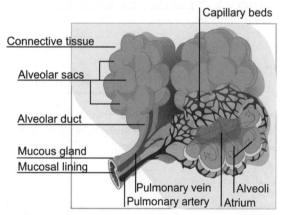

FIGURE 1.13 Structure of lung alveoli. *Source: Pixabay.com, http://pixabay.com/en/science-diagram-illustration-41510/.*

the alveoli surface is coated with a thin mucous coating, enabling the dissolution of the lung gases to occur.

The connective tissue of the chest wall determines the minimum and maximum volume of the chest cavity, but does not control the minimum or maximum lung volume. The connective tissue of the lung is primarily elastic and tends to collapse. There is some stiffness from connective tissue. Pathologies that increase this stiffness lead to difficulty in breathing.

There are two main laws of physics associated with lung function:

- Boyle's law (P.V = K): In a container filled with gas, if you decrease the volume, the pressure will correspondingly increase, and vice versa.
- Dalton's law: In a mixture of gases, each gas behaves as if it were on its own. It exerts a *partial pressure* that is independent of that exerted by other gases in the mixture.

Immuno function in lungs is important because the lungs have close contact with ambient air. Lung lymphoid tissue synthesizes immunoglobulins (predominantly IgA). Mucous secretions are the first line of defense, and filter gaseous microbubbles.

Lungs have prodigious biochemical processing capabilities, mainly peptides (e.g., angiotensin, bradykinin, vasopressin), amines (serotonin, histamine, dopamine, norepinephrine), and prostaglandins.

INTRODUCTION TO BLOOD

Blood is made up of cells, proteins, carbohydrates, lipids, ions and water. Table 1.5 shows the specific functions of each. It has nonformed and formed elements (cells), whose characteristics are described in Table 1.6.

The major role of red blood cells (erythrocytes) within the circulation is:

- Transport of O_2 to tissues
- Transport of CO_2 from tissues

Platelets (thrombocytes) are used for:

- Clotting: Hemostatic plug, surface for reactions to occur
- Store for bioactive compounds (chemotactic factors, growth factors, enzymes, etc.)

White blood cells (leukocytes) are made up of the following:

- Lymphocyte: Produce antibodies
- Neutrophil: Phagocytose bacteria (first line defense)
- Monocyte: Phagocytose bacteria (second line defense), major component of inflammatory response (become macrophages)
- Basophil/eosinophil: Phagocytosis

In addition to cells, there are many protein systems within blood, including:

- Clotting system: Thrombosis and hemostasis
- Fibrinolytic system: Destruction of clots

TABLE 1.5 Functions of the Different Constituents of Blood

	Specific Functions
Cells	Oxygenation
	Clotting
	Host defense
Proteins	Mainly regulation
Carbohydrates	Energy
Lipids	Building blocks
Ions	Isotonic balance
	Cell regulation
Water	Volume

TABLE 1.6 Concentrations of the Different Cell Types in Blood

	Normal Cell Concentrations × 10^6/ml	Diameter (μm)
Red cells	5000	10
Platelets	150–400	2
White cells	4–10	
Neutrophils	60%	14
Lymphocytes	30%	8
Monocytes	6%	17
Eosinophils	3%	15
Basophils	0.5%	15

- Complement system: Immune defense
- Immunoglobulins: Five subclasses, highly specific
- Protein inhibitors: Negative feedback
- Transport proteins: Waste disposal, etc.

- Surfaces are provided by platelets
- All active factors are serine proteases (except fXIII), cleaving following factor
- All active factors can be inhibited by plasma inhibitors

THROMBOSIS, HEMOSTASIS, AND INFLAMMATION

Blood comprises a cellular component (platelets, red blood cells, white blood cells) and a noncellular component. The noncellular components interact to allow the body to maintain a leak-free circuit that does clot internally (Figure 1.14):

- Coagulation cascade (clotting)
  - Intrinsic pathway, important in biomaterials
  - Extrinsic pathway, prevents hemorrhage
- Fibrinolytic system
- System of inhibitors

When a blood vessel is injured, there are a number of stages of action that prevent hemorrhage, as shown in Table 1.7.

The blood coagulation system is controlled and perpetuated by a system of serine proteases (Figure 1.15). There a number of common themes to all the reactions:

- Surfaces are required for many of the complexes to form: Activation of fXII, fXI, fX, fII (prothrombin)

TABLE 1.7 Blood Vessel Injuries and Response

Action	Response
Injury to vessel	Spasm of vessel
Platelets stick to exposed collagen	Causes release of platelet granules
ADP stimulates other platelets	Platelet aggregate plugs hole
Tissue factor released into blood	Extrinsic cascade activated
Blood around injury clots	Red blood cells get caught up in fibrin
Endothelium releases tPA	Clot dissolves

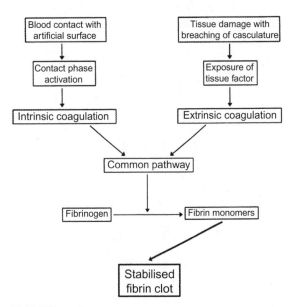

FIGURE 1.15 Basic representation of the clotting cascade. *Source: Nicholas P. Rhodes.*

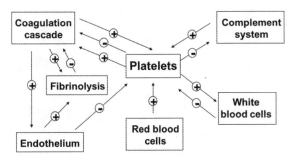

FIGURE 1.14 Interaction of cells and protein systems within blood. *Source: Nicholas P. Rhodes.*

HOMEOSTASIS AND REGULATION

The general principles of homeostasis are that actions are performed to ensure the maintenance of the status quo. Figure 1.16 shows an example.

The principal physiological systems that are regulated are:

- Muscular
- Skeletal
- Skin
- Respiratory
- Digestive
- Circulatory
- Immune
- Excretory
- Nervous
- Endocrine
- Reproductive

Systemic calcium levels are maintained by using the skeleton as a reservoir (Figure 1.17).

In the circulation, blood pressure, O_2/CO_2 balance, pH, and salt balance (Na+ , K+ , etc.) are regulated. This is achieved using specialized receptors and sensors:

- Baroreceptors: Blood pressure
- Chemoreceptors: O_2/CO_2 balance and pH
- Osmoreceptors: Salt concentration

Baroreceptors measure blood pressure and are found in the walls of the large arteries of the neck, particularly in the carotid sinus, the base of the internal carotid artery, and the aortic arch. They are sensitive to changes in pressure and fire off a greater rate of signals when the pressure builds, signaling to the cardioregulatory and vasomotor centers of the brain (medulla oblongata).

- Cardioregulatory center: Increases/ decreases parasympathetic stimulation of the SA node in the heart
- Vasomotor center: Increases/decreases vasodilation

Short-term control of heart rate is by parasympathetic stimulation and vasodilation by sympathetic stimulation. These regulate blood pressure within seconds, and the effects last seconds to minutes. This comes into effect when pressure drops dramatically, for example, when you stand up.

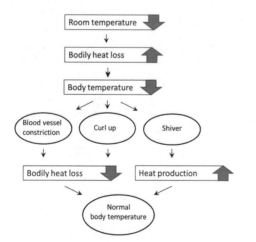

FIGURE 1.16 Homeostasis relative to temperature regulation. *Source: Nicholas P. Rhodes.*

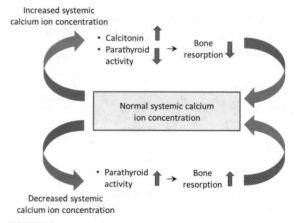

FIGURE 1.17 Control of systemic calcium levels. *Source: Nicholas P. Rhodes.*

Long-term regulation mechanisms for regulation of blood pressure over a span of hours occur from the following:

- Kidneys: Release renin from juxtaglomerular apparatus and release aldosterone from the adrenal cortex (Figure 1.18)
- Capillaries: Fluid movement into/out of tissues
- Blood vessels: Mechanical stretching leads to vasodilation
- Baroreceptors: Stimulate posterior pituitary gland, leads to release of ADH (antidiuretic hormone), and causes the kidneys to resorb more water (Figure 1.19)

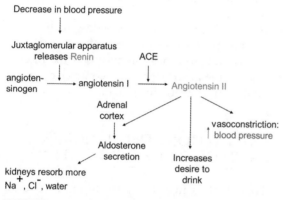

FIGURE 1.18 Mechanism of regulation in kidneys. *Source: Nicholas P. Rhodes.*

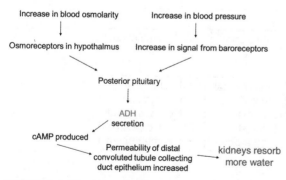

FIGURE 1.19 ADH (vasopressin) mechanism of regulation. *Source: Nicholas P. Rhodes.*

- Heart atrial cells: Mechanical stretching of these cells leads them to release atrial natriuretic hormone, causes kidneys to increase urine volume

Control of blood volume by regulation of kidneys ensures the correct isotonic balance: at 50 mm Hg blood pressure the urine produced is zero times the normal urine volume, urine produced at 200 mm Hg blood pressure is eight times the normal volume. The effects last minutes to hours and correct the gross mismatching of volume.

The are two chemocenters in the brain (medulla oblongata), with the following characteristics:

- Detect changes in chemistry: pH, O_2, CO_2
- Two sites of detection:
  - Vascular system (carotid/aortic bodies)
  - Brain (medulla oblongata)
- Analogous to baroreceptors
- Stimulate the same neural pathways

The chemoreceptors that exist in the medulla oblongata only function during a central nervous system ischemic response: when blood pressure is less than 50 mm Hg, extreme concentrations of H^+ and CO_2 build up.

The circulation regulates the core temperature of the body. When the hypothalamus detects changes in core temperature, it causes constriction or dilation of blood vessels in skin. Decreases in skin temperature below a critical value causes dilation of skin blood vessels to prevent frost bite.

Gross mechanical trauma leads to rapid vasoconstriction. Extreme vascular shock (loss of blood pressure) due to mechanical trauma or anaphylotoxins leads to a reduction of circulation in the least important organs.

RENAL PHYSIOLOGY AND HOMEOSTASIS

The renal system (kidneys) controls the blood content of a number of important solutes and electrolytes. It controls blood osmolarity (concentration), acid-base balance, and volume (hyper/hypovolemic), and generally works by osmosis.

Regulation is principally achieved through hormones, the most important being:

- Renin–angiotensin–aldosterone axis: Absorption of NaCl and H_2O
- ADH (antidiuretic hormone): Absorption of free H_2O

The most important regulation is of blood volume, generally too low rather than too high. Kidneys try to conserve volume and solutes. The major high volume effect is in too much water. After that, volume is controlled by osmolarity. Control of high sodium is achieved by reducing adsorption.

In the kidneys, the molecules that are reabsorbed in the proximal tubule are:

- Glucose
- Amino acids
- Vitamins
- Chloride
- Potassium (K^+)
- Biocarbonate (HCO_3^-)

Following this, the molecules are absorbed in the Loop of Henle, known as a countercurrent multiplication system:

- Only water escapes on the descending limb (by osmosis)
- Only salt escapes on the ascending limb (by active pump)

Molecules adsorbed after the loop, in the distal tubule are (by osmosis):

- Ammonia
- Urea
- H^+

Water balance is maintained by permeability of the collecting duct:

- If the blood is too dilute, collecting ducts become impermeable and water goes out to the bladder (up to eight times the normal urine rate)
- If blood is too viscous, collecting ducts become permeable and water is reabsorbed (down to zero times the normal urine rate)
- The tubule absorbances are controlled by ADH (antidiuretic hormone)

NUTRITION, THE PANCREAS, AND GLUCOSE REGULATION

Food that is ingested goes through several processes as shown in Figure 1.20.

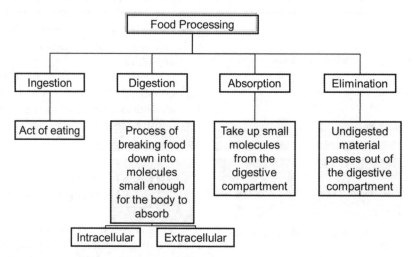

FIGURE 1.20   Food processing. *Source: Nicholas P. Rhodes.*

Food is made up of protein, carbohydrates, and lipids, all of which have different nutritional values (Figure 1.21). Proteins are digested into amino acids; carbohydrates undergo glycolysis; and fats are degraded into fatty acids and the glycerol backbone (Figure 1.22). All of these components are processed ultimately into the Krebs cycle and the electron transport chain.

Regulation of appetite occurs through hormonal feedback (see Figure 1.23).

Leptin is produced by adipose (fat) tissue. Leptin suppresses appetite as its level increases. When body fat decreases, leptin levels fall, and appetite increases. The hormone PYY is secreted by the small intestine after meals, and acts as an appetite suppressant that counters the appetite

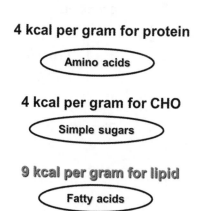

FIGURE 1.21 Net energy values of different food groups. *Source: Nicholas P. Rhodes.*

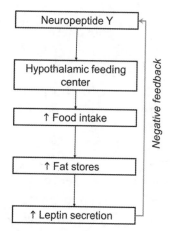

FIGURE 1.23 Hormonal regulation of appetite. *Source: Nicholas P. Rhodes.*

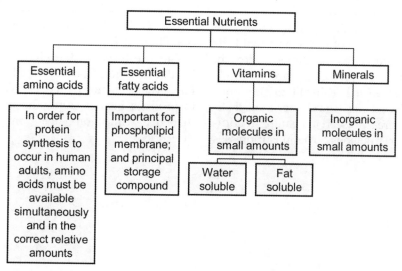

FIGURE 1.22 Physiological uses of the different food groups. *Source: Nicholas P. Rhodes.*

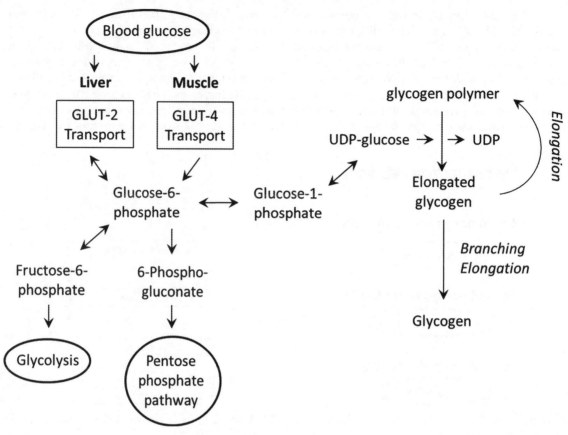

FIGURE 1.24 Energy storage in the body. *Source: Nicholas P. Rhodes.*

stimulant ghrelin. Ghrelin is secreted by the stomach wall and is one of the signals that triggers feelings of hunger as mealtimes approach. In dieters who lose weight, ghrelin levels increase, one reason it is difficult to stay on a diet. A rise in blood sugar level after a meal stimulates the pancreas to secrete insulin. In addition to its other functions, insulin suppresses appetite by acting on the brain.

Energy is stored in the body as glycogen (see Figure 1.24).

Glucose levels in the blood are dictated by insulin secretion from the β-cells of the islets of Langerhans in the pancreas. When the ability to produce insulin stops, the patient becomes diabetic. Glucagon acts in the reverse direction.

2

Research Methodology

Azzam Taktak

Royal Liverpool University Hospital, Liverpool, U.K.

STUDY DESIGN

Before embarking on a research study, it is very important to carefully consider all the issues and potential pitfalls that can make the study fail or, worse still, result in wrong conclusions. At the focus of study design should be the final objective (or objectives); what is the question we would like to answer?. The question should not be "How do I analyze my data?" but rather "How do I prove or disprove a certain theory?" or "How do I find out if events A and B are somehow related?" The answer to the last two questions will determine how to analyze the data and interpret the results.

Broadly speaking, there are two types of studies, observational and experimental (Altman, 1991). In observational studies, we collect data on one or more groups of subjects purely from an observer's point of view. That is, we do not interfere with the clinical management of these subjects. An example would be to compare the survival rate of infants with low birth weights compared with those with average birth weight. Another example is to look at the prevalence of heart disease in groups of subjects from the general population with different socio-economic status. Data for these studies can either come from clinical records or from surveys. Experimental studies, on the other hand, require the researcher to deliberately influence the clinical management of the subjects to investigate the outcome. Typical examples of these types of studies include drug trials.

There are two types of observational studies: case-control studies and cohort studies. In case-control studies, a number of subjects with the disease in question (cases) are identified and compared with a group of subjects without the disease but who are otherwise comparable (controls). The past history of these groups is examined to determine their exposure to a particular risk. In cohort studies, two groups are identified, one exposed and one not exposed to a particular risk. The groups are followed up over time and the occurrence of the disease in question in each group is identified.

In both designs you can sometimes have more than one case group. For example, if we are studying the association between smoking and lung cancer, we might have two case groups: present smokers and those who have smoked in the past but stopped smoking prior to being recruited for the study. We might go on further to divide the present smokers group into heavy smokers and light smokers (measured in a unit called pack-years).

The advantage of cohort studies is that they do not rely on the accuracy of medical records which can sometimes contain errors or be incomplete. The disadvantage is that if the disease in question is rare, it will need a large number of subjects to be recruited and may take years, which can be costly. Another problem with cohort studies is that subjects sometimes drop out of the study. They might, for example, stop smoking halfway through the study, or refuse to take part or move house or die of an unrelated disease. These problems are known as *loss to follow-up*. Another problem that can occur in both types of studies is that certain aspects can change over time. Clinical practice might change over time, certain risk factors might affect older subjects more than younger ones, and so on. Moreover, there are issues related to feasibility and ethics to consider with cohort studies. Consider, for example, a study looking at association between car accidents and drivers being under the influence of alcohol. Here, a case-control study is the only feasible option. As blood samples are always taken from drivers who have been involved in a crash and analyzed for alcohol, reliable data should be possible to obtain.

A serious problem that some clinical studies can experience is the effect of confounders. A confounder is a variable that has not been taken into account that can completely skew your results. A well-known example from the literature is a study by Charig et al. (1986) on the effectiveness of keyhole surgery on the treatment of kidney stones. In this study, 350 subjects treated with keyhole surgery (cases) were compared with another 350 subjects treated with the more traditional open surgery

(controls). They concluded that keyhole surgery had a higher success rate than open surgery. Suppose, however, we separated the subjects according to the size of the stone. It is extremely likely that those with smaller stones (<2 cm diameter) were more likely to undergo keyhole surgery than open surgery. They also had better chance of removal of the stone due to its small size. The size of the stone is a confounder. Results of the two groups separately can show an association in the opposite direction, with open surgery proving to be more successful in both groups.

A term that is often heard associated with clinical trials is *randomization*. Randomization is a process designed to eliminate or reduce errors due to bias. For example, in a drug trial, if we decided to give the first 100 subjects the new drug and the next 100 subjects the existing drug or placebo, we might introduce some bias if, for example, clinical settings that could influence outcome have changed in due course. The best way to eliminate this bias is to allocate the subjects to the cases or controls groups at random. To do this, we need a random sequence of numbers, which we can obtain from software packages or statistical tables. Let us consider the following random sequence:

91470387540015331276

If we decide that any number in the range 0 to 4 will be allocated to the cases group (N) and 5 to 9 allocated to the controls (C) group, we will have the following sequence:

CNNCNNCCCNNNNCNNNNCC

So the first subject is allocated to the controls group, the second to cases, and so on. Here we notice that 8 subjects were allocated to the controls group and 12 to the cases. If the numbers were large enough, we should see a split that is very close to 50:50.

Supposing we are comparing the performance of 3 blood pressure devices on 10 normal subjects to see if the devices produce similar results. Since the subjects are normal healthy volunteers of a limited age group, say 20 to 30 years old, we are not expecting any significant variation between subjects. The only bias to consider here are the order these measurements are taken. It is hypothetically possible that there is an upward trend in the measurements due to subject fatigue, for example. We therefore would want to randomize the order in which the measurements are taken. There are 6 possible combinations to take these measurements:

1. ABC
2. ACB
3. BAC
4. BCA
5. CAB
6. CBA

Using the above sequence, we first need to eliminate any numbers above 5 and add 1 (since we are starting from 0). This will give us the following sequence:

2514651126

In this case, not all combinations were chosen equally. For example, number 3 was not selected, whereas number 1 was selected three times. This is a feature of randomness and is also a reflection on the small sample size for the number of trials compared with the number of possible combinations. In a way, it is not that important that balance was not reached in the above setup. The most important thing is that we have randomized the order of taking measurements, thereby reducing the chance of bias. If it was a requirement to balance the above design then we would need to look into what is known as a block design. This is a complex topic and is beyond the scope of this chapter.

Bias can also occur in some clinical trials due to the observer or the subject themselves. They might subconsciously affect the outcome of the trial by somehow manipulating the results. For example, the observer might look for reasons to discard a particular observation if it did not agree with his or her own hypothesis. It is desirable, therefore, if the observer was unaware of the conditions of the experiment to reduce the possibility of bias, a process known as *blinding*. The subject might also influence the outcome if he or she behaved differently under different conditions. If the observer and the subject were both unaware of the conditions, this is known as *double blinding*. This is not always feasible, however. For example, if a trial is being conducted to investigate the efficacy of surgery against other forms of treatment, blinding would not be an option. Sham surgery is sometimes carried out in these situations (subject to an ethical committee approval) to blind the subject but not the observer. In such cases, it might be possible for someone else other than the surgeon to conduct the analysis without knowing which treatment the subject has received in any particular session.

HYPOTHESIS GENERATION AND TESTING

Let us consider a simple case of a drug trial where we are trying to assess the efficacy of a new drug. Suppose we have two groups of subjects, those treated with the new drug, which we call group T (for treatment), and those who are receiving an existing drug or placebo, which we call group C (for control). What we are often interested in is to show that the proportion of people who get better taking the new drug (call them θ_T) is significantly larger than that in the control group (call them θ_C). Obviously, if all subjects in group T are cured ($\theta_T = 1$) and none in group C are cured ($\theta_C = 0$) then we can safely conclude that the drug is hugely successful. Often, though, this is not the case. Both θ_T and θ_C will be somewhat effective. It might turn out, for example, that $\theta_T = 0.45$ and $\theta_C = 0.4$. Is the difference here significant or is it purely due to chance? In other words, if we repeat the experiment, will we see similar results or will θ_T be closer to or even lower than θ_C?

We need statistics to answer this question. Statistics does this by first assuming that the drug is ineffective:

$$\theta_T = \theta_C$$

This is our null hypothesis. If the evidence from the data suggests that there is very small probability that this is true, this gives some evidence in favour of rejecting the null hypothesis. In some cases, we might not know whether the difference is positive or negative. In the above example, this means that we don't know whether the new drug is better or worse than the old one. The alternative hypothesis can therefore be expressed as:

$$\theta_T \neq \theta_C$$

This is called a 2-sided alternative since θ_T can be greater than or less than θ_C. Sometimes the association can only go one way. For example, if we conduct a study to assess the efficacy of a sleeping pill, we are interested in determining whether the pill increases the length of sleep or not. Here we make the prior assumption that the drug cannot decrease the length of sleep. Such assumptions are often not possible to make in real life with any certainty. If we measure the average difference of the length of sleep in a number of volunteers and call that D, our null hypothesis becomes

$$D = 0$$

and the alternative hypothesis in this case is 1-sided:

$$D > 0$$

One-sided tests are rarely used and the above example was shown for illustrative purposes only.

APPLICATION AND INTERPRETATION OF STATISTICAL TECHNIQUES

There are many statistical packages that can carry out statistical analysis. Examples of these include SPSS, SAS, Minitab, GenStat, R, MATLAB (Statistics Toolbox), and so on. Even Microsoft Excel, which is primarily a spreadsheet tool, can carry out a number of sophisticated statistical analyses but only after installing the data analysis add-on. There are also nowadays many online packages that carry out the analysis, but the user must take care that they trust them first before using them. In this section, we will demonstrate some statistical analysis using the following website, developed by the author: *http://clin-engnhs.liv.ac.uk/MedStats/MedStats_Demos.htm*

We saw earlier that to prove the effectiveness of a drug or treatment or a device or any other intervention, we need to set a hypothesis first that the intervention is not effective and seek to disprove this hypothesis. A statistical tool or family of tools to carry out this type of analysis is generally called a *test of significance*. The significance probability is denoted as p and is often called the p *value*. The p value is the probability of getting data as extreme as or more extreme than that observed, given that the null hypothesis is true. Very small p values that are < 0.01 provide strong evidence against the null hypothesis. On the other hand, p values that are > 0.1 show very little evidence against the null hypothesis. Values in between are an indication of marginal evidence and should be treated with caution. To measure the p value, we need a test statistic, which can be estimated from the data. We now look at two common types of test tests.

The T-Test, ANOVA and the Z-Test

The t-test is used when the data can be reasonably modeled by a normal distribution, such as, for example, taking mean blood pressure readings from 100 normal subjects. A histogram is a very quick method to check the distribution of the data, although there are some pitfalls when using histograms. Another tool to asses the normality of the data visually is the normal probability plot that most statistical software packages offer. This tool plots the ordered values of the variable against normal scores from the standard normal distribution. More formally, there are statistical tests to assess the normality of the data such as the Lilliefors test or the chi-squared test. If the data appear to be skewed, we can apply some transforms such as the log transform or the square root to make it look more normal.

Here is an interesting example on the assessment of normality in data. The data set X has been sampled from a normal distribution with a mean of 60 and standard deviation of 10. This might represent, for example, weight in kg of 25 high-school students. The data set is shown below:

{60.5, 45.1, 74.3, 53.1, 51.4, 55.1, 54.6, 53.5, 57.2, 40.3, 63.6, 48.1, 48.3, 73.8, 54.2, 64.7, 75.9, 50.7, 65.6, 58.2, 47.1, 55.3, 71.5, 71.6, 74.8}

A histogram of the data does not give good indication that the data are normally distributed (Figure 2.1(a)). A normal probability plot on the other hand looks more promising as the data lie roughly along a straight line but with a slight curvature (Figure 2.1(b)). A Lilliefors test reveals that we cannot reject the null hypothesis that the data are normally distributed ($p = 0.139$). Now let us perform the following transform on the data: square the values and divide by 100. The new data set might now represent weights of a younger population such as elementary school pupils, for example. Now let us repeat the above analysis for a test of normality. The histogram and the normal probability plot are not very informative (Figure 2.2(a) and (b)). The evidence

(a)
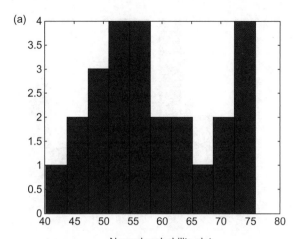

(b)
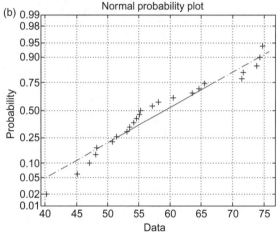

FIGURE 2.1 (a) Histogram of data drawn from a normal distribution. It is not apparent from the histogram that the normal distribution model is a good fit to the data. (b) Normal probability plot for the same data showing that the data lie roughly along a straight line.

from the Lilliefors test however is marginal ($p = 0.049$). Taking a square root transform would help satisfy us here that the data can be reasonably modeled with a normal distribution. These tests, as well as data sampling, were all carried out using the MATLAB Statistics Toolbox.

Let us look at an example of a statistical test. A study was conducted to investigate the

(a)
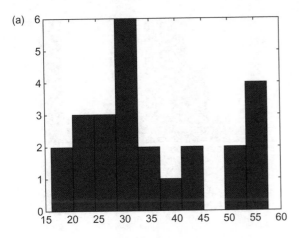

(b)
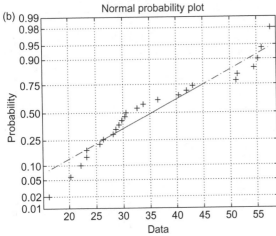

FIGURE 2.2　(a) Histogram of the square transform of the above data. (b) Normal probability plot of the square transformed data.

association between birth weight and death in infants with severe idiopathic respiratory distress syndrome (SIRDS) (Van Vliet and Gupta, 1973). A group of 50 infants with SIRDS were recruited, 27 died and 23 survived. The average weight of the survivor group was 2.21 kg compared to an average of 1.86 kg in the deceased group. We wish to know whether the difference in weight is significant or if it is due to chance only.

The appropriate test to do here is a 2-sample t-test for difference in means. The test requires that three assumptions are met:

1. The data are normally distributed.
2. The samples are independent.
3. The two groups have equal variances (a rule of thumb can be applied here that the two variances do not differ by a factor of more than 3).

Once we are satisfied that the above assumptions are met, we can proceed to analyze the data. Go to the previously mentioned website and click on Student's 2-sample t-test. Click the View button next to About This Program line. This will open a window that describes the test. Here, you will also be able to download the SIRDS data. Upload the data as described and then click Evaluate. The program displays a t statistic of -2.2538 and a p value of 0.029 (2-tailed). There is therefore moderate evidence from the data that the difference in weight is statistically significant but we probably need to collect more data to be sure.

If there are more than two groups to compare, a family of statistical tests called ANOVA (Analysis of Variance) are used. ANOVA is a very wide and complex topic and we will only cover the basics of it here. For more information on the subject, the reader is referred to the list of recommended books at the end of this chapter.

We will now visit some of the basic aspects of ANOVA using a hypothetical example. Let us suppose we collected data from a number of subjects using 10 different instruments. A 1-way ANOVA test tells us whether the instruments produce similar results or not. The null hypothesis is that the distributions of results between instruments are the same. If the p value is small (p < 0.01) this provides evidence against this null hypothesis, that is, there is a difference somewhere in the measurements. ANOVA does not tell you where the difference is. If we want to find out where the difference

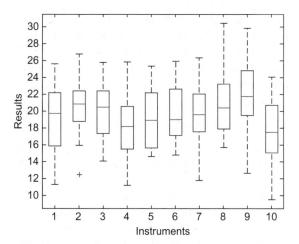

FIGURE 2.3 Box-whisker plot of results of 10 instruments on independent samples.

is, we must carry out more analysis such as box-whisker plots to visually analyze these differences. A box-whisker plot of this hypothetical data set is shown in Figure 2.3. The *p* value for this data set is 0.013 indicating moderate evidence against the null hypothesis. On close examination of Figure 2.3, we notice that instrument number 10 produced slightly lower results on average than some other instruments, which is the most likely contributing factor to the slightly low *p* value. In fact, all measurements were sampled from the same normal distribution with random noise. Notice that the number of samples does not have to be equal for each instrument for this analysis. However, the 3 assumptions needed for the t-test above are also required for ANOVA.

An alternative to the t-test is the z-test. This test is usually used for other types of data that are not normally distributed but can be approximated by a normal distribution when certain conditions are met. Examples of such data include proportion data, which can be modeled using binomial distribution, and count data, which can be modeled using Poisson distribution. Both these types of data can be approximated by a normal distribution under certain conditions.

Before we go on any further, it is important to understand the concept of *confidence intervals*. The confidence interval is a term used when calculating a random variable. In estimating this variable, instead of quoting a single value (point estimator), we acknowledge the fact that there is some amount of uncertainty in our estimation, and we call this the confidence interval. The 95% confidence interval is often quoted but the meaning of this term is sometimes misunderstood. If we are trying to estimate a parameter θ and we obtain a 95% confidence interval $(\widehat{\theta}+, \widehat{\theta}-)$, the interpretation of this interval is this: if we repeat the experiment a large number of times then the true value of θ would be included in this interval in 95% of the experiments. Of course, most often, we carry out an experiment only once so the implication is that there is a 5% probability that our interval misses out θ completely.

In the previous example where $\theta_T = 0.45$ and $\theta_C = 0.4$, suppose we had 100 subjects in each group. Using the same link as before, we now click on Significance Test for Difference in Proportions. We enter 100 in the number of samples in groups 1 and 2 boxes. We enter 45 successes for group 1 and 40 for group 2 (the order does not matter here). The program returns the following values:

$\theta_T = 0.45$ (95% C.I. 0.35−0.55)
$\theta_C = 0.4$ (95% C.I. 0.3−0.5)

As can be seen, there is a significant overlap in the two confidence intervals so we cannot rule out the possibility that the two proportions are similar. The program also returns the z-statistic and the *p* value, which in this case are 0.716 and 0.474, respectively. This is a high *p* value indicating that there is little evidence against the null hypothesis that the two proportions are the same.

Now let us suppose that we based our estimation of θ from a sample of 1000 in each group. We now enter 1000 in the number of samples in each group and 450 and 400 in the number of successes. We get a different picture:

$\theta_T = 0.45$ (95% C.I. 0.42−0.48)
$\theta_C = 0.4$ (95% C.I. 0.37−0.43)

The z-statistic is much higher now with a value of 2.265 and the p value is much smaller at 0.024 indicating moderate evidence against the null hypothesis of equal proportions. We can keep going like this and will notice that the evidence gets stronger with more data.

Nonparametric Tests

These tests do not make any assumptions about the distribution of the data as they perform the analysis on the ranks of the data rather than the absolute values themselves. Examples of such data may include comparing responses to a questionnaire from two groups whereby responses are graded as: 1 − Excellent, 2 − Good, 3 − Average, 4 − Poor, 5 − Diabolical. There is a clear trend in the sequence of the above numbers but the distances between them are not defined. The test to do in this case is called a Mann−Whitney test.

Although nonparametric tests are more convenient in that they do not make any explicit assumption on the distribution of the data, they are less powerful than parametric tests since they ignore absolute values. For example, for the SIRDS data set above, had we been tempted to use the Mann−Whitney test, we would obtain a p value of 0.076 (2-tailed), which provides only weak evidence against the null hypothesis that the two distributions are the same. If we then apply a threshold of 0.05 for the p value as is common practice in medical literature, we would reject the null hypothesis under the 2-sample t-test and not reject it under the Mann−Whitney test.

Knowing something about the data is very important in making a judgment regarding the distribution of the data. Data representing weight, height, and blood pressure in a normal population should be adequately modeled by a normal distribution. Responses to questionnaires, on the other hand, are very unlikely to be normally distributed. Age is likely to have some right skew.

Correlation and Regression

The two terms *correlation* and *regression* are often used synonymously, but there is a subtle difference between the two. Correlation refers to the fact that knowing something about one variable tells you something about the other. Regression is a mathematical equation that allows you to predict the value of one variable (known as the response variable) from another (known as the explanatory variable). We can see why the two terms are often quoted together since if the two variables are not well correlated, it is meaningless to try to generate a regression model for these variables. The simplest form of a regression model is the linear regression model. If the explanatory variable is represented by x and the response variable by y, the linear regression model describing the relationship between the two can be modeled by the equation of a straight line:

$$y = mx + c$$

Note that if we came across the following relationship

$$y = mx^2 + c$$

this is still considered a linear model since x^2 can be easily replaced by another variable, say t. This is also true for any of the following:

$$y = me^x + c$$
$$y = m\log(x) + c$$
$$y = 1/x$$

If we have a set of x and y of continuous measurements on a number of samples and we wish to see how they are correlated, the first step is to do a scatter plot on the data. We could do this task quite easily on any software such as Microsoft Excel. Figure 2.4 shows a plot of a hypothetical set of values. We then perform linear regression analysis on the data and again we can do this with Excel or any other similar package. This is shown as the solid line in Figure 2.4. We can use the equation of the best fit line to predict values of y for given x. The r^2 value that the software calculated is known as

the *coefficient of determination*. It tells us how much of the variation in the data can be explained by the best fit line with the rest of the variation being random noise. The square root of this value (i.e., r) is called Pearson's correlation coefficient. It takes an absolute value of 1 if the correlation was perfect and 0 for no correlation. Most statistical packages will give you a p value or a confidence interval with the r value and it is good scientific practice to quote these as well as the r value itself.

It is often said that correlation does not imply causation. Just because x and y correlate strongly does not mean one causes the other. It might be that the correlation we find is due to a third factor that we have not considered that is also correlated with these two variables and is the true causation. To determine causality, we need to ask ourselves, does the association make scientific sense? Is it consistent with current knowledge and can it be repeated under different settings? (Greenhalgh, 2010).

Table 2.1 provides a summary of statistical tests, with some clinical examples. Some of the

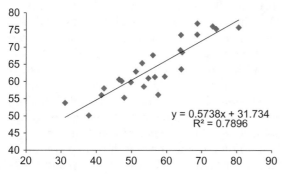

FIGURE 2.4   Best fit line to a set of hypothetical data with the coefficient of determination r^2.

TABLE 2.1   Summary of Statistical Tests with Some Clinical Examples

Purpose	Example	Parametric	Nonparametric
Compare paired samples	Taking heart rate measurement before and after exercise on a number of healthy volunteers	1-sample t-test	Wilcoxon signed-rank test
Compare two unrelated samples	Measuring birth weight of infants with SIRDS and comparing the survived against deceased groups	2-sample t-test	Mann–Whitney test
Compare more than two sets of observations on the same sample	Taking measurements on a number of subjects using different instruments to assess the differences between devices (but not differences between samples)	One-way ANOVA	Kruskall–Wallis
Compare more than two sets of observations on a single sample under different conditions	Different operators making measurements on a number of samples using different methods of preparing the samples	Two-way ANOVA	Friedman
Investigate correlation between two continuous variables	Correlation between systolic and diastolic blood pressure in a group of patients with hypertension	Pearson correlation	Spearman correlation
Investigate correlation between two categorical variables	Correlation between smoking and lung cancer	None	χ^2
Investigate correlation between two ordinal variables	Agreement between disease severity and QALY score	None	Kappa

tests referred to in this table have not been covered here. Interested readers are referred to the list of recommended books for more details.

LITERATURE SEARCHING AND REFERENCING

Searching medical literature has never been easier with the availability of online tools such as PubMed, Scopus, Web of Knowledge, and so on. PubMed is the tool most widely used by clinicians, healthcare scientists, and other healthcare professionals. It is a free resource that is developed and maintained by the National Centre for Biotechnology Information (NCBI) at the U.S. National Library of Medicine (NLM), located at the National Institutes of Health (NIH). PubMed comprises millions of citations for biomedical literature from Index Medicus and MEDLINE. On the PubMed website[1] there is an online tutorial and a link to a YouTube demo on how to use the resource.

To demonstrate how PubMed works, let us assume we want to do a search for publications on Sudden Infant Death Syndrome (SIDS). We type the first word "sudden" and we immediately get a drop-down box with some suggestions. We select "sudden infant death syndrome" from the list and press select. We get a list of nearly 10,000 publications dating back to 1945. We also get a histogram showing how many articles were published each year since 1945. Now let us refine our search a little bit by looking at publications in the last decade, that is, starting from 2000. Click the "Custom range" option on the left side and specify the date range from the January 1, 2000 to the present date. We now get around 3500 publications. Let us refine the search even further by searching for "Clinical Trials" and "Randomized Controlled Trials"

only. We click on these two links on the left side and we now get a much smaller number near 100.

Now let us suppose we are doing research looking for any association between SIDS and breast-feeding. We click "Meta-Analysis" on the left and we find that there is an article by K.L. McVea et al. published in the *Journal of Human Lactation* in February 2000 on this subject (McVea et al., 2000). An examination of the abstract tells us that the study is a summary of 23 cohort and case-control studies and the combined evidence shows that infants who were bottle-fed were twice as likely to die from SIDS than those who were breast-fed. It is very important, however, not to jump to conclusions here. Remember: correlation does not imply causation. In fact, the paper could not rule out the presence of confounders.

If you are doing the search from a computer that is connected to an educational institution's network, chances are you will be able to read the full article taking advantage of your institution's library subscription with the publisher, sometimes through a third party. The paper tells us that the analysis was conducted by searching the MEDLINE database between 1966 and 1997. The search included a number of MeSH (Medical Subject Headings) terms such as *sudden infant death*, *cot death*, *crib death*, *breast-feeding*, and *infant nutrition*.

In biomedical engineering, we are probably primarily interested in reviewing articles that assess a particular technology. For example, Lisboa and Taktak published an article on the use of artificial neural networks in cancer (Lisboa and Taktak, 2006). In the period between 1994 and 2003, there were 396 studies published with only 27 being either clinical trials or randomized controlled trials. The majority of these studies showed an increased benefit to healthcare in the use of this technology. The uptake of this technology in the manufacturing of medical devices remains

[1] *www.ncbi.nlm.nih.gov/pubmed*

low, sadly, with only a handful of devices utilizing the technology to date.

There are other powerful online literature search engines besides PubMed. One such engine is called Scopus, which is also free to access via most academic networks. From a clinical engineering point of view, Scopus has the advantage over PubMed in that it also searches for scientific and technical journals and books that are not included in MEDLINE.

Another useful tool is Google Scholar. Although it is slightly less structured than PubMed and Scopus, it has the advantage that it can be accessed from anywhere and it trawls the whole Internet to find matches to your query. Sifting through the results, however, can be time consuming and you cannot limit your search to clinical trials or meta-analysis only, for example. Another major drawback with Google Scholar is that it finds publications that have not gone through a peer-review process as well as those that have, so use it with caution.

If you are embarking on a literature search from new, it is a good idea to build yourself a database if you haven't got one already. There are a number of bibliography software packages available such as Reference Manager, EndNote, and so on. These packages link to word processing software such as Microsoft Word which helps a great deal in taking care of citations and generating a reference list when writing a scientific paper.

Let us look at an example of how to import references from PubMed into EndNote and linking it to a document in Word. First revisit the PubMed site with the search for meta-analysis studies in SIDS since 2000. Select three studies that relate to SIDS and breast-feeding. Click "Send to:", select "File" and select "MEDLINE" as the format, and save the file to the hard disk. Next, open EndNote, create a new library, and choose import from the File menu. In the Import dialogue box choose the MEDLINE filter as the import function and

select the file you have just downloaded. You should see all three references you have just selected appear in your library.

Next open Microsoft Word. There should be an EndNote menu item in the menu bar. You can insert references in your document in many different ways. If EndNote is still open, you can highlight the reference you want to insert and click Insert Selected Citation from the menu. Alternatively, type the name of one of the authors (e.g., McVea) and click Insert Citation. If there is more than one reference for this author you will be presented with a list that you can choose from. Once you have finished typing your document, you will want to format your references in the style of the journal you are submitting to. In the Style drop-down box you will notice numerous styles such as Harvard, Vancouver, or other styles that are more specific to certain journals such as the *BMJ*.

References

Altman, D.G., 1991. Practical Statistics For Medical Research. Chapman & Hall/CRC.

Charig, C.R., Webb, D.R., Payne, S.R., Wickham, J.E., 1986. Comparison of treatment of renal calculi by open surgery, percutaneous nephrolithotomy, and extracorporeal shockwave lithotripsy. Br. Med. J. (Clin. Res. Ed.). 292, 879–882.

Greenhalgh, T., 2010. How to Read a Paper: The Basics of Evidence-Based Medicine. Wiley-Blackwell.

Lisboa, P.J., Taktak, A.F.G., 2006. The use of artificial neural networks in decision support in cancer: a systematic review. Neural Netw. 19, 408–415.

McVea, K.L., Turner, P.D., Peppler, D.K., 2000. The role of breastfeeding in sudden infant death syndrome. J. Hum. Lact. 16, 13–20.

Van Vliet, P.K., Gupta, J.M., 1973. THAM v. sodium bicarbonate in idiopathic respiratory distress syndrome. Arch. Dis. Child. 48, 249–255.

Further Reading

Armitage, P., 2000. Statistical methods in medical research. Blackwell Scientific.

Cohen, L.H., 1996. K.M.E. Practical Statistics for Students: An Introductory Text. Sage Publications Ltd.

Harrell Jr., F.E., 2006. Regression Modeling Strategies: With Applications to Linear Models, Logistic Regression, and Survival Analysis. Springer.

Peat, J.B., Elliott, E., B, 2008. Statistics Workbook for Evidence-based Healthcare. Wiley-Blackwell.

Van Belle, G., Heagerty, P.J., Fisher, L.D., Lumley, T.S., 2004. Biostatistics: A Methodology For the Health Sciences. Wiley.

3

Good Clinical Practice

Anthony Scott Brown

Royal Cornwall Hospitals NHS Trust, Truro, U.K.

BACKGROUND

The standards of clinical research have developed over many decades during which there have been many trials that today we would frown on or even be horrified by for being unscientific, unethical, or both. There was the Public Health Service syphilis study conducted in Tuskegee, Alabama. This study researched the natural effects of untreated syphilis in black males between 1932 and 1972; it involved 600 participants, of which 399 who had the disease were not treated for its effects. Furthermore there was no informed consent for this study (CDC, 2011). Another early disaster was a trial of the drug thalidomide developed by German scientists during the Second World War. It was prescribed to pregnant mothers as a treatment for morning sickness and resulted in the birth of babies without arms or legs (Foggo, 2009). Thankfully clinical research has moved on considerably and the standards of both clinical and ethical practice have improved immeasurably. That said, unfortunate incidents in clinical trials do still occur.

The risks of clinical trials were brought to a fore by the media in 2006 with the TGB1412 disaster. This drug was developed by the German biopharmaceutical company TeGenero and the trial was run by the contract research organization (CRO) PAREXEL International Limited, at the Northwick Park Hospital in the United Kingdom. Eight healthy volunteers entered into the phase 1 trial of the anti-CD28 monoclonal antibody TGN1412. Two of the volunteers received a placebo and the remaining six who received the new drug became seriously ill within minutes and were admitted to intensive care (Saunders, 2006; Brown, 2011). Although these occurrences are rare, they do raise the awareness of the public, healthcare providers, and the manufacturers to the potential harmful consequences of clinical trials.

PHASES OF CLINICAL RESEARCH

Clinical research is undertaken in a series of phases, and with each subsequent phase there is a reduced risk to patient safety. For clinical trials of investigational medicinal products (CTIMPs) there a four phases, whereas for clinical investigations of medical devices there are

TABLE 3.1　Comparison of the Phases of Drug and Medical Device Trials

Clinical Trial of Investigational Medicinal Product	Details and Purpose	Clinical Investigation of Medical Devices	Details and Purpose
Phase I	First time in man (FTIM), healthy volunteers Collection of tolerability data Pilot dose findings and investigation of pharmacokinetic (PK) and pharmacodynamic (PD) profiles	Premarket approval study	To obtain CE marking Includes pilot studies
Phase II	Therapeutic pilot study Demonstration of pharmacological activity assessment of short-term tolerability profile		
Phase III	Larger scale study of subjects with target disease Comparison of efficacy with current treatments		
Phase IV	General population Long-term safety data	Evaluation with a CE marked device	Registry and audit-type studies

just two phases. A comparison of the two types is given in Table 3.1.

STANDARDS IN CLINICAL RESEARCH

Probably the first standard to be introduced in clinical research was the Nuremberg Code of 1947. This came about following the legal trials of military war crimes, and one of the most significant requirements set down in the code was that the voluntary consent of the human subject is absolutely essential (Nuremberg Military Tribunals, 1949).

A major step forward was the development of the Declaration of Helsinki (1964) which evolved out of the Nuremberg Code. There have been six amendments to this over the years, the latest being in 2008. The 2000 revision was controversial and the 2008 is still disliked, therefore in Europe clinical research protocols still refer to the 1996 revision and in the United States the declaration is not used at all. The first Good Clinical Practice (GCP)

standard was published in America in 1977 as one of the U.S. Food and Drug Administration (FDA) regulations.

The problems of differing regulations and standards in countries across Europe meant there was considerable repetition of trials and that delayed new pharmaceutical products getting into the marketplace, which consequently put more participants potentially at risk. Results of trials in one country were not accepted in another country because the trial was conducted under a different set of standards.

To provide some standardization, a series of conferences with representation from regulatory authorities was organized: the International Conference on Harmonisation of Technical Requirements for Registration of Pharmaceuticals for Human Use (ICH). The cosponsors of the ICH are:

- European Commission (EC), European Union
- Food and Drug Administration (FDA), United States
- Ministry of Health, Labour and Welfare, Japan

- European Federation of Pharmaceutical Industries Associations (EFPIA)
- Pharmaceutical Research and Manufacturers of America (PhRMA)
- Japan Pharmaceutical Manufacturers Association (JPMA)

The aim of ICH is to harmonize the processes within clinical research, focusing on quality. There are a range of guidelines divided into four key areas: quality, safety, efficacy, and multidisciplinary. The ICH E6 Guideline for Good Clinical Practice is widely available.

There are a range of regulations that have been developed over the years which have built on the Declaration of Helsinki and ICH GCP. The two European directives are 2001/20/EC and 2005/28/EC, which each country needs to transpose into law. In the United Kingdom the main regulation is the Medicine for Human Use (Clinical Trials) Regulations 2004 which was implemented into law through Statutory Instrument 2004/1031. The

statutory instruments in the United Kingdom can be visualized as a wall with the Declaration of Helsinki being the foundation; it is this "wall" that protects research participants and ensures the research is conducted to the highest standards (Figure 3.1). Similar arrangements are in place in other European countries.

GOOD CLINICAL PRACTICE

Clinical trials of investigational medicinal products (CTIMPs) are important in developing both the safety and efficacy data around a new drug before it is granted a product license (Brown, 2011, p. 104). Good Clinical Practice (GCP) is a set of internationally recognized ethical and scientific quality guidelines that should be followed to ensure that the rights and well-being of participants are protected and that the data produced from the research are valid and reliable. To put it simply, GCP ensures that the research is conducted to high standards of ethical and scientific integrity. The ICH GCP E 6 document covers the following elements of clinical trials:

- Design
- Conduct
- Performance
- Monitoring
- Auditing
- Recording
- Analysis
- Reporting

Central to the principles of GCP is informed consent and this comprises 21 elements which are briefly outlined in Figure 3.2. Although these principles are specific to clinical trials involving investigational medicinal products, they underpin all research activity including investigations involving medical devices.

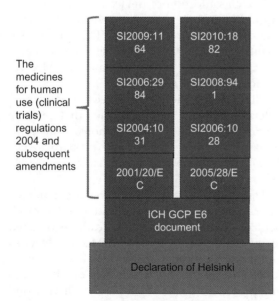

The medicines for human use (clinical trials) regulations 2004 and subsequent amendments

SI2009:1164	SI2010:1882
SI2006:2984	SI2008:941
SI2004:1031	SI2006:1028
2001/20/EC	2005/28/EC

ICH GCP E6 document

Declaration of Helsinki

FIGURE 3.1 The wall of clinical trials regulations.

1. Trial involves research
2. Purpose of the trial
3. Trial treatments; probability for random assignment
4. Trial procedures, including invasive procedures
5. Subject's responsibilities
6. Experimental aspects of the trial
7. Reasonable forseeable risks and inconveniences
8. Alternative (available) procedures and treatment(s)
9. Compensation
10. Anticipated prorated payment
11. Anticipated expenses
12. Subject's participation is voluntary throughout
13. Monitors, auditors, inspectors: access to notes
14. Confidentiality
15. Pledge to inform subject on new information
16. Contacts: information, trial-related injury
17. Forseeable circumstances of termination
18. Duration of subject's participation in trial
19. Approximate number of subjects in the trial
20. Permission to inform the subject's GP/family doctor; the use of tissues, organs, samples, and DNA during and after the trial; data protection 95/46/EC

FIGURE 3.2   The elements of informed consent.

CLINICAL INVESTIGATIONS FOR MEDICAL DEVICES

Although clinical trials for medicinal products (drugs) have been a requirement for many years, it was not until 2010 that the equivalent was required for medical devices. The requirement was introduced in the revision to the Medical Devices Directive 2007/47/EC and this had to be transposed into law in the member states by December 21, 2008; it finally came into force March 21, 2010. The changes can be briefly summarized as follows:

- Clinical data are required for all classes of medical device (i.e., Class I, IIa, IIb, or III) irrespective of whether they are already CE marked or not.
- All serious adverse events (SAEs) must be reported to the competent authority (CA; in the United Kingdom the CA is the MHRA, in the United States it is the FDA).

Classification of Medical Devices

The classification of devices was introduced in the Medical Devices Directive (93/42/EEC). The new Directive 2007/47/EC has subsequently amended this earlier directive and also incorporates the active implantable Medical Devices Directive (90/385/EEC).

The classification of a medical device is determined by a set of 18 rules that provide statements relating to situations, functions, parts of the body treated, and properties (EC, 2010). An overview of the classifications, risk, and rules is given in Table 3.2.

The rules are designed to allow the manufacturer to determine the classification of the product. For detailed guidance the reader is directed to the European Commission Medical Devices Guidance Document (MEDDEV 2.4/1 Rev. 9, June 2010; EC, 2010) which sets out the rules and provides a series of simple flowcharts to aid manufacturers in determining the classification

TABLE 3.2 Classification of Medical Devices

Classification	Risk	Rules	Application
I	Low	1–4	Noninvasive
IIa	Medium	5–8	Invasive, transient, or short term*
IIb	Medium	9–12	Additional rules for active devices
III	High	13–18	Miscellaneous rules

Short-term use >60 mins and <30 days.

of their device. Brief details are also given in the MHRA Bulletin No. 10, The Classification Rules (MHRA, 2011). Examples are:

Class I: Hospital beds/hoists, noninvasive electrodes (e.g., ECG or EEG), and plasters
Class IIa: Fixed denture prosthesis, reusable surgical instruments, and tracheal tubes
Class IIb: Radiological equipment, surgical diathermy, and anaesthetic machines
Class III: Bone cement, biological heart valves, and contraceptive diaphragms

Accessories are classified in their own right separate from the device.

While this classification holds true for countries in the European Union, the Unites States, whose competent authority is the Food and Drug Administration, has a different system for classification. The FDA has established classifications for approximately 1700 different generic types of devices and grouped them into 16 medical specialties referred to as panels. Each of these generic types of devices is assigned to one of three regulatory classes based on the level of control necessary to assure the safety and effectiveness of the device. The three device classes and the requirements (regulatory controls) that apply to them are:

1. Class I: General Controls
   a. With exemptions
   b. Without exemptions

2. Class II: General Controls and Special Controls
   a. With exemptions
   b. Without exemptions
3. Class III: General Controls and Premarket Approval.

COMPARING CLINICAL TRIALS AND CLINICAL INVESTIGATIONS

It is anticipated that readers of this book are more likely to be involved in clinical investigations of medical devices, however it is useful to have an understanding of the phases of clinical trials as underpinning knowledge. Table 3.1 compares the phases of clinical trials against the stages of a clinical investigation for medical devices.

The key difference is that there are only two phases in a medical device trial. The first phase is a premarket approval study, the purpose of which is to obtain CE marking. Once CE marking has been obtained, the next phase is an evaluation while in clinical use. This second phase tends to be either registry or audit-type studies to monitor long-term safety.

ISO 14155 STANDARD

The international standard for medical device trials is ISO 14155:2003; it is the equivalent and similar in many ways to ICH GCP, the standard for clinical trials of investigational medicinal products (CTIMP). ISO 14155:2003 is published in two parts:

1. Part 1 General Requirements: Defines procedures for the conduct and performance of clinical investigations of medical devices
2. Part 2 Clinical Investigation Plans: Provides the requirements for the preparation of a Clinical Investigation Plan (CIP) for the clinical investigation of medical devices

The aim of these standards is essentially threefold. Firstly, to ensure that human subjects are protected and understand the foreseeable risks and potential benefits (if any), and having understood this freely give informed consent to participate in the study. Furthermore to ensure that the trial is scientifically well designed and that its conduct will establish the performance of the medical device by providing clinical data that are both valid and reliable; that are reproducible. Finally, it also acts as a reference document for sponsors, monitors, investigators, ethics committees, and regulatory authorities.

For healthcare organizations the responsibilities of the sponsor and clinical investigator will be important as well as the monitor. It may be that some of the duties of the sponsor can be delegated to a clinical trials unit or a contract research organization (CRO) and this should only be undertaken when a contract or a written agreement clearly specifies the duties that have been delegated.

Sponsor Responsibilities

The key responsibilities of a sponsor are outlined in the following list, however readers are advised to refer to the ISO 14155 standard for definitive guidance. The standard lists 15 key responsibilities (ISO, 2003, p. 10−11):

1. Selection of an appropriate clinical investigator (CI) and investigation site
2. Appointment of an appropriate monitor to oversee its conduct
3. Prepare and keep current the clinical investigator's brochure
4. Provide the CI with the Clinical Investigation Plan (CIP) and subsequent amendments
5. Sign the CIP

6. Supply the medical devices as specified in the CIP
7. Ensure the CI is provided with the appropriate training to use the device in accordance with the CIP
8. Ensure all deviations from the CIP are reviewed and reported
9. Appropriate recording and review of all adverse events
10. Inform all principal investigators (PIs) about serious adverse events (SAEs) and all serious adverse device effects during clinical investigations
11. Inform the CI when the clinical investigation is prematurely terminated or suspended and the relevant bodies
12. Inform the CI of the developmental status of the device
13. Review and approve any deviation from the CIP taking appropriate actions as necessary
14. Collect, store, and keep secure all relevant documentation
15. Ensure accurate device accountability and traceability systems

Note that for a device trial the term *clinical investigator* is used, whereas for a drug trial *chief investigator* is the normal nomenclature.

Clinical Investigator Responsibilities

For those involved in clinical engineering design it is likely that they may at some time become a clinical investigator (CI) for that medical device; the following outlines the responsibilities entrusted to the role.

The clinical investigator must be appropriately qualified, experienced in the field of application, and familiar with the investigation methodology. Paramount is training in informed consent and this could be obtained through attending ICH GCP training. There are 21 elements of informed consent for clinical trials of investigational medicinal products and in general these are also relevant for device trials; these were shown in Figure 3.2.

The clinical investigator is responsible for the day-to-day conduct of the clinical investigation as well as for the safety and well-being of the human subjects involved in the clinical investigation (ISO, 2003, p. 12). Full details are given in the ISO standard ISO 14155:2003.

CLINICAL INVESTIGATION PLAN

The Clinical Investigation Plan (CIP) is the key document in device trials; it is effectively the equivalent of the protocol in a clinical trial. The CIP is defined as follows (ISO, 2003, p. 6):

> The CIP shall be a document developed by the sponsor and the clinical investigator(s). The CIP shall be designed in such a way as to optimise the scientific validity and reproducibility of the results of the study in accordance with current clinical knowledge and practice so as to fulfil the objectives of the investigation.

Key elements of the CIP include:

- General information to include a comprehensive list of all the CIs, PIs, coordinating CIs, and investigations centers/sites, name and address of the sponsor, monitoring arrangements, data and quality management, an overall synopsis of the clinical investigation, approval and agreement to the CIP
- Identification and description of the medical device to be investigated
- Preliminary investigations and justification of the study including literature review, preclinical testing, previous clinical experience, and device risk analysis and assessment (this process is described in EN ISO 14971:2007 which is explored in Chapter 6 "Risk Management")

- Objectives of the clinical investigation
- Design of the clinical investigation
- Statistical considerations
- Deviations from the CIP and how they are handled and recorded
- Amendments to the CIP (ISO, 2003, p. 6–11)
- Adverse events (AEs) and adverse device effects
- Early termination or suspension of the investigation
- Publication policy
- Case report forms (CRF) (the means by which data are captured during the investigation)

APPROVALS TO UNDERTAKE RESEARCH

There are generally two regulatory approvals needed prior to undertaking research; namely ethical approval and competent authority (CA) approval. Ethical approval in the United Kingdom is given through the National Research Ethics Service (NRES); other countries have similar bodies, sometimes called ethical review boards (ERBs). The proposed research is reviewed by a research ethics committee (REC) which comprises expert and lay members to give an opinion on whether the rights and well-being of the participants are suitably protected. There are various timelines within which a response from the ethics committee is required and this will vary from one country to another.

The competent authority is concerned with the science of the research and the safety of the patient. In the United Kingdom the competent authority is the Medicines and Healthcare Products Regulatory Agency (MHRA). Table 3.3 and Table 3.4 give examples of competent authorities.

TABLE 3.3 Examples of Competent Authorities for Drug Trials

Country	Competent Authorities for Drug Trials
France	Agence française de sécurité sanitaire des produits de santé (AFSSAPS)
Germany	Bundesinstitut für Arzneimittel und Medizinprodukte (BfArM)
Italy	Agenzia Italiana del Farmaco (AIFA)
Luxemburg	Division de la Pharmacie et des Médicaments
Norway	Statens Legemiddelverk
Sweden	Läkemedelsverket
United Kingdom	Medicines and Healthcare Products Regulatory Agency (MHRA)

TABLE 3.4 Examples of Competent Authorities for Medical Device Clinical Investigations

Country	Competent Authorities for Medical Device Clinical Investigations
France	Agence française de sécurité sanitaire des produits de santé (AFSSAPS)
Germany	Zentralstelle der Länder für Gesundheitsschutz bei Arzneimitteln und Medizinprodukten (ZLG)
Italy	—
Luxemburg	Ministère de la Santé
Norway	Sosial- og helsedirektoratet
Spain	Ministerio Sanidad y Consumo Agencia, Española de Medicamentos y Productos
Sweden	Läkemedlesverket
United Kingdom	Medicines and Healthcare Products Regulatory Agency (MHRA)

References

Brown, A.S., 2011. Clinical trials risk: a new risk assessment tool. Clin. Governance. 16 (2), 103–110.

Centers for Disease Control and Prevention (CDC), 2011. U.S. Public Health Service Syphilis Study at Tuskegee: The Tuskegee timeline. Available from: <http://www.cdc.gov/tuskegee/timeline.htm> (accessed 22.10.11).

European Commission (EC), 2010. Medical DEVICES: Guidance document – Classification of medical devices (MEDDEV 2.4/1 Rev. 9). Available from: <http://ec.europa.eu/health/medical-devices/files/meddev/2_4_1_rev_9_classification_en.pdf> (accessed 25.10.11).

European Parliament and the Council of the European Union, 2007. Directive 2007/47/EC. Official J. Eur. Union L247: pp. 21–55.

Foggo, D., 2009. Thalidomide 'was developed by the Nazis': The damaging drug may have been developed as an antidote to nerve gas. Sunday Times. February 8, 2009.

International Organization for Standardization (ISO), 2003. ISO 14155 Clinical investigation of medical devices for human subjects.

Medicines and Healthcare Products Regulatory Agency (MHRA), 2011. Competent Authority (U.K.) Bulletin No. 10 The Classification Rules. London: Medicines and Healthcare Products Regulatory Agency.

Nuremberg Military Tribunals, 1949. Trials of War Criminals before the Nuremberg Military Tribunals under Control Council Law No. 10 Washington D.C., U.S. Government Printing Office. vol 2: pp. 181–182.

Saunders, S., 2006. Post-mortem on the TGN1412 Disaster, Institute of Science and Technology. Available from: <http://www.i-sis.org.uk/PMOTTD.php> (accessed 10.01.10).

Further Reading

World Medical Association (WMA), 1996. Declaration of Helsinki: Recommendations Guiding Medical Doctors in Biomedical Research Involving Human Subjects. Available from: <http://www.jcto.co.uk/Documents/Training/Declaration_of_Helsinki_1996_version.pdf> (accessed 25.09.11).

Useful Websites

National sInstitute for Health Research, <*www.nihr.ac.uk*>.
National Research Ethics Service, <*www.nres.nhs.uk*>.
Medicines and Healthcare Products Regulatory Agency, <*www.mhra.gov.uk*>.

Glossary

Clinical investigation The systematic testing of medical devices following an ethical and scientifically approved Clinical Investigation Plan.

Clinical investigator The person responsible for the day-to-day conduct of the clinical investigation as well as for the safety and well-being of the human subjects involved in the clinical investigation.

Clinical trial The systematic testing of investigational medicinal products (drugs) following an ethically and scientifically approved protocol to determine the efficacy and safety of a new medicinal product.

Good Clinical Practice (GCP) A set of internationally recognized ethical and scientific quality guidelines which should be followed to ensure that the rights and well-being of participants are protected and that the data produced from the research are valid and reliable.

Medical device Apparatus or instrument used for the diagnosis, prevention, monitoring, and treatment of disease.

4

Health Technology Management

Justin P. McCarthy*, Richard Scott[†], Paul Blackett**, John Amoore^{††}, and Fran J. Hegarty***

*Clin Eng Consulting Ltd, Cardiff, U.K., [†]Sherwood Forest Hospitals NHS Foundation Trust, Nottinghamshire, U.K., **Lancashire Teaching Hospitals NHS Foundation Trust, Lancashire, U.K., ^{††}Department of Medical Physics, NHS Ayrshire and Arran, Scotland, U.K., ***Medical Physics & Bioengineering Department, St James's Hospital, Dublin, Ireland

INTRODUCTION

The strategic plans of a healthcare organization determine its current and future operations. Strategic plans are subject to change, evolving both incrementally to meet challenges and opportunities, and in a more profound way to take account of fundamental changes of direction or spheres of activity. Healthcare technologies available will also affect strategic developments. The evolution and development of healthcare technologies trigger developments in clinical care. Over the past decades technologies have transformed healthcare, making what was once impossible now routine. As the delivery of healthcare has become progressively more dependent on the use of sophisticated healthcare technology there has emerged a corresponding requirement for the strategic management of this technology so its use can support the organization's strategic objectives. Today the management of healthcare technology within the health institution must be informed by, and remain responsive to, the changing strategic plans of that organization and external regulatory frameworks.

The term *healthcare technology management* (HTM) describes the role that embraces, but is not limited to, a focus on scientific and technical support of electromedical devices and clinical information technologies, including their financial stewardship (AAMI, 2011). This includes management of healthcare technologies and computer systems that are highly integrated and interoperable. The healthcare organization should develop a holistic healthcare technology management system that can support the deployment of these assets. Such a system will have processes for supporting devices from the strategic acquisition and deployment of technology down to the day-to-day support requirements (MHRA, 2006).

An HTM system is complex and to implement fully requires a number of different processes to run concurrently and in an inter-connected way. In the model described here the strategic management is realised by having a corporate-level HTM policy that brings into being a multidisciplinary group called the *medical device committee* (MDC). This committee develops the organization's response and implements it at the corporate level by writing and reviewing an annual HTM plan (AHTM plan).

The tactical management of medical devices is developed by specific departments within the organization who are mandated by the MDC to realise the AHTM plan. The clinical engineering department would be one such department which would usually have an organizational-wide responsibility. The roles and responsibilities of any group tasked with managing devices will be set out within the AHTM plan; however each department must develop a specific program that takes into account the particular devices in its charge and also the context within which they are used. Each department charged with realising the AHTM plan will therefore develop an HTM program for the devices they look after. This program in turn will consist of specific day-to-day operational support plans for groups of devices. These equipment support plans are likely to be numerous and concurrent. The strategic objects of the organization contribute to and are informed by the specific equipment support plans by ensuring that the three layers of the system are interconnected and responsive to each other. Each layer can be run within a quality improvement cycle with regular reporting built into the system between each layer.

THE STRATEGIC HEALTHCARE TECHNOLOGY MANAGEMENT SYSTEM

Managing healthcare technology and the risks involved should be governed by a strategic, organization-wide policy that will inform and guide those who use and manage the technology. Many stakeholders can contribute to such a policy—board members, clinicians, general managers, finance managers, clinical users of healthcare technology—but it is only clinical engineers that have this area of work as a central feature of their role profile. The systems engineering skills of clinical engineers can contribute significantly to the development and review of the hospital-wide HTM system and policy. (The term *hospital* is used to mean a corporate healthcare delivery organization rather than a particular set of buildings.)

The HTM policy will outline the processes by which responsibility for the ongoing management of particular devices or technologies will be assigned to departments within the organization. In many health institutions, a medical device committee (MDC) is established as the means of realising the HTM policy. The MDC should analyse the deployment of devices and systems and review their associated risks and benefits, to ensure their implementation supports the organization's clinical, corporate, and financial goals.

One way of doing this is to develop and regularly review a plan for how the policy will be implemented. Such a plan might be reviewed and altered annually in response to changes in the organization or the governance within which it operates. Such a plan will be referred to as the *annual healthcare technology management plan* (AHTM plan). The MDC committee should ensure that the plan is applied throughout the health institution.

In the AHTM plan the MDC should assign authority and responsibility for the ongoing equipment management of medical devices and systems to the appropriate departments within the organization. The MDC acts in a coordinating capacity in this regard ensuring all devices and systems are assigned to appropriate departments. The MDC does not develop the specific support solutions; that is done within each department. However, each department will report back to the MDC on the effectiveness of the solutions it develops and delivers. In this way the MDC can assure that appropriate solutions are being applied across the organization and the board has visibility of this.

The AHTM plan should also identify the requirements for new acquisitions of devices either to support service development or as part of a planned replacement program. Where acquisitions are sanctioned, the MDC should initiate an acquisition project to plan and manage the procurement and commissioning of the new devices.

The ongoing support of devices is realised by the departments charged with developing specific programs of scientific and engineering activity, which are in turn tailored to support specific groups of devices in particular clinical environments.

Implementing the policy through developing and delivering an AHTM plan is the process that will maximise the benefit to the patient and to the organization, and reduce and control risk. While the policy and its implementation involves many stakeholders it is only clinical engineers that contribute to all elements of the policy. Clinical engineers contribute significantly to the development and review of the hospital-wide strategic policy. The clinical engineering department's links with senior management of the organization and the lead professionals for the different clinical specialties will support the necessary dialogue to ensure effective equipment prioritisation decisions, whether these involve the transfer of existing equipment or procurement of additional or replacement equipment. The involvement of the clinical engineering department and individual clinical engineers is collaborative and interdisciplinary. It is the role of the clinical engineer to apply engineering principles and practice, incorporating their practical experience of supporting healthcare technologies to create a framework in which

the risks associated with the acquisition and use of medical devices are minimised. Clinical engineers have a commanding expertise in the development and delivery of healthcare technology management programs. So, it is the clinical engineer who is uniquely placed to develop the overarching management systems. In short, clinical engineering led healthcare technology management services, working in partnership with managers and clinicians, are essential in delivering robust patient focused healthcare and corporate success.

Clinical engineers will often be the drivers of the healthcare technology management programs within the hospital and develop processes by which the existing medical device infrastructure is assessed annually to ensure it is efficient and effective. This is a good example of how the advancing care and equipment management roles are complementary and integrated. To be able to advise effectively, clinical engineers should know both the current state of the existing technology deployed in the hospital and be able to provide an overview of new developments, and then be able to articulate how implementing a new technology will affect the hospital's ability to meet its corporate goals.

IMPLEMENTING THE HEALTHCARE TECHNOLOGY MANAGEMENT PROGRAM

When it comes to implementing the policy through realisation of the AHTM plan and developing and managing the healthcare technology management programs, the MDC will assign this to one of the departments with direct responsibility for equipment management. Usually this will be the clinical engineering department (CED). The CED must then develop its HTM program which sets out the day-to-day workings of how clinical engineers will manage the medical devices, ensuring that devices are maintained in a satisfactory condition and kept available for use.

Equipment management can be achieved for any particular equipment or system by considering its management over its operational life a beginning, a middle, and an end phase. Publicly Available Specification PAS 55, developed by the Institute of Asset Management and published by the British Standards Institution (BSI), provides a comprehensive guide to these activities (IAM, 2008). It is the basis for the development of an ISO standard in this field. In this section we will look at the middle phase of the life-cycle model and the ongoing processes necessary to support the device over its working life. The HTM program can be considered a cyclical process within the AHTM plan. It is used to develop and regularly review the operational support that individual departments put in place to support their healthcare technology.

The challenge for any clinical engineering department is how to best deploy resources to optimally manage the diverse and complex range of devices for which it has responsibility. Usually this has to be achieved with finite resources, which rarely allow for an ideal solution to be put in place. Consequently some form of risk evaluation is necessary to identify items that carry a higher requirement for management to control the corporate, clinical, and financial risk. These risks are independent of each other and each needs careful consideration during the planning stage.

If one imagines the HTM program as a cyclical process, then the starting point of that cycle is the equipment entering clinical service once the previously discussed acquisition project is completed. Once commissioned, there is a need for a formal plan to be put in place to support its use. This is referred to as the *equipment support plan* (ESP). When a device is first put into use, the equipment support plan for the first year will be established. However, over time the support requirements may

change and so at regular intervals the equipment support plan may need to be altered. For the purposes of illustration in this chapter we will assume the equipment support plan will be reviewed annually. So the HTM program as a whole is an ongoing process that consists of a series of equipment support plans that are reviewed and monitored within a quality management system.

At any one time, a clinical engineering department will be implementing an HTM program that supports all the devices it has responsibility for. This will include many different devices, each in a different point in its own life cycle and each used within a particular clinical context. The HTM program will therefore consist of a number of equipment support plans each tailored for specific groups of devices and all running concurrently. It is through the development and delivery of the device-specific equipment support plans that the HTM programs can focus on the particular requirement of the vast range of devices supported, while also ensuring the HTM program as a whole meets corporate, clinical, and financial objectives.

The support processes provided by the clinical engineering department that make up a healthcare technology program are detailed in Figure 4.1. Once the MDC defines the overall AHTM plan, and clearly defines the responsibility, authority, and resources, the clinical engineering department implements the processes detailed within the "Implement HTM Programme" box.

Defining the Aims, Objectives, and Scope of the Program

Most healthcare establishments will have overarching corporate objectives, goals, and targets for short, medium, and long-term periods. Usually, these become more detailed and expanded as these objectives travel down

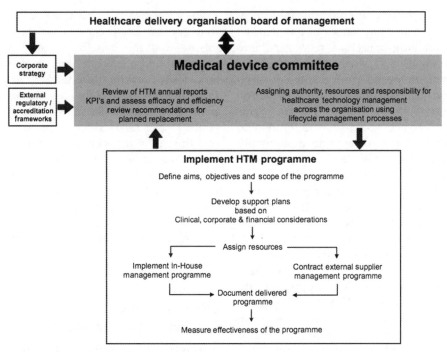

FIGURE 4.1 Implementing a healthcare technology management program.

through the management hierarchy until they arrive at the CED door hopefully communicating the local objectives and explaining how they fit into the overall "big picture." Review of these corporate goals should inform the development of the program. By setting out the aims, objectives, and scope of the program, the CED can incorporate the corporate overview into its own plan.

Healthcare establishments, whether privately or publicly owned, are usually subject to standards set by various bodies. Clinical associations and groups, together with professional institutes and numerous government agencies and accreditation bodies, dictate minimum levels of care, safety, and service to be provided by the establishment. The support provided by the CED will be influenced by these standards and the CED is strategically placed to provide evidence to support the establishment's claim that it is complying with

the standards. Equally the CED has an obligation to bring to the attention of any other departments who deliver healthcare technology management programs, and the various committees and groups within the establishment, any shortcomings and failures that need to be addressed to ensure future compliance with standards.

As with any product design or service delivery, understanding the customer's requirement is an essential ingredient. The delivery of healthcare technology management is an interdisciplinary endeavor: responsibilities will be shared between the clinical engineering department and those involved with device procurement and use, with all stakeholders having an active part to play. The most successful clinical engineering services are to be found where the service users have had a say in the initial service design. In this way, a real sense of shared purpose is achieved. As with any engineering endeavor, it is best practice to build on what has worked successfully, taking objective responses into account, refining, and evolving services for customer benefit. Be adaptable and prepared to change—if that cherished idea has evidently not worked then start again! This plan–do–check–act model will be familiar to those who have worked with formal quality management systems (ISO, 2005).

Defining the aims, objectives, and scope for the healthcare technology management program provides a clear overview at the outset of what support the CED is expected to provide. Although this can, in some cases, take the form of legal agreements between organizations, it is usually less formal, being discussed, documented, and agreed internally. This support specification for the CED can, and should, be reviewed annually considering any changes in the strategic direction of the establishment or requests from individual clinical services. A typical support specification would include:

- An acceptance service for new devices
- Response times to breakdown requests
- Planned maintenance completion time scales and targets
- Repair time targets
- Equipment availability targets
- Provision of medical device management advice
- Safe disposal of devices
- Maintenance and access to a medical equipment inventory
- Delivery of device-specific training
- Addressing medical device alerts and investigating incidents
- User Support

As we start to focus on discussions relating to equipment management it is important to remember that a key role for the clinical engineering department is to identify, evaluate, and appropriately address risks. In some cases it may be possible to eliminate a risk, however, there will always be risks associated with the delivery of healthcare. The challenge is to ensure that the benefits of any treatment or intervention outweigh any potential harm. So careful consideration needs to be given as to how devices covered by an equipment support plan are used clinically. The CED can contribute to risk reduction not only by implementing maintenance programs but also by protecting time to provide meaningful user support and training. Where risks are identified through participation in maintenance activity, the CED should act to control those risks or, if that is beyond the scope of the equipment support plan, escalate the risk up to the MDC for consideration at the corporate level.

Developing the Device-Specific Equipment Support Plans

Having defined the goal of the HTM program, the CED must now organise itself to deliver the program. Given the diversity of devices and clinical environments supported it is not surprising that a "one size fits all" approach will not work. The CED will most likely develop a number of equipment support plans for different devices or clinical areas, which define in more detail how the service support will be delivered for each.

In doing so, the CED should carefully consider and balance how to meet the clinical, corporate, and financial requirements of the organisation. Ideally the plan will ensure the efficacy of the equipment and allow the clinical engineers to play a complete role in supporting the clinicians. However, financial constraints rarely allow the ideal to be delivered. It is unlikely that the clinical engineering department will become expert in the support of all medical devices. The monetary or staffing resources needed to train and maintain competence on a one-off particular model of equipment can become unreasonable and the

CED may choose to outsource the support to the manufacturer or an external service supplier. The CED should research how to best support all devices through a mix of in-house and external service support.

The ongoing technical maintenance of healthcare technologies has three key components. These are scheduled maintenance, performance verification, and unscheduled maintenance. Scheduled maintenance consists of all proactive activities whose purpose is to minimise the likelihood of failure of the device in service. Performance verification includes all proactive processes that assure devices that appear to be working are working optimally. Unscheduled maintenance covers all reactive actions that are initiated as a result of a reported real or suspected fault or failure of a device or system. Clinical engineers use risk management methodologies to optimise the hospital resources to deliver a healthcare technology management program that is focused on ensuring that the clinical work of the hospital can be delivered optimally. These technology maintenance functions can be delivered by in-house teams of clinical engineers, by a contracted third-party maintenance organisation or can be contracted to the service engineers of the equipment manufacturers, or a combination of all three. One of the key roles of the lead professional clinical engineer is to determine the optimum mix of services to meet the organisation's need. Management of the devices takes place within a business model and so clinical engineers work closely with the finance and procurement departments to ensure that not only is the technical equipment management effective but the support arrangements are also an efficient use of financial resources. Consequently, the clinical engineering role includes the development of methodologies for the support of healthcare equipment, management and regular review of these methodologies, and the management of both fiscal resources and personnel.

Clinical engineers understand the devices and the clinical context in which they are used, and in developing the individual support they often imaginatively foresee problems that could arise and develop the support strategies to mitigate them.

Assigning the Resources

It is axiomatic that funding for the CED is necessary to meet its agreed service specification targets. Staff must be recruited; tools, test equipment, and spare parts be bought; and service contracts placed. The hospital should allocate a staff and revenue budget that allows the CED to deliver the HTM program. Departments that are fully funded to maintain their establishment's medical devices to manufacturers' specifications and frequency are rare! Pressures of finance, labour, expertise, and materials brought to bear on the CED meet head-on the requirement to maintain equipment in a safe and operational condition to meet the needs of the clinician and patient. Where allocated resources are insufficient to deliver a comprehensive solution, the head of CED must first optimise the use of available resources and must then communicate the residual risk to the CEO and board. Usually, labor and parts are associated with work undertaken by in-house clinical engineers and their support staff, while external work can comprise equipment on service contracts and those items of equipment sent away for repair. The CED needs to implement some form of financial control so that an accurate position of finances can be maintained. Working within a larger organisation it is likely that other departments such as procurement and finance will have introduced ways to do just this, but smaller or independent departments will need to start from scratch.

A useful and important tool to assist in the classification of medical devices for risk assessment is to place the devices into a risk category. ECRI Institute[1] has long suggested devices are placed into low, medium, and high-risk categories and usually a record of the category is held on local clinical engineering databases. The categorisation can be used to inform decisions regarding maintenance strategy (ECRI Institute, 2007). The choice of risk category however is up to the clinical engineering department in consultation with clinical users but general definitions are given. For example, high-risk devices would include ventilators and defibrillators—devices whose failure would reasonably be expected to cause immediate patient harm. Medium-risk devices would include ECG recorders and BP monitors which could possibly cause delay in treatment if failed, and low-risk devices are those that are unlikely to lead to any serious complications. The risk category itself does not dictate what maintenance a device requires but it can be a useful component of a risk assessment. Also it is important to realise some items of the same type might be put in different risk categories because of different clinical area or clinical use to which they are put.

Implementing an In-House Maintenance Program

Once defined in the equipment support plan and the resources assigned, teams are set into action. The scheduling of work is an important task for the team leader. He or she has to develop and implement a program that includes routine scheduled maintenance and performance verification processes, while also having sufficient resources to respond to breakdowns or unscheduled maintenance events. Also remember that ongoing clinical user support and training might be as

[1]www.ecri.org/Pages/default.aspx or www.ecri.org.uk/index.html

important to clinical effectiveness as routine performance verification.

Typically, equipment requires some form of inspection on an annual basis with some devices requiring this more frequently. This performance verification activity is intended to assure that the device is performing to specification and is safe. It is considered good practice for a device to be seen at least once per year, and manufacturers usually provide a checklist to detail what should be checked and how often this should be done. CEDs use manufacturers' guidelines but are also as influenced by local experience gained in supporting devices. It should also be noted that some countries have particular requirements for regular electrical testing of "portable appliances."

Additionally, some devices may require scheduled maintenance, that is, routine replacement of parts such as filters to prevent failures. While these may be replaced at a set frequency, sometimes parts require changing after a number of operating hours, which introduces an additional complexity. Others yet may require parts such as batteries to be replaced at a differing frequency to an inspection. Any system introduced by the CED needs to carefully manage the peculiarities of the medical devices within its care.

For particular groups of devices it makes sense for the performance verification and scheduled maintenance activities to be aligned and undertaken at the same time. For example, the team that verifies the performance of all the defibrillators in the hospital will usually manage the planned replacement of the defibrillators' batteries and change the battery in the clinical area during one of its scheduled visits. The planning and monitoring of these proactive maintenance actions is best carried out in conjunction with the medical equipment database. This database will usually automatically generate reminders for scheduled work, sometimes a set period before it being due.

This gives the clinical engineer time to contact the user and arrange a time for the maintenance to be carried out. Likewise, should the device not be available, or not found, a procedure must be in place to advise the user that it has not been maintained.

Corrective maintenance, or repair, is the process of restoring a medical device to a safe, functional, normal condition (see the definition in the draft second edition of IEC 62353; IEC, 2007). Corrective maintenance is unpredictable and the department has to put a system into place to handle difficult situations with varying demands on a whole range of devices where specialist knowledge may not be available and conflicting requests made. Clinical engineers with appropriate training and experience are expected to diagnose and rectify these problems, with reference to technical documentation and the manufacturer if required. The process that gets the device back into use must be done quickly as it is possible that diagnosis or therapy to patients may be delayed due to the device being out of service. All corrective actions will also be documented in the medical equipment database so that it contains a full service history for each device.

Contracting Out External Maintenance Programs

The contracting of maintenance to an external supplier requires careful consideration. The transfer of work to an outside contractor does not remove liability from the healthcare organization nor responsibility from the CED. Due diligence needs to be exercised that the contractors are suitable. This might entail requesting training records and evidence of knowledge, experience, and skills, examining insurance cover, and considering business continuity issues. These are particularly important when the contractor being considered is not an authorised service agent of the manufacturer, sometimes called a "third-party contractor." Even if a contractor is approved by the manufacturer it is still appropriate to request and review its responses.

With the choice of contractor having been made there is a decision on the level of cover required. Typically this ranges from fully comprehensive cover, which includes all the required inspections and any repairs necessary, to basic cover, which only includes the recommended service. Many variations exist in-between, parts excluded, onsite or offsite servicing, accidental damage included, parts included to a monetary value, only one repair visit included, and so on. It is usually the CED's decision as to which level of cover is best suited to the organization. If the equipment is new it will probably be reliable so the decision may be taken to only cover the equipment with a basic contract; if it is unreliable it may be financially attractive to have fully comprehensive cover.

One particular approach is to have a contract with the external contractor, yet have the in-house clinical engineering department perform "first-line fault finding." These or similar contracts are often called partnership contracts. With such a contract when a fault is reported the CED takes the initiative and looks to see what the fault is and liaises with the contractor. If repairs can be undertaken by the CED then this may done, or at the least, the contractor knows the general condition of the equipment before they attend onsite. This solution allows the CED to retain involvement with the equipment while having technical backup easily at hand.

Whichever method of external contract is made, records need to be kept of the contract and period of cover, along with service reports. The scheduled and unscheduled maintenance delivery by the external service supplier should also be documented within the medical equipment database.

Documenting the Delivered Program, the Role of the Medical Equipment Management Database

An essential component of any healthcare technology management program is the existence of a register of all medical devices being managed by the CED. The department cannot manage what it does not know about, therefore a requirement to establish a medical device inventory is immediately identified. It is important to consider what devices should be included in the database. In many healthcare organizations there are several technical departments that manage specific types of devices. Renal dialysis is a typical example but there are others such as radiotherapy engineering, rehabilitation engineering, and pathology which may have maintenance teams dedicated to their specific equipment. While the CED will probably have the greatest quantity of equipment, a decision will need to be taken whether to include the devices from these other departments or not. The significant advantage of doing so is that the organization has a single point of reference for its medical devices, but perhaps at the detriment of suitability and data quality. It is unlikely that one database design will suit everyone's technical purposes.

The management of the HTM program relies heavily on the upkeep of the medical equipment database. Over the years CEDs have approached this from different directions. Some departments have designed their own database, tailoring it to their own needs, while others have opted to buy a ready-made database. There are advantages and disadvantages in both cases but the major consideration is one of support. In the case where the database has been developed locally, support is usually provided by one person with an interest in information technology. If they leave, retire, or become ill, there is likely to be no contingency plan to handle this. On the other hand, a commercially available database will be offered with service support albeit at a cost. Therefore, it is quite justifiable to have such a crucial part of the department supported professionally.

The medical equipment database is the repository for the medical equipment asset register and the service histories of these assets. The database may also support scheduling of routine work across the different teams. Reports generated from mining the data are useful in the day-to-day management of the delivery of the plans and in measuring the efficiency and effectiveness of the program as a whole. Data and information from this source will also be part of the significant contribution from the CED to the organization's annual HTM plan.

As the medical device management system (MDMS) equipment database touches on all aspects of the CED it is likely to be the repository and hub of all stored and accumulated data. It is therefore vital that efforts are made to maintain the accuracy of the data. The established database needs regular housekeeping which will highlight areas that have become outdated and records that have not been completed correctly. Although computer systems have many ways of validating inputs it is surprising how small errors can appear, and of course no amount of validation can keep clinical departments from changing their names or moving equipment around! Regular refresher training can remind clinical engineers of the correct ways of using the database and can avert any buildup of poor practice.

Financial Control

All the actions identified as part of the equipment support plan need to be resourced, set into action, and controlled. Controlling costs is an important part of the program and so some means of measuring resource utilisation needs to be established.

The optimum mix of in-house or externally contracted maintenance services needs to be decided and justified. Often the decision to manage devices in-house is made where the CED has the capacity and capability to do so. However, it is important to remember that the HTM program must also support the hospital's financial goals. Regardless of the service support mix chosen there is a need to control the costs associated with it. Where the service solution is fully outsourced the costs associated with external contractors will be readily available as service contracts will be procured and their cost clearly identified. However, in such situations it is not uncommon for the CED to also provide some degree of front-line support. As a minimum this usually includes managing communications between the clinical staff and the service company, but may extend to having a quick first look to rule out user error or difficulty with operation situations. Where service is delivered in-house it can be difficult to assess all the costs associated with its provision. It should include an estimate of staff costs, noncontract external service, spare parts, and cost of staff training as a minimum. However, other overhead costs such as workshop space, tools, test equipment, and energy costs can be less visible in the hospital context. Regardless, the CED should make all reasonable attempts to construct a complete financial analysis of its function and include financial analysis in the decision making when developing the HTM program.

Clinical engineering departments may analyse costs in different ways. A top-down approach might be to divide the overall department budget including overheads such as lighting, heating, and IT on a pro-rata basis to each device supported. A bottom-up approach might be to track resource utilisation as part of recording each maintenance action within the MDMS database and then aggregate

the costs up. The actual method developed should be implemented in conjunction with the hospital's finance department to ensure compliance with the corporate financial policy.

Being able to estimate the total cost of each equipment support plan enables analysis of how cost effective each plan is. An estimate of the cost of the plan can be achieved by looking at the internal and external costs associated with each element. Figure 4.2 shows a simple template that can be used to get an estimate of the support costs associated with an equipment support plan.

Evaluating Effectiveness

As well as these financial controls, a set of key performance indicators (KPIs) developed for the work undertaken by the CED is a valuable and clear way in which the general performance of the CED can be monitored. The management of the organization and the members of the department will have a keen interest in how well the CED is working and it is usually accepted that KPIs are reported on a monthly basis.

In developing KPIs it is useful to adopt a balanced scorecard approach, where several aspects of the department are taken into account rather than presenting an overwhelming list of technical data. There are no fixed rules as to what should be included but reference to the service specification agreed on would be a good start in developing these. Perhaps KPIs that have been grouped into four categories—service user, internal management, continuous improvement, and financial—would be appropriate. Such a set of KPIs prove a useful tool when looking at trends over a long period of time. Changes in work practice and external influences on the department can be quantified.

Another way of assessing the effectiveness of the service is to ask the user of that service

Maintenance action	CED Internal Costs		CED External Costs	
	Staff cost	Spares	Contract	Noncontract
End-user training				
Unscheduled maintenance				
Scheduled maintenance				
Performance verification				
Subtotals				
Cost of internal support				
Cost of external support				
Total cost				

FIGURE 4.2   Table used to estimate the total cost of an equipment support plan.

what they think of it. Many ways of obtaining feedback from users of services exist; some are more suitable than others to the healthcare environment. For example, it is unlikely that focus groups would be successful as these would take clinical staff away from patient care and therefore would be costly. Sometimes the simpler solutions are the best and a straightforward customer survey will yield good results.

Closing the Circle

The most effective improvements to service delivery are likely to come from those committed to delivering it. Clinical engineers committed to excellence will strive to improve the service. The annual review provides a structured way of doing this. The effectiveness of the established program will be assessed by the clinical engineering department. This should be done objectively, based on evidence and available data. Assessment of how the department is performing against its targets can be achieved through analysis of the KPIs. Qualitative data from customer feedback surveys and a review of the complaints received also inform the review. Finally, critical thinking from all members of the clinical engineering department can help identify ways in which the service can be improved.

As part of the review process the clinical engineering department will write a quality improvement plan. A quality improvement plan (QIP) is an annual, detailed plan that describes the improvement actions identified during the annual review. The usual form of a QIP is a table with actions to address a particular improvement listed alongside other key information.

No.	Objective	Action	Due date	Person responsible	Review	Completed date
1	Reduce response time for breakdown requests to within 24 working hours.	Introduce named engineer Rota to deal with incoming requests.	October 2nd	John Smith	Weekly until complete	

FIGURE 4.3 Quality improvement plan template.

A simple QIP format is shown in Figure 4.3.

Additional columns could be included to list resources required or challenges to be overcome. The QIP, however, should be understandable by all staff as it is a team plan and everyone needs to play their part in the improvement of quality. A useful acronym to aid creation of QIP actions is "SMART" (Meyer, 2003).

Actions should be:

- Specific: Not vague or easily misunderstood
- Measurable, in some form: How do you know the action has been achieved?
- Achievable: Must have the agreement of all parties
- Realistic: Need to be attainable within availability of resources
- Timely: Should be deliverable within some agreed time scale

Where items on the QIP are deliverable within the capability of the department, they will be progressed. Usually these will take the form of changes to specific equipment support plans. Through writing and implementing the QIP the clinical engineering department closes the quality circle and improves the program.

As part of the annual review, the clinical engineering department may identify improvements that are beyond its capability. We have already seen that review of the HTM program may identify changes to the AHTM plan so as to better control a risk. Perhaps the clinical engineering department may suggest that the MDC consider managing a group of devices using an equipment library approach rather that devices being owned by an individual department. Through critical review of the HTM program, clinical engineers, informed by working with end users close to the point of care, can often identify actions that improve practice and control risk. We have also seen that the annual review of the HTM program can identify equipment for planned replacement. Since many of the risk issues and planned replacement projects are outside the scope of the HTM program, to close the quality circle the clinical engineering department should report and escalate them up to the MDC as part of the annual report.

SUMMARY OF THE HEALTHCARE TECHNOLOGY MANAGEMENT PROGRAM

In summary, we can define the healthcare technology management program as the concurrent implementation of equipment support plans for different groups of devices. These plans are developed, delivered, and reviewed as part of a quality cycle. The plans may indicate that components of each are delivered by in-house teams or are contracted to third-party or supplier maintenance organizations. Regardless of which, the documentation of the plans and the work undertaken is recorded in a database used to provide statistics for the annual review, and, as required during the year, to provide KPIs that guide the management of the plan's implementation.

The operational support process outlined takes up a significant amount of the working hours within a clinical engineering department. Much of it is routine and involves equipment maintenance. However, the additional engineering and technical support that a clinical engineering department can provide to a hospital, together with the high-level strategic input described earlier, make up effective HTM and contribute significantly to the safety of patients, to clinical outcomes, and to good governance within the hospital.

It is worth highlighting that all the processes within Figure 4.4 are developed, delivered, and managed by the clinical engineering department. Consequently, clinical engineers need to develop their management knowledge and skills. Looking at Figure 4.4 it is clear why the American College of Clinical Engineering definition of a clinical engineer places equal weight on the application of management and engineering skills when defining the role of the clinical engineer (ACCE, 1992). It is also worth stressing that since the equipment management practice is operating within a quality cycle and the review process responds to application issues from the clinicians and to changes in the hospital strategy, the practice of healthcare technology management is also supporting and advancing patient care.

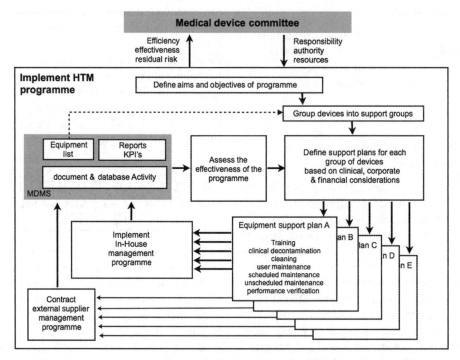

FIGURE 4.4 Healthcare technology management (HTM) program as a quality management cycle.

References

American College of Clinical Engineering (ACCE), 1992. Clinical Engineer (defined), [Online]. <www.accenet.org> (accessed 09.05.13).

Association for the Advancement of Medical Instrumentation (AAMI), Forum Participants Move Forward with Name, Vision for Field, 28.09.2011. [Online]. Available: <www.aami.org/news/2011/092811.press.future.forum.html> (accessed 08.05.13).

ECRI Institute, 2007. ECRI-AIMS Overview, [Online]. Available: <www.ecri.org.uk/documents/ECRI-AIMS_Overview.pdf> (accessed 09.05.13).

International Electrotechnical Commission (IEC), 2007. IEC 62353: Medical Electrical Equipment - Recurrent Test and Test After Repair of Medical Electrical Equipment. IECIEC), 2007, Geneva.

International Organization for Standardization (ISO), 2005. ISO 9000:2005 Quality Management Systems — Fundamentals and Vocabulary. ISOISO), 2005, Geneva.

Institute of Asset Management (IAM), 2008. PAS 55-1:2008 Asset management. Specification for the Optimised Management of Physical Assets. BSIIAM), 2008, London.

Medicines and Healthcare Products Regulatory Agency (MHRA), 2006. DB2006(05) — Managing Medical Devices, [Online]. Available: <www.mhra.gov.uk/Publications/Safetyguidance/DeviceBulletins/CON2025142> (accessed 09.05.13).

Meyer, P.J., 2003. What would you do if you knew you couldn't fail? Creating S.M.A.R.T. goals. In: Attitude Is Everything; ISBN 978-0-89811-304-4, The Meyer Resource Group Incorporated.

Glossary

Annual healthcare technology management plan (AHTM plan) An annual plan developed by the MDC which outlines how the HTM policy will be implemented at corporate level in the current year.

Equipment support plan A document that sets out the range and scope of maintenance and support action to be delivered by the clinical engineering department for a specific device or group of devices. Also includes details of the information sources on which the plan is based and the means by which the plan's effectiveness can be assessed.

Healthcare technology management policy (HTM policy) A hospital-wide policy that identifies the corporate requirements for the deployment and ongoing support of healthcare technology and sets out a course of action by which the corporate objectives are met. The HTM policy should be issued by the hospital board and reviewed at corporate level as part of the regular management of the governance structures within the organization.

Healthcare technology management program (HTM program) A process designed to deliver appropriate scientific, engineering, and technical support to ensure healthcare technology remains safe and effective in clinical practice.

Medical device committee (MDC) A multidisciplinary committee established at corporate level within an organization charged with responsibility for the development, regular review, and implementation of the organization's AHTM plan.

5

Leadership

Merlin Walberg

Phoenix Consultancy USA, Inc.

INTRODUCTION

One afternoon, I was sitting in the lobby of a very luxurious hotel in North Carolina, waiting for a client meeting to begin. A member of staff walked through the lobby, noticed a small piece of paper on the carpet, bent down, picked it up, and carried on his way.

This tiny behavior made a big impression on me. It demonstrated true, DNA-centered leadership! We are certain that nowhere in his job description did it say, "responsible to always look for bits of paper on the lobby floor."

In essence, this was a true (if small) demonstration of leadership: seeing a need, deciding that you are responsible, determining the outcome wanted, taking initiative, changing something, modeling the behavior you wish to see in others. *Walking the talk*.

This chapter is to introduce you to *leadership* as it applies to you, a *scientist*. Our goal, outcome wanted, #EndInMind is for you to make the connection that it is not only your skills as a scientist that determine your effectiveness and ability to make a difference, it is those skills, that mindset and inner rigor, combined with DNA-centered leadership that will enable you to achieve your dreams. We are talking about self-leadership (as demonstrated in the little vignette above) and leadership of others—bringing people together around a common goal and achieving it.

By way of further introduction, we invite you to read everything here with two different pairs of spectacles: professional spectacles and personal spectacles. What we describe in this chapter applies to being successful in work and in life overall. Amazing and true; see if you agree. By trying out what is proposed here you will be able to tell if it is true and works, just like with your scientific training. Learn, try out, practice, reflect on what worked and what did not, revise your practice, try some more, and get better and better at making things work.

There is a fundamental difference between learning leadership and everything else you learned to be a scientist. As a scientist you start with a big picture and look at possibilities, then narrow down those possibilities with the goal of finding the one right answer. It is very exact and specific. As a leader, you look at a very specific situation and back up from that issue to broaden your picture, find out who else is or needs to be involved, explore possible ways forward, and then make a plan and enact it. There are always different approaches possible and never one right answer, which is what makes it tricky.

So what is leadership? Hundreds of books are written each year in an attempt to answer that question.

EXERCISE

Start by thinking for a moment about a great leader that you have known in your life. A person who, according to you, demonstrated leadership as part of his or her way of being, had the skills, qualities, abilities, and attributes that caused other people to follow voluntarily. This last point is important, as we really want to be talking about positive forms of leadership, not the kind that provides heavy penalties for *not* following (though that can be very effective in the short term).

What did he or she do? How did they behave? What were (are) they like? How did (do) they make you feel? Scribble down a list.

The list probably includes the following:

- Has vision
- Communicates clearly and often
- Listens
- Is strategic
- Is approachable
- Values others
- Has a certain confidence, gravitas
- Is knowledgeable
- Sees the big picture
- Good at delegating, negotiating, influencing, and making decisions
- Can deal with conflict and differences
- Engenders trust and respect
- Is consistent and fair
- Vales and empowers others
- Is flexible
- Makes me feel valued

The list indicates who a leader is as well as what a leader does. We depict these two aspects as *inner* and *outer* (Figure 5.1).

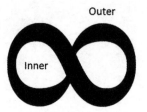

FIGURE 5.1   Inner and outer loops.

As a scientist, you will be very familiar with the symbol shown in Figure 5.1. In terms of leadership however, the symbol demonstrates the ever-present relationship between what is going on *inside you*, all invisible and hugely powerful in your experience of the world:

- Your thoughts
- Your health
- Beliefs about yourself and others
- Emotions
- Interpretation of the behavior of others and previous experiences
- Attitudes
- Values

and that which is visible *in the world*:

- Behaviors
- Skills
- Interactions with others
- Words you and others speak
- Your abilities
- How you spend your time
- How others view you

Why does this matter? Think about it this way: A tennis player learns a set of skills and strategies and rules, gets good tools, and goes out to play the game. What affects how that game is played? Does the particular opponent matter? Does the player's view of court surface matter? Does the history of matches played between the two on the court matter? Does the crisis at home just before the game matter?

What matters the most is our tennis player's view of these things, his or her ability to focus on what is important, to manage things that are not important, and be totally present to what is needed in this situation right now.

The ability to be self-aware, learning to notice the impact of what is happening inside of ourselves and others and learning to manage all of that enables the leader to have desired outcomes more frequently; and with more positive impact.

It is very easy to see the experienced high-level tennis player working with this knowledge. Much time is spent training for the *inner game* as well as on the outer game (Gallwey, 1986).

EXERCISE

Think of a situation that happened recently which did not go as you would have liked. Can you find anything that was happening in your inner world that contributed to that?

Think of a situation that happened recently that did go as you wished. Can you notice anything in your inner world that contributed to that? What are the conditions that enabled that? How can you create more of those conditions in the future?

LEADERSHIP AND MANAGEMENT

Before we go any further, let us be clear about these two inter interrelated concepts. *Leadership*, by definition, indicates that one is going somewhere, hopefully forward! If you are leading yourself, you are both leader and follower. If you are leading a team, as we have already said, you want followers to join you on the journey. When you are in leadership mode you are thinking about an end goal and how whatever you are about to say or to do is moving you closer or further away from that goal. *Management* is about handling what currently exists and improving it, making it more efficient, smoothing out the wrinkles, and so on.

Everyone needs to do both, lead and manage. Sometimes at the same time. So how can you tell what you need to do when?

If you are delegating a new task to someone with the intention of helping that person to develop new skills, you are leading. If you are delegating just to find a more efficient way to divide the work load, you are managing. Get the difference?

If you are a scientist with no line management responsibilities, these both still apply to you. For example:

- Ensuring your work is done in time to pass it on to the next link in the chain: management
- Recommending a new link in the chain to create a better service: leadership
- Keeping up to date with scientific journals: self-management
- Training for new skills: self-leadership
- Prioritizing work delegated to you: management
- Offering to join a project team: leadership

In his classic article "Managing the Dream: Leadership in the 21st Century" Warren Bennis clarifies the distinction between leadership and management. The following list is from that article (Bennis, 1989):

- The manager administers; the leader innovates
- The manager is a copy; the leader is an original
- The manager maintains; the leader develops
- The manager focuses on systems and structure; the leader focuses on people
- The manager relies on control; the leader inspires trust

- The manager has a short-range view; the leader has a long-range perspective
- The manager asks how and when; the leader asks what and why
- The manager has his eye on the bottom line; the leader has his eye on the horizon
- The manager accepts the status quo; the leader challenges it
- The manager is the classic good soldier; the leader is his own person
- The manager does things right; the leader does the right thing

EXERCISE

Think about your workload. How much time do you spend managing things? How much time do you spend leading? Is there more scope to develop either? Both?

What is one aspect of your work that is calling out for more leadership on your part? Will you make a commitment to respond to that challenge? Even if no one asks you to do so, or even notices?

A tremendous amount of self-awareness, clarity about values, and personal goals gives one the strength and insight to tackle such things.

EMOTIONAL INTELLIGENCE

This brings us to the subject of emotional intelligence. This is often described as the key to outstanding performance, particularly at senior levels.

Here is a working definition: Emotional intelligence is the ability to understand one's own emotions and those of others and their impact on behavior, and then to use them intelligently.

EXERCISE

Think for a moment about the role emotions play in everyday work situations. What do they affect? Be specific. What sort of impact do they have? How much do they affect performance? Morale? Productivity?

The previous exercise demonstrates how emotions play a huge part in almost everything. Positive emotions can be contagious. Negative emotions too. And yet, emotions are rarely discussed or valued as part of the way we conduct business. We do, however, speak indirectly with others about the emotions that impact the way of working offline.

That being said, it becomes easier to understand just why having high emotional intelligence is considered *the* differentiator between good and great performance. Daniel Goleman's research of outstanding leaders showed that intellect was a driver of outstanding performance, but emotional intelligence (EQ) proved twice as important for jobs at all levels (Goleman, 1996). And the higher one gets, the more important it is. Technical knowledge is not a differentiator at senior levels; influencing, team effectiveness, motivation, inspiring others are. All of these require high EQ.

How can that be true? It is because the technical skill to perform your job is expected as the baseline. What makes the difference is how you put those skills to work with other people. The good news is that emotional intelligence can be learned; that is, you can increase your EQ, unlike your IQ which stays the same throughout life.

Daniel Goleman is the man who popularized emotional intelligence. He describes the key components in great detail. In the following we touch on just a few of the first characteristics of these components.

Self-Awareness

- Know which emotions you are feeling and why
- Realize the links between your feelings and what you think, do, and say

- Recognize how your feelings affect your performance
- Have a guiding awareness of your values and goals

Once you are aware of these things, the next step is to manage or regulate them. This does not mean denying your feelings, rather managing the way they are used instead of having them unconsciously determine your behavior. For example, speaking about being angry rather than shouting because you are angry. See the difference?

Self-Regulation or Self-Management

- Manage impulsive feelings and distressing emotions well
- Stay composed, positive, and unflappable even in trying moments
- Think clearly and stay focused under pressure

Self-Motivation

Self-motivation is about taking responsibility for your behavior, goals, and about persevering despite obstacles, which is a very important part of self-management.

Social Awareness

Social awareness is fundamentally about intending to understand the moods and needs of others. Let's focus again on the first aspect, empathy. Emotionally intelligent people share the following traits:

- Attentive to emotional cues of others and listen well
- Show sensitivity and understand others' perspectives
- Help out based on understanding other people's needs and feelings

You can see that if you include other people's frame of reference, lots of things change. You will then be looking for opportunities to meet the needs of others, ways to ensure diverse backgrounds and viewpoints can flourish, and have a coaching attitude, where one intentionally develops the skills and abilities of others.

Social Skills and Relationship Management

This is about managing relationships through applying the other skills and connecting more with people to improve how people respond to you. Emotionally intelligent people share the following:

- Skilled at persuasion
- Deal with difficult issues straightforwardly
- Foster open communication and stay receptive to bad news as well as good
- Fine-tune what they say to appeal to the listener

EXERCISE

Do a bit of research. Study yourself and your emotional reactions to things. Begin to develop your awareness of what affects you, why, how your behavior changes when your emotions are getting the best of you. This awareness raising is the first step in developing your emotional intelligence further than it already is. As a leader, the more you can develop this, the more you can choose responses and behaviors that will encourage others to follow your lead.

Emotional intelligence begins with listening. Listening to yourself and then listening to others. In fact, when I go to a new client organization and ask people that work there to describe the leaders, they usually say "he/she listens." *Daniel Goleman, 1996*

Why is listening important? Listening is the key skill that enables us to understand each other. Unlike speaking, in most educational systems we are never taught to listen. It is often assumed that if we are fortunate enough to have two ears that work well, we listen well. This is not necessarily so. You have probably heard it said that the single most important human need is the need to be loved. One can go further and say that the most important need we have is the need to be understood. A key component of being understood is having others who are willing to give their attention to you and to listen. This is a hugely valuable gift that you have to offer in every interaction. It costs you nothing, except a bit of your time.

Listening to understand what motivates and has meaning for others offers the single most powerful tool to influence others. Choosing to influence others is leadership. Listening is an essential ingredient in a wide range of work activities: understanding and solving problems, dealing with enquiries, responding to customer needs, attending meetings, staff supervision and development, team work, conflict resolution, effective negotiation, delegation, building cooperative alliances, and leading and implementing change. The list could go on. Table 5.1 outlines the consequences of not listening from the point of view of the speaker, the person who is not listening, and an organization as a whole.

The consequences of poor listening are severe and dramatic. Not listening includes half-listening: reading, writing, or thinking about other things while someone is speaking to us. In an organization where the predominant culture is one of not listening or only half-listening, staff contribution decreases hugely over time, morale suffers, and absenteeism increases.

Listening does not necessarily mean agreeing. We are really only talking about listening and giving attention to someone.

Note: Active listening is only possible for short periods of time. In fact, we spend much of our time cutting out noises and distractions. However, it is critical to know when to listen, and to have the skill and motivation to do it.

LISTENING ON THREE LEVELS

To really understand what someone is saying, we need to learn to listen to the whole person: not just the words that are being said, but also what lies between and behind the words. This is described as listening on three levels: to the content, the feelings, and the intentions.

1. *Content*: This is what we usually listen for: the facts, information, the details, the story line. However, as we think at about 500 words per minute, and speak at about

TABLE 5.1 The Consequences of Not Listening

Speaker	Non-Listener	Organization
Feels:	Loses respect	Loses commitment
Undervalued	Looks foolish	Increases poor performance
Demotivated	Gets avoided	Wastes resources
Irritated	Misses opportunities	Deals with increased mistakes
Frustrated	Wastes time	Has higher costs
Anger	Is considered rude	Has reduced quality

125 words per minute, there is a lot of time for our mind to wander to similar experiences we have had, preparation for a counter argument, or the day's shopping list!

Developing the capacity to listen accurately to content is helped by trying to be as objective as possible. This means holding back our own feelings about what we are hearing, resisting thinking about our own experiences, and trying to capture the speaker's words.

2. *Feelings*: By listening to the feelings we can discover the relationship between the speaker and the "story." We listen between the lines of a perfectly rational story to hear feelings of resentment, frustration, excitement, hope, and so on.

It is important to hear these feelings, because they can linger far longer than the events to which they are related, and will have a tremendous impact on future interactions.

Developing the capacity to listen accurately to feelings is helped by holding back our own feelings, likes and dislikes, and trying to develop empathy. Empathy allows us to pick up the feelings of the speaker, rather than our own. This is done by listening to the words people choose, the tone of voice used, and looking at facial expression and changes in body language.

3. *Intentions*: Listening to the "will" of the speaker will enable us to find their motivation, commitment, and direction. This information is vital in negotiations and agreements, to know what the prospects for implementation and support are.

Developing the capacity to listen accurately for the intentions of the speaker requires that we hold back our own wishes, suggestions, and advice, and that we are interested in the outcome for the speaker. Despite the fact that intentions are often buried and unconscious it is possible to

hear them by listening to the emphasis given, the amount of detail, the first and the last thing said, and the energy used to describe different aspects. It can be very helpful to the speaker if you are able to hear their intentions and reflect them back.

One more important thing about listening: You will note that in all cases, to listen to some-one else's story to hear the content, feelings, or intentions requires that your inner space is clear of your own story. If you really want to listen to someone else, do not start talking about the time the same thing happened to you, how you would feel if it were happening to you, or giv-ing them advice that is based on what you would do if you were them.

Remember: The goal is to understand them better, not to get them to understand you.

EXERCISE

Try it out. In a one-on-one situation, see if you can hear and repeat the story without embellishment. You can stop the speaker by saying "Hang on a minute, I just want to check that I got what you have just been saying."

To capture feelings, listen to the words they choose, any change in the tone of their voice, and any changes in body language that might be indicating feelings. Ask about those, rather than assume you know what they mean. "Can I just check? You said you are fine with this idea only the way you said 'fine' so emphatically makes me want to ask if there is anything more you might tell me to help me understand more of how you feel about this."

Listening for someone's will is actually pretty easy. Does it sound like they are really committed to taking action? Or just like it's a good idea? Or does it sound like they are say-ing "yes" just to get you to finish? You might say "Thank you for agreeing to do this. When can I expect the first step?" or "I hear that you think this is a good idea, I am just not hearing that you are really committed to acting on it. Is that so? What can I do to help get you more committed? Is there something still concerning you?" or "I hear you speaking a bit about the two possible options and you describe option one in great detail and option two only briefly. It sounds to me like option one is the one you prefer. Is that so?"

If you are going to a meeting with a colleague, why not speak with that colleague before the meeting and ask him or her to notice what is actually said, how important it is, what actually applies to you and your team, and what will really happen as a result. Or something similar. You get the idea.

Making Listening Visible

For the speaker to feel valued, motivated, worthwhile, and encouraged they need to know that they are being listened to. Therefore it is important to avoid doing things like doodling and shuffling through papers. Instead:

- Give the speaker your full attention, even if it is only for long enough to say that you are unable to listen at the moment and to arrange another time to talk.
- Keep eye contact with the speaker while being sure to avoid staring.
- Sit or stand reasonably still; fidgeting indicates impatience, doing other activities indicates disinterest.
- Periodically summarize and reflect back what you have heard. This helps both you and the speaker to keep track of what's being said. Do not change subjects!
- Allow silence to help you communicate patience and to enable the speaker to draw more out of themselves.

Remember: The thing that will most indicate you are listening is giving your full, relaxed attention and concentration to the speaker.

Creating the Right Environment

- Find a quiet space: wherever possible ensure an atmosphere of privacy.
- Eliminate distractions: divert phone calls, put up a "do not disturb" notice, put your work aside.
- Eliminate barriers: come out from behind the desk, be at the same height level, use understandable language.

Create the right "inner" environment by clearing your mind, so as to make a space for what the speaker has to say.

Remember: What you think you are hearing on the three different levels must never be assumed to be correct. Test it out with the speaker by reflecting back what you have heard and asking if it is right.

Table 5.2 summarizes the challenges of listening on three levels.

SEVEN HABITS OF EFFECTIVE LEADERS

In addition to active, empathetic listening, what else are the key behaviors of excellent leaders? Stephen Covey's excellent book *The Seven Habits of Highly Effective People* describes these behaviors perfectly (Covey, 2004).

Briefly, Covey researched what successful people had in common and discovered that there were seven habits consistently present. Habits are behaviors that people use regularly, so much so, that they do not require conscious thought (eventually!). Habits are things that can be learned and changed so *anyone* who wishes to adopt these habits can do so. We translate into meaning that excellent leadership can be learned, which is good news. That's not to say that everyone can be a chief executive or a leader of a social movement or a prime minister; all of these things require more than good habits. However, good habits are required for

TABLE 5.2 Challenges of Listening on Three Levels

		LISTENING ON THREE LEVELS				Challenge for the listener
		How are we listening?	How is it expressed?	What are we listening for	Where is it coming from?	
CONTENT	P A S T	Objective Level Most common, surface	Ideas Information	What is said	Head	To be open minded
FEELINGS	P R E S E N T	Subjective Level Between the lines	Values Attitudes	How it is said	Heart	To be Empathetic
INTENTIONS	F U T U R E	Operational Level What is behind, or underneath	Motivation Commitment	What is intended	Hands & feet	To retain interest in the speaker

© *Merlin Walberg Phoenix Consultancy USA, Inc. 1988.*

all of them. As described earlier, there is a baseline of skill and knowledge that is expected for senior positions, it is emotional intelligence, and having excellent self-discipline and interpersonal habits, that make the difference between good and great (Collins, 2001).

Here are the habits:

Habit 1: Be Proactive

This means to fundamentally accept that you are responsible for your own behavior.

"What? Of course I am," you might say. Really? Do you ever think or say: "It wasn't my fault because," "I didn't have...," "He makes me angry," "I have to...," "What can *I* do?" "If only...then I would...," "I don't have time," "I can't."

You can see how easy it might be to say one of those, right? What lies behind each of those statements is someone saying "I am not responsible for my feelings, behaviors, or thoughts."

How different would it feel to say: "This is not what I would have wanted. Given this situation, I will," "The big decision is out of my control and so what I think is important for us to focus on is...," "My time is very limited, I will look at my priorities," "I prefer...," "I choose...," "I feel angry when..."

Proactivity means choosing your response to whatever comes your way, rather than reacting unconsciously. Make sense? So, on what basis do you make decisions about those choices?

Another key way to determine how proactive you are is to think about how you spend you time and energy. If you are focused on things you can influence, make decisions about, and achieve, you are spending your time in your *circle of influence*. Here is where positive energy rises and you feel a sense of making things happen and contributing. Are you thinking and focusing on things you cannot influence? If so, you are in the arena that Covey calls the *circle of concern*. This is where your energy gets drained, because you are not able to impact these things, and indeed you are less effective in your circle of influence as you waste energy that could be better used elsewhere.

Things in the circle of influence might be very important, such as decisions about policy, funding, the skills of others on whom you rely, for example. Yet, if they are outside your ability to influence, the proactive response is to accept them for what they are and deal with *that* reality rather than the one you wish were true—even if you are right! Think about it.

EXERCISE

Reflect on the conversations you have with others over and over again. Are they about things in your circle of concern? Is that why you keep having the same conversation, because there is nothing you can do about it? Do you know people who do that? How does it feel?

Habit 2: Begin with the End in Mind

This is the habit of personal leadership. It means thinking about what you want the outcome to be and having that as your guiding light for all decisions. Covey talks about having a personal mission statement comprised of your values and things that are important to you.

EXERCISE

You are attending your retirement party. What would you like to hear people say about you? About what you stand for? What you would not stand for? The impact you had on them? The workplace? Your profession? The answers to those questions will lead you toward your core values, which, no matter how challenging or restricted you find the situation you are in, will give you peace of mind, knowing that you have done the best you can from your own point of view.

Begin with the end in mind is useful for small things as well. Someone comes to your office, or the lab, and starts speaking to you. You are wondering what on earth they are talking about. You can say: "I only have a few minutes at the moment. What is it you would like from this conversation?" "What do you need from me?" or "How can I help?"

As important, when you are about to speak with someone, think first of your ends in mind for that conversation. Have you ever walked out of a meeting with your supervisor knowing that it did not go well, that you did not get what you wanted? Not sure quite what happened? This is often because you were not clear before you went in about what you wanted.

There are always two ends in mind:

- Content of issue outcome
- Relationship or interpersonal outcome

Thinking about having two ends in mind will certainly have an impact on how you say what you say, bearing in mind that you want to have the relationship continue or get better.

Habit 3: Put First Things First

Now that you have decided that you are responsible for your own behavior and have more clarity about what you are responsible to achieve, this habit, the habit of self-management, will help you to get there. There is a quote of Goethe's that is the essence of this habit: "The things which matter most must never be at the mercy of the things which matter least."

This habit is asking you to actively think about how you choose to spend your time and to prioritize important things. Table 5.3 illustrates this beautifully.

EXERCISE

For one week keep track of how you spend your time. You can use a model of these four quadrants to log your activities. How much of your time falls into the different quadrants? Be honest with yourself. What changes would you like to make? How might you make one or two to give you some more quadrant II time?

The first three habits are what Stephen Covey calls the "private victory." That means no one need know you are doing them. Doing them is your own personal victory. Others will certainly notice the results of those habits, and they do not need to participate for you to succeed with them.

The next three habits are referred to as the "public victory." These habits do indeed involve others in a more obvious way. You will see that they provide a very powerful role model about the way to engage with individuals and teams. In fact, all of us would like to be treated in the way these habits prescribe. We just don't always remember to do it for others!

TABLE 5.3 Prioritizing Your Activities

	Quadrant I. Activities Important and urgent	Quadrant II. Activities Important not urgent
Important	Crises • Pressing problems • Deadline-driven projects	• Crisis prevention • Values clarification • Preparation & planning • Relationship building • Renewal & evaluation
	Quadrant III. Activities Urgent not important	Quadrant IV. Activities Not urgent and not important
	• Interruptions, some phone calls • Mail, some reports • Some meetings • Some pressing matters • Many popular activities	• Trivia, busy work • Some mail • Some phone calls • Time wasters • Many pleasant activities
	Urgent	

Habit 4: Think Win-Win

This is a very interesting habit. The habit is called *think* win-win, not win-win. Pedantic? Maybe, but maybe that one word is central. The habit means that in all interactions with others it is your job to *always* be thinking about yourself and the other and how the situation in which you are both involved can have mutual benefit. In fact, if I were naming the habit, I would call it "think mutual benefit." Mutual benefit implies the ever-present intention to look for positive outcomes, no matter how difficult and different the points of view. This is the way to build long-term relationships. Win-lose or lose-win are always a setup for someone to try to get even; not a good prescription for long-term sustainability and certain implementation of agreements. Think about it: You feel screwed into the ground by someone who thinks they have just negotiated a great deal. How likely are you to work really hard to make sure the arrangement works perfectly?

Mutual benefit even applies if you are saying "no" to someone or telling someone they no longer have a job in your department. How is that possible? The way you give the information can either enable someone to walk out of the door with their head held high or crawling. Very different.

Habit 5: Seek First to Understand, Then to Be Understood

This is the habit of empathetic listening. We have already discussed listening in great detail. The key to the effective use of this habit is that you do it *before* giving your point of view. The idea is that if you listen first and speak second, you will have information that can help you pitch what you want to say in a way that makes it as meaningful and accessible as possible. After all, that is what you want isn't it? To be listened to? So this habit, as with the others, asks you to model it first. It is guaranteed to save you time

by helping you to only have to say it once to be understood. It will help you to not have to go back and ask again, as you will have been listening the first time. Also, as you have been listening on three levels you really have the full story. It is hugely motivating for someone to feel heard and you may gain the added benefit of having someone listen to you as well. Nonetheless, whether they do or not, it is your job to continue to do so. Eventually they will too.

Habit 6: Synergize

This is the habit of creative cooperation or teamwork. Covey calls it the highest form of human interaction. Have you ever experienced real synergy? When you and another person or group of people are all focused on something creative, interesting, and engaging and are working away at it? No point scoring, no worrying about who speaks more or less, only interest in the topic itself and everyone contributing what they have to offer. Suddenly two or three hours have passed.

What are the conditions necessary for this to possibly happen? To lead the possibility of synergy, your job is to enable diversity of opinion and create a culture where it is okay to express opinions, even if they turn out not to be the ones chosen for implementation; where it is safe to try and have a go. Of course, we are not suggesting that this is suitable for everything. Scientific tests, for example, should not be done creatively. They must be exact and follow strict protocols. However, there are many real synergies in science. Indeed it is often synergy that creates breakthroughs in science. It is the leader's job to determine whether precision or divergent thinking is needed and to be flexible enough to foster both.

Habit 7: Sharpen the Saw

Last of all, and perhaps the most important of all, is this final habit of self-care. It is the habit of taking care of yourself so you are in the best shape to carry out all of the other habits. This requires courage to take breaks, nourish oneself with good food, and create a life with spiritual, social, and intellectual nourishment, for you to be your best. Great leaders know that burnout or operating at 50% capacity is just not going to yield top-notch results all the time. It takes some strength to resist "working so hard" that others may perceive you as really committed. Here you are asked, again, to be the role model for the way of life that actually yields the highest returns.

EXAMPLES

"So what is all this Leadership stuff? Surely it is just common sense dressed up with a few long words and some catchphrases. The real problem is that **they** don't listen to the real experts..."

Sound familiar? It is something I still hear from many scientific colleagues, and I would admit to having shared many of the underlying beliefs myself. Having undergone some leadership training, and even more powerfully, having had to lead scientists across a wide range of specialisms without formal hierarchical authority, my views are now radically different. Here are a few examples:

Emotional Intelligence

Working in safety critical areas like radiotherapy physics means getting the right balance between essential checking and safety procedures, and providing prompt personalised radiation dose treatment for patients. When someone has this balance wrong, engaging in purely objective discussions about safety checks and QA frequency may only embed their views, leading to conflict. You need to understand their (and your) emotional

commitment to these views, and the values which underpin them. Only then can you reach a consensus which is right for the patient.

The 7 Step Process of Change

Given that scientists are committed to rational evidence based decisions, once new evidence is available and tested, of course we will all happily change to a new procedure or process...except we don't. Scientists are human too, and we need to go through the same adaptions to change as any other group. When merging (and reorganising) medical physics and clinical engineering colleagues working in the same organisation in related but historically separate groups, the process took longer than needed and was more distressing than it should have been because we didn't acknowledge the need for people to undergo change processes. A more structured approach, which would have acknowledged the concerns, hopes and fears of the individuals, would also have been more efficient.

Asking Questions Rather than Giving Answers — Empowering Others

Scientists are good at questions, after all we are professional sceptics, and we are trained to spot the flaws in arguments and to expose them forensically. That approach is fine for purely scientific discourse, but is terrible training for dealing with people that we want to motivate. People give of their best (and give much more) when engaged and empowered. Wanting a junior colleague to take responsibility for a new area of work in the safe use of medical equipment, motivation came from discussing the scope of the clinical problem with them, and asking for their ideas on how scientists could contribute, how best to engage with nursing and medical colleagues, etc. The outcome was an improved clinical service as well as a motivated individual aware of her capabilities to take on bigger tasks in future.

Taking Responsibility to Lead, even When it is not your Job

The more we learn, the more there is to learn, and we love our personal area of scientific expertise, so we go deeper rather than broader. Fine, and sometimes necessary, but also limiting. Being the expert in a specialised field doesn't preclude us from being good in other fields as well. When the opportunity came to move out of my specialist area of science, to a role which encompassed all the applications of science in medicine, my first reaction was to question whether anyone could be an expert in all these areas. Once you realise that no-one can be, then the issues become clearer: who is best placed to lead the experts who do work in these diverse scientific disciplines? And if not us, then who?

Using Leadership Tools in a Scientific Environment

The value of leadership tools has been to develop personal awareness and listening skills, and I have also had the opportunity to put the skills I have learned into practice when implementing a 7 day service within a multi disciplinary team. Previously our team only provided a five day service, so moving to working over seven days was quite different.

When faced with a challenge or difficult situation I have learned to listen to my inner self, developing my self confidence and being able to project that confidence to my outer self which is the face that others see. A valuable skill to develop in becoming a good leader; enabling you to deal with your fears and face challenges.

When introducing anything new to an established service it is crucial to recognise that the staff in that service will have some resistance to that change.

Possession of leadership tools puts you in a strong position to resolve this resistance to change and find solutions that empower others.

I found it important to use emotional intelligence and the skill of listening on 3 levels, as so little of our communication is verbal, it was through the actions and non verbal communication of the team that I was able to identify particular issues, including concerns about how a seven day rota would operate and the impact on individual team members.

By being aware of the 7 Stage Process of Change I was able to anticipate and understand the process the staff were going through. The change process was discussed as a team and everyone in the team was able to voice their ideas and any concerns.

Initially there was staff resistance, some staff stated problems with childcare, and this was followed closely by an awareness and recognition of the need to change, to offer a better service, increased results and outcomes for patients. Staff quickly realised that there might be benefits for them as well, the 'what's in it for me?' Staff decided that they would like to have a day off on the week in lieu of weekend working.

By thinking win win the strategies that resulted in mutual benefits and solutions were employed. I was able to outline the benefits to the patients and service. As a leader I value and respect others, I like to put myself in their place, and it was important to allow the staff to realise the benefits themselves without pressurising them.

The next stage in the process was to have a mental try out of the process—what will the benefits be if we do this? Staff, patients, unit, we could have better results and outcomes.

- Improve the quality of care
- Offer greater patient choice

- Reduce cancelled treatment cycles
- Reduce risk
- Reduce complaints
- Service expansion and development
- Increase success rates
- Increase business
- Increase service profile
- Generate additional income

This was quickly followed by a real world try out as a pilot study, with a planned review after 6 months. The cost of offering weekend services was not as much as anticipated and offset by additional income generation.

The staff gave their commitment to the arrangement and after the trial period we had evidence based improved outcomes, and staff took pride in offering the best options possible to the patients, enjoying the increased success.

After the six month pilot the weekend working was integrated and became part of normal service, with staff finding additional things to do on a Saturday morning, including extra patient clinics.

The successful implementation was due to the use of the leadership tools, using the appropriate tool at the right time.

Emotional Intelligence/Listening on Three Levels

Background

I had recently taken over line management responsibility for a member of staff who had been working in the department for a couple of years, although only just been rotated into my laboratory. Within a few months of this rotation, the annual appraisal of this person was due. I had been very pleased with the progress they had made and was impressed by the quality of their work and obvious commitment to the job. I expected the appraisal to be straight forward.

Initially the appraisal went well. The discussion of recent achievements was straightforward and the feedback I was giving was very positive. Appropriate responses were being received to the questions I was posing yet I began to feel that something was out of synch. Whilst the content of the conversation was stating everything was fine, the feelings that I was picking up on were telling me the exact opposite. At the time I couldn't quite put my finger on what it was but in retrospect I think it was the lack of visible emotion displayed, as if the person was simply going through the motions. The appraisal ended with an opportunity for the member of staff to raise any additional concerns they may have. I was told that everything was fine and there was nothing else they wished to raise. The expression on their face was not in keeping with their words so I pushed the point several times but their answer did not change. I ended the appraisal with an offer for them to come and have a chat with me at any time.

I wrote up the appraisal paperwork that afternoon and was still convinced that something was very wrong. Having only known this person for a relatively short period of time, and in a professional capacity, I had no idea what it could be and whether it was even work related. I was also unsure what the best way to approach them with my concerns would be. In the end I chose to go with my gut instinct and the following day I asked to speak to them again. At this point the person became extremely upset and confided in me.

Acting on what I heard and using my emotional intelligence to offer a gentle and repeated opportunity for the member of staff enabled this person to trust me enough to speak so that we could work together on the issue at hand.

Power

One more thing for us to look at that is essential to understand as a leader is the use of power in organizations and how people get things done. It is not as simple as it might seem.

The understanding of power and how it works among people is central to achieving impactful leadership. Often one thinks about power in terms of hierarchy in organizations, and that those who are the most senior have the most power. In fact, we usually think of power as a weapon to use to get your own way.

We admire it in men and are very wary of it in women. There is much to be done in our understanding of power to enable power to have the same value for both genders. Here is something that might help. Let us define power as "the ability to make things happen."

How does that definition make sense in terms of leadership within the scientific professions and in the professional world in general? Can one have too much ability to make things happen? Can it be that making things stable and consistent is powerful? That enabling exceptional teamwork is powerful? That joining together with colleagues to solve complex problems is powerful? If so, would it be hierarchical power that was most effective? Let's have a look.

Positional Power—Formal Authority

This type of power is invested in a title or position. Whoever occupies that position has that authority. True, it takes some doing to get into that position in the first place, but the amount of power is the same for anyone in that position. Some examples are CEOs, police officers, parents, athletic coaches, head teachers, heads of departments, clinical directors, and so on.

The use of this type of power is *very* effective for quick short-term results. Those responding to this power do so because the person in the powerful position has the ability (either perceived or real) to make things difficult or unpleasant for the follower.

"Put away your toys or you will not be allowed to. . ."

"Follow my instructions or I will have to find someone else who will."

"This is the way we do it on this team."

"The decision has been made to change this protocol. You are required to change the way you do it. Today."

Positional power is highly respected to get someone to change their behavior really short term, perhaps in emergency situations, to avoid an accident, move a group of people quickly, or make a quick decision as in changing direction on a sailboat. It is highly valued in the military.

However, if positional power is overused in ordinary situations, the followers feel invisible, devalued, and unable to contribute. One consequence of this is that people often do what they are asked, but do as little as is possible to be safe.

Research shows that there is 30% discretionary effort that employees have at their disposal to offer their workplaces. Not one single percent of that will go to a leader who simply uses his or her position to get people to behave in the way they want. So what else can one do to influence people?

Expertise

We have heard it said that knowledge is power, and this is true! Think about meetings you have attended where the chairperson is a very senior person who has an idea or plan for making a change. You, the subject expert, in front of a room full of senior people might be in a position to say: "That is an interesting idea and unfortunately, it will not work. The reason is. . ." In such situations expertise is considerably more powerful than position. Of course, it makes sense.

The fewer people that have the knowledge, the more powerful expertise is. This is also a source of power for people who are not very high up in the hierarchy. Think of the person who schedules holidays or takes care of the payroll. An important side note here: Sometimes, when someone's experience of their power is tied to their area of expertise (like being the only one who knows the method for running the payroll), you might try to offer help so that they do not feel "too burdened to take a holiday." This offer would be to have them train someone else to do what they do. It is your job as a leader to ensure that the person in charge of payroll has a way to contribute, feel valued, and experience themselves as powerful if you plan to take away their unique expertise.

EXERCISE

Take two minutes to think about and jot down all of the areas in which you have expertise and knowledge. Think broadly. These could include computer skills, knowledge about processes, about people, about what is happening in your field, and more.

Resource Control

We know that holding the key to resources is powerful. The finance team, the CFO, the auditors; all powerful people. What other resources that you control are a further source of power for you? Your time; the time of those on your team; information; perhaps some machinery or equipment? As with expertise, resources can be highly valued, especially if they are rare. There is a great opportunity here to build relationships and expand your influence: You can trade resources!

You can offer to help another colleague with a short-term deadline and build up "credit" with that colleague for future needs you may have. This can be adapted and bring great benefit to multiple parties over long periods of time. Look for opportunities to offer to help, no matter how busy you are. It makes you visible, gets you to be viewed as a positive contributor, shows what you are capable of, and builds up knowledge of your expertise and resources so that your power is magnified.

EXERCISE

Think about and jot down as many resources that are within your gift as you can. Think about those around you who you have easy access to as well as yourself.

Interpersonal Skills

Have you ever known a brilliant professor who was such a boring speaker that his or her brilliant ideas and knowledge just didn't get heard? Or someone who was extremely capable, got lots of things done and done well, and could have done so much more if they were able to delegate, share, and communicate better with others?

What about someone who seems to be able to build relationships with anyone, get people on board easily with new ideas, who is a great listener who makes others really feel heard? A person who others are happy to work with and who somehow finds ways to cooperate even when there are differences?

This type of power trumps everything we have discussed so far. The people who possess it know that interpersonal skills are by far the biggest source of personal power that anyone can have. They have the skills and awareness needed to make others feel valued and important: the ability to listen, communicate clearly and gently, understand the impact of emotions, and work with others positively. These include the skills of self-awareness and self-management and knowing that asking questions is a much better way to engage people in a conversation than making statements about what you want.

EXERCISE

How good are your interpersonal skills? Which ones need some attention? Which ones serve you really well and can be built upon and expanded?

EXERCISE

The Power Net

To get a visual representation of your political situation you can draw yourself and your relationships and the power that flows between you in both directions.

1. Draw a circle in the center of a page to represent yourself.
2. Draw circles to represent specific people with whom you interact. Draw these circles closer to you for people with whom you spend a lot of time and further away for those who have an important effect on your work though they may not spend much time interacting with you (patients, suppliers, commissioners, for example).
3. Draw lines to connect yourself with all of the people. On the lines write what type of power each person in your net has in relation to you. Do they have the right to decide what you do? What information do you need? What is the basis of their power? What sort of power do you have in relation to them?

You can also think of the power net in terms of dependence: who depends on you for what? And who do you depend on for what?

Information

Information is one of the most important political resources, and politically useful information cannot be gained through formal channels only. Tuning in to informal information networks is a vital part of understanding organizational politics and hence of being able to manage power.

Change and Opportunity

Managing power not only involves being able to analyze the political framework in which you find yourself, but also recognizing when this changes and taking the opportunity that this may present. The ability to act assertively (rather than passively or aggressively) may be important if you are to make the most of such opportunities.

CONCLUSION

To be a leader requires a paradigm shift: to feel deep inside that you are responsible. It is about getting it in your DNA; "I matter, what I see matters, what I do matters, and how I do it matters most of all." Every day you interact with people, processes, and problems that will benefit greatly from your commitment to thinking about what you can do to make things and relationships with people better. To think from the other person's perspective: why would he or she do, think, or say that? Then to pitch what you want to say in language that they understand, always being aware that there are two ends in mind in every situation: the content of what you want to achieve and the relationship/interpersonal outcome that you want. Keeping the relationship intact makes it possible to work on the content further. Forgetting the relationship and only working on the content may mean that the content gets sabotaged, because the other party feels hard done by.

At all levels of your scientific career you can be a leader, if you choose to do so. The decision is yours. Enjoy the journey.

References

Bennis, W.G., 1989. Managing the dream: leadership in the 21st century. J. Organ. Change Manag. 2 (1), 6−10.

Collins, J., 2001. Good to Great. Random House.

Covey, S.R., 1989. The Seven Habits of Highly Effective People. Simon & Schuster.

Gallwey, W.T., 1986. The Inner Game of Tennis. Pan Books.

Goleman, D., 1996. Emotional Intelligence: Why it Can Matter More Than IQ. Bloomsbury.

6

Risk Management

Anthony Scott Brown

Royal Cornwall Hospitals Trust, Truro, U.K.

INTRODUCTION

Preventable mistakes are common in healthcare. The risks associated with healthcare are being highlighted, and hospitals and healthcare organizations are taking steps to manage these through active risk management. An acceptable framework should provide a clear definition of "reasonable risk" and an account of normative justification for this definition. This is largely determined by the risk appetite of the organization—how much risk the organization is prepared to accept. In addition, it should elucidate a set of operational criteria or markers that delineate in an operationally useful way the parameters or boundaries that separate reasonable from excessive risk. Finally, it should articulate practical tests that deliberators can use to determine whether or not these operational criteria have been met in a particular case (London, 2006; Brown, 2011). This chapter is structured loosely in a similar manner to the International Standard ISO 14971:2007 which relates to the application of risk management to medical devices.

DEFINITION

In any discussion about risk management we must first establish what we mean by risk. Risk can be defined as "a situation involving exposure to danger [noun] or exposure (someone or something valued) to danger, harm or loss" (Oxford Dictionaries online). Other definitions include the following.

Rosa (2003) defines risk as "a situation or event where something of human value (including humans themselves) is at stake and where the outcome is uncertain." The Royal Society (1992) uses the definition "the probability that a particular adverse event occurs during a stated time period or results from a particular challenge."

It is accepted that the concept of risk has two components (BSI, 2007, p. V):

1. The probability of occurrence of harm
2. The consequences of that harm, that is, how severe it might be

(*Note*: The European Standard EN ISO 14971:2007 has the status of a British Standard.)

The Office of Government Commerce (OGC) define risk as "an uncertain event or set of events which, should it occur, will have an effect on the achievement of objectives" (OGC, 2007, p. 1).

CONSEQUENCES AND LIKELIHOOD

Clearly the definition is open to interpretation, but essentially to make it useful in managing risk we must be able to "quantify" it. The basic equation is:

Risk = Consequences × Likelihood
[Likelihood is sometimes replaced by probability]

THIRD DIMENSION OF RISK

Most risk assessment systems consider risk in terms of consequences and likelihood (sometimes referred to as probability and impact matrices; APM, 2010), however in some circumstances there is a third dimension: time (Figure 6.1). The time factor is principally used in project management either for a stage or the overall project (OGC, 2009). Hence there is proximity (i.e., the time before the risk is apparent) and also a duration (i.e., the time period over which the risk is present).

The overrunning of a stage or project can have major implications. According to the OGC (OGC, 2007, p. 26), impacts on an organizational activity are usually considered in terms of the organizational objectives and hence examine the impact on:

- Costs
- Timescale
- Quality/requirements

The OGC guide also introduces the concept of a time dimension which impacts and feeds

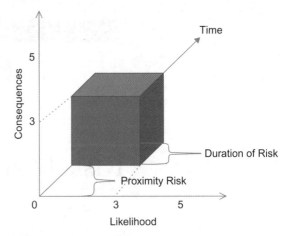

FIGURE 6.1   Three-dimensional model of risk.

into the risk assessment. The concept of time-dependent risk is not usually considered in many industries including healthcare, however it may be appropriate for specific risks, for example, a new surgical procedure, a licensed drug being used "off label," or a clinical investigation of a new medical device.

In reality, medical device risk is time dependent. For example, using a syringe driver to deliver an infusion to a patient. At a later time the same device may be used again either on the same patient to deliver a further infusion or a different patient. The second infusion might be a different drug and the syringe driver may be operated by a different member of staff with more or less experience and training. In areas such as intensive therapy units (ITU) there may be multiple devices connected to a patient. In this instance the risk consequences and likelihood may be different. This is shown in Figure 6.2 which shows two risks separated by a period of time. The first risk has a consequence of 3 and a likelihood of 3, whereas the second risk has a consequence of 2 and a likelihood of 4.

In reality, of course, while this may be true for a single medical device, there are several hundred similar devices in an acute care

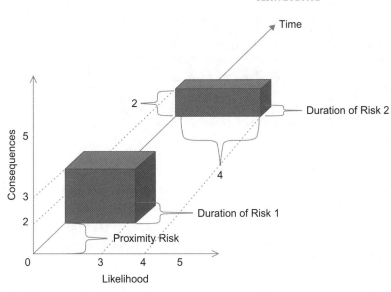

FIGURE 6.2 Three-dimensional model of risk showing two distinct events.

hospital so the risk is continuous. In areas such as intensive care or the operating theaters there may be multiple devices connected to the patient. In many ways we should consider the number of simultaneous infusions so we can aggregate the risk, but as this is not possible to estimate let alone measure, the risk is treated as though it were continuous.

Although the overall or average risk over time would be lower, it would not be practical to undertake a formal risk assessment prior to using every medical device. However, it would probably be useful to undertake a further reassessment should there be any significant change of staffing or for a change in the environment.

RISK APPETITE

One area of risk not yet mentioned is the concept of risk appetite, or the capacity of the organization; this is usually determined or set by the board. According to the Office of Government Commerce, "Risk appetite plays a vital role in supporting an organization's

objectives and orchestrating risk management activities" (OGC, 2007; p. 20). Each organization will take a view on what level of risk they are willing to take or accept; it may well be dependent on how the organization would be able to manage a risk. For example, an organization with substantial financial reserves and capital levels would be better placed to live with financial risk than other organizations with less capacity. Another way of looking at risk capacity is the burden it places on the organization and how it enacts on the way the organization operates. It may, for instance, influence decisions about changing the business direction.

For clinical engineering departments involved in major equipping or re-equipping programs an understanding of corporate risk appetite is essential so that a risk management strategy and escalation rules are appropriately defined (Sowden, 2011, p. 138). Properly defined and communicated risk appetite helps to insulate a program from unwelcome surprises and provides it and its projects with clear tolerances in which to operate.

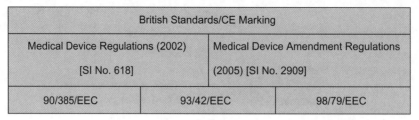

FIGURE 6.3   Legislation wall of safety.

RISK ACCEPTABILITY

Linked to risk appetite is risk acceptability: the extent to which the organization will accept a given level or risk. Tolerance thresholds may be specified and probably the two simplest to specify are time and cost.

In some countries, a good example being the United Kingdom, healthcare organizations can "pool" risk. The National Health Service Litigation Authority (NHSLA) in the United Kingdom has the Risk Pooling Scheme for Trusts (RPST) whereby meeting governance standards in individual organizations will entitle them to reduced insurance premiums.

MEDICAL DEVICES DIRECTIVES

There are a plethora of technical standards relating to medical devices, designed to ensure consistency in the safety of a device throughout its life. These must be seen as the starting point in the risk profile. Underpinning these are the Medical Devices Directives (Figure 6.3). There are three European Directives concerning medical devices:

- Active implantable Medical Device Directive (90/385/EEC)[1]
- Medical Devices Directive (93/42/EEC)*

[1]These standards were incorporated into U.K. law in a new standard 2007/47/EC on March 21, 2010.

- In vitro diagnostic Medical Devices Directive (98/79/EEC)

The EU Directive 2007/47/EC (which amends the Medical Devices Directive 93/42/EEC) specifies:

- Clinical data will be required on all classes of medical devices (even if already CE marked)
- All serious adverse effects (SAEs) must be reported to the competent authority (CA). The competent authority in the United Kingdom is the MHRA. (European Parliament and the Council of the European Union, 2007)

The requirements for Good Clinical Practice in medical device clinical trials are set out in BS EN ISO 14155:2003, Clinical Investigation of Medical Devices for Human Subjects, and was explored in Chapter 3 "Good Clinical Practice."

The Medical Devices Directives were implemented into U.K. legislation by the Medical Device Regulations 2002 (S.I. No. 618). This statutory instrument consolidates all existing medical device regulations into a single piece of information that came into force on June 13, 2002. Subsequently the Medical Device (Amendment) Regulations 2005 (S.I. 2005 No. 2909) were introduced. In other countries the directive would be implemented through their own legislative processes into regulations. The regulations place obligations on manufacturers to ensure that their devices are safe and fit for their intended purpose. In addition, the majority of medical devices need to be CE marked before being placed on the market.

There are further safeguards in the manufacturing process through British Standards such as the ubiquitous BS EN ISO 60601-1-6:2010, Medical Electrical equipment. General Requirements for Basic Safety and Essential Performance, which is further supported by the specific standards for types of devices such as surgical diathermy. This legislation and regulations are easily conceptualized as a brick wall to protect the safety of patients and staff.

The safety of medical devices is overseen by the competent authority (CA) in each country whose remit is the regulation of medicines and medical devices and equipment used in healthcare and the investigation of harmful incidents. Their guidance on managing medical devices recommends that healthcare organizations should strive to ensure that the service departments that maintain their equipment should be registered to BS EN ISO 9001 or similar. BS EN ISO 9001:2008 is a general quality management standard that is not specifically targeted at medical devices but is more concerned with having processes and checks in place to ensure a consistent quality and is underpinned by the ethos of continuous improvement. Increasingly, in-house electrobiomedical engineering (EBME) departments are becoming registered to this standard. There is tacit knowledge surrounding the rationale for routine maintenance, and in particular maintenance schedules are hotly debated.

Healthcare organizations, their staff, and patients take on faith that medical devices are intrinsically safe based on the safeguards put in place through the legislation. In the United Kingdom the safety of medical devices is overseen by the Medicines and Healthcare Products Regulatory Agency (MHRA), an executive agency of the U.K. Department of Health, which aims to "Take all reasonable steps to protect the public and safeguard the interests of patients and users by ensuring that medical devices and equipment meet appropriate standards of safety, quality and performance and that they comply with the relevant Directive of the European Union" (MHRA, 2003, p. 1).

BS ISO 31000:2009

This is an international standard that outlines the principles and generic guidelines on the principles and practice of risk management within an organization. A useful accompanying standard is BS ISO 31100, the code of practice that provides guidance on the implementation of BS ISO 31000. The standard 31000 sets out a framework to embed risk management in any organization. The framework comprises seven elements:

1. Understanding the organization and its context
2. Risk management policy
3. Embedding risk management into processes within the organization
4. Identifying lines of accountability
5. Recognition of the need and identification of resources
6. Internal reporting mechanisms, such as risk registers, and communication channels
7. External communication and reporting (if required)

Overall the risk management framework must not be seen as a one-off task, but must be continually revisited to ensure that it remains fit for purpose. The basic review cycle is shown in Figure 6.4.

Risk Management Policy

Like all policies the risk management policy must be subject to periodic review. To ensure risk management is core to the business it must link in to other policies and the objectives

FIGURE 6.4   Risk management review process.

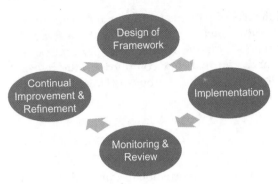

of the organization. For organizations involved in medical devices, whether that be manufacturing or in use, one of the key objectives must be patient safety. The organization's risk appetite must be clearly articulated in the policy and this may be informed by legal and financial aspects.

The policy must also clearly define the roles and responsibilities for managing risks; often in large organizations there is a risk manager who oversees the risk management framework and processes. It is usual for each risk to have a named risk owner. The risk owner is responsible for the identification of the hazard, the evaluation and grading of the risk, and subsequent control measures. There needs to be in place a regular review process whereby each risk is reviewed to ensure that the control measures are effective and that the residual risk is correctly graded.

The policy must set out the process, methods, and tools used to manage the risks within the organization. Typically this will comprise a risk matrix to grade the consequences and likelihood and then the recording of risks, traditionally by means of a risk register. There can be different levels of risk register, for example, in a hospital the ward or department would hold a risk register, and risks that cannot be managed at this level would sit on a higher level risk register at the directorate or division. Finally, there would be an organization risk register held by the board.

AS/NZS 4360:2004

The AS/NZS 4360 risk management standard (now in its third revision in 2004 from the original 1995 standard) is the basis of many risk management standards across the world and underpins the new International Standard ISO 31000. The latest revision places more emphasis on embedding risk management within the organizational culture rather than merely "quantifying risk." Furthermore it considers risk as an opportunity in addition to risk as a threat. This represents a sea change in risk management which was once linked closely with the blame culture and has now matured into a fair and just culture.

Essentially the process comprises four steps:

1. Risk identification
2. Risk analysis
3. Evaluation of risk
4. Risk control

Like any management process, communication and monitoring is essential at every stage.

ISO 14971:2007

This international standard on the application of risk management to medical devices is aimed at the manufacturing industry and outlines in some detail a process to identify the hazards associated with all medical devices. It describes a method of risk assessments, identification of control measures, and review. This standard should be read in conjunction with BS EN ISO 60601-1-6:2010, Medical Electrical Equipment − Part 1: General Requirements for Basic Safety and Essential Performance, and any specific ISO 60601 standards for the device in question. It is strongly recommended that the implementation of a risk management system for medical devices becomes part of an overall quality management system (QMS); the associated international standard is ISO 9001:2008, Quality Management Systems − Requirements.

The ISO 14971 standard states (BSI, 2007, p. 5):

> The manufacturer shall establish, document and maintain throughout the life-cycle an ongoing process for identifying hazards associated with a medical device, estimating and evaluating the associated risks, controlling these risks, and monitoring the effectiveness of the controls.

The process is divided into four parts:

1. Risk analysis: The identification of the hazards and estimating the potential risk that they pose
2. Risk evaluation: Determining if the risk is acceptable or if further reduction is necessary
3. Risk control: Identification and analysis of possible control measure options and their implementation
4. Product and post-production information: A system to collect and review information about the medical device during the production stage in manufacturing but also subsequently in use during installation, maintenance, or operation

Failures "in the field" may well be reported back to manufacturers through the competent authority (CA) and will result in an incident investigation. Readers are directed to the International Standard ISO 13485:2003, Medical Devices − Quality Management Systems − Requirements for Regulatory Purposes, for additional guidance.

To demonstrate that a device has been designed to be safe, manufacturers are required to undertake risk assessments and compile a risk management file. BS EN ISO 14971:2007, Medical Devices, Application of Risk Management to Medical Devices, provides a framework to systematically assess the risks. Any EBME department undertaking the design and manufacture of devices for use either in their hospital or to be placed "on the market" should therefore undertake a systematic estimate and evaluate the associated risks and ensure that this is documented appropriately.

The resultant risk management file can be used to inform the basis of subsequent reliability centered maintenance. Readers are directed to Moubray (1992), which gives a detailed insight into reliability centered maintenance.

The risk management process as outlined in BS EN ISO 14971 includes a subset of risk assessment. There are six phases to the risk management process: risk analysis, risk evaluation, risk control, residual risk evaluation, risk management report, and production/post-production information.

Risk Analysis

Theoretical perspectives on risk will not be elaborated on here as they have already been discussed in a recent Institute of Physics and Engineering in Medicine (IPEM) report entitled Risk Management and its Application to Medical Device Management (Brown and Robbins, 2007). Risk analysis requires the identification of key safety characteristics of the device, and identification of the hazards. For each of the hazards an estimation of the risk should be made. Essentially this is about considering the consequences (severity) and likelihood (probability) of a hazard becoming a risk. This is a systems approach to failure as conceptualized by Reason's (2000) "Swiss cheese" model. "The systems approach concentrates on the conditions under which individuals work and tries to build defenses to avert errors or mitigate their effects" (Reason, 2000, p. 768).

To ensure consistency, each of these terms should be defined or quantified in the risk management file. Typical definitions are given in Table 6.1 and Table 6.2.

When considering medical device design, a risk may be the failure of a safety critical component, in which case the likelihood may be estimated in mean time to failure (MTTF), for risks where is not possible to estimate the likelihood, for example, of software failure, or malicious tampering the worst-case scenario should be used. ISO 14971 gives some useful examples of sources of information or data for estimating risks (BSI, 2007, p. 10):

- Published standards
- Scientific technical data
- Field data from similar devices already in use, including published reported incidents
- Usability tests employing typical users
- Clinical evidence
- Results of appropriate investigations
- Expert opinion
- External quality assessment schemes

TABLE 6.1   Typical Definitions for Consequences

Negligible (none)	Treatment /diagnosis error resulting in minimal or no detectable effect to patient condition
Minor	Minimum harm/injury to patient with short-term consequences, e.g., additional monitoring or minor treatment
Serious (moderate)	Significant but not permanent harm with medium-term consequences and requiring further treatment
Critical (major)	Significant injury with long-term/permanent consequences
Catastrophic	Resulting in patient death

TABLE 6.2   Typical Definitions for Likelihood

Improbable (rare)	Rarely occurs, >1–5 years
Remote (unlikely)	Not expected to happen more than yearly
Occasional (possible)	May reoccur occasionally, >6 monthly
Probable (likely)	Likely to reoccur >monthly
Frequent (almost certain)	Frequently reoccurs >weekly

Risk Evaluation

Having estimated the risk for each hazardous situation, the risk must be evaluated.

For most practical purposes this will be "measured" using a risk matrix. Most healthcare organizations use a 5×5 risk matrix based on the Australian Standards AS/4360:2004 where the two axes correspond to scales for consequences, sometimes call severity and likelihood. A typical example is shown in Table 6.3.

Risk Control

Following on from risk evaluation we must now put in place processes to control the risks. The introduction of control measures (barriers) can reduce the consequences (severity) or reduce the likelihood (probability), either way this will diminish the risk. There is a hierarchy for controlling risks. The first and most preferable is elimination by using a different approach to achieve the same goal. Where elimination is not feasible then control

TABLE 6.3 Example of 5 × 5 Risk Matrix

Consequences	Likelihood				
	Improbable	Remote	Occasional	Probable	Frequent
Negligible	1	2	3	4	5
Minor	2	4	6	8	10
Serious	3	6	9	12	15
Critical	4	8	12	16	20
Catastrophic	5	10	15	20	25

TABLE 6.4 Risk Rating Numbers

Risk Rating	Risk Rating Number
Very low	1–7
Low	8–12
Medium	13–15
High	16–25

measures must be put in place. The hierarchy of control is (CIEH, 2005, Figure 6.5):

1. Elimination or avoidance
2. Substitution
3. Controlling risks at source
4. Separation and isolation
5. Safe working procedures
6. Training, instruction, and supervision
7. Personal protection
8. Other considerations: welfare facilities, first aid, emergency procedures

Lack of control can compromise risk, however careful attention to manufacturing process can minimize the risk. Techniques such as hazard and operability (HAZOP) studies, which originated in the British chemical industry, could prove useful.

Risk Management Report

Having put in place systems to either minimize or eliminate risk, there now remains only overall residual risk. It is therefore necessary to re-evaluate the risks to determine whether the overall residual risk posed by the medical device is acceptable. Inevitably there will be some risks that cannot realistically be eliminated. There are two possible outcomes from the evaluation of the overall residual risk:

1. Risk unacceptable: This will require the collation of further evidence/literature to determine if the clinical benefits outweigh the overall risk. If judged acceptable the evidence to justify this decision should be included in the risk management file; if this is not the case the risk remains unacceptable.
2. Risk acceptable: The manufacturer should decide what evidence is pertinent to include in the accompanying document in the risk management file.

According to ISO 14971 "Compliance is checked by inspection of the risk management file and the accompanying documents" (BSI, 2007, p. 13).

Risk Control

Least Effective

Most Effective

Other considerations/Em. Procedures

Personal Protection

Training, Instruction & Supervision

Safe Working Procedures

Separation or Isolation

Control at Source

Substitution

Elimination or Avoidance

- For any risks identified control measures should be put in place to mitigate the risk
- "Hierarchy of Controls"

FIGURE 6.5   Hierarchy of risk control. *Source:* CIEH, 2005.

Production and Post-Production Information

The risk management report is an important document in the risk management file and acts as a check or quality control which provides an assurance that the risk management plan has been implemented correctly, the overall residual risk is acceptable, and that mechanisms are in place for the compilation of production and post-production information. This information can be drawn on for the issuing of Field Safety Corrective Actions (FSCA) through the MHRA, and, where necessary, Medical Device Alerts (MDA).

One way of ensuring that processes are consistently followed is through the introduction of a quality system that is externally certified, such as BS EN ISO 9001:2001.

RISK TOOLS

Monte Carlo Modeling

Monte Carlo analysis is a widely used qualitative risk analysis technique. It is a powerful technique but does require considerable preparatory work and judgment to ensure it models the risk appropriately. Although traditionally used in project management, it could be successfully used in clinical engineering for a major re-equipping project or a new operating theater suite.

Monte Carlo simulation is a complex stochastic technique used to solve a wide range of mathematical problems. Monte Carlo methods randomly select values from a given distribution to create multiple scenarios of a problem. Each time a value is randomly selected, it forms one possible scenario and solution to the problem (Cummins et al., 2009).

A key feature of this technique is that it offers an approach that comprises the assessment of any number of individual risk events with the analysis that concerns overall project risk. Essentially there are three elements (APM, 2010, p. 153):

- Risk estimates
- Monte Carlo modeling
- Simulation and analysis of results

Risk Estimation

The first step is risk estimation and is based on the assumption that it is possible to estimate the outcome of a risk event on a continuous probability distribution. Cost and time are common parameters in clinical engineering, though patient safety may be a more useful parameter. The *Project Risk Analysis and Management Guide* (APM, 2010, p. 148–150) suggests five factors to be considered when estimating risk:

1. Understand the implications of the probability distribution shape (e.g., Beta PERT, triangular, and general triangular).
2. Consider all relevant sources of uncertainty.
3. Follow a process designed to avoid bias and unrealistic low income variance.
4. Avoid assumptions that the most likely value is the one that is planned.
5. Adopt a lessons-learned approach to compare estimate against actual incomes.

Hazard and Operability Study

A hazard and operability (HAZOP) study is a systematic process for examining a system to identify potential hazards and potential operability problems. Hence HAZOP can be used to inform risk assessments.

HAZOP studies require more detail than other techniques, such as failure mode and effects analysis (FMEA) and fault tree analysis (FTA), regarding the systems under consideration, but produce more comprehensive information on hazards and errors in the system design (IEC 61882:2001). The international standard IEC 61882 was prepared by the International Electrotechnical Commission (IEC) Technical Committee 56. Readers are directed to this standard for further guidance and specific examples of the study process.

The HAZOP process is designed to be used by multidisciplinary teams, therefore in the healthcare sector this could be the clinical engineer in conjunction with a clinical colleague. The composition of the clinical team (e.g., surgeons, physicians, nurses, physiotherapists, and radiographers) will very much depend on the type of system being examined. Alternatively if this is part of a design or development process for a new medical device the team would comprise the designer of the electronic circuitry, the mechanical engineer who specified the enclosure and interface ports, for example, a clinical engineer, and a maintenance technician.

The HAZOP study procedure has four key steps and is shown as a flowchart in Figure 6.6.

The following four subsections look at each of the steps in more detail.

Definition

Before a study is started both the scope and objectives must be clearly defined. This is paramount for a successful study and will ensure that "the system boundaries and its interfaces with other systems and the environment are clearly defined" (BSI, 2001, p. 15). The objectives of a study can be best described as stepping stones toward the overall aim, and these must be specific and measurable.

An example of a team in the healthcare sector was described earlier; each member of the team must be assigned a role. The key roles include the study leader, a recorder, designer, user, specialist, and a representative from the maintenance staff.

Preparation

The study leader has a pivotal role and must be trained in HAZOP techniques. This person assumes the responsibility for all of the preparatory work including obtaining and formatting the information, and organizing the meetings. In many ways, for a HAZOP study, the leader will act as the project manager and use similar techniques to the PRINCE 2® project management methodology.

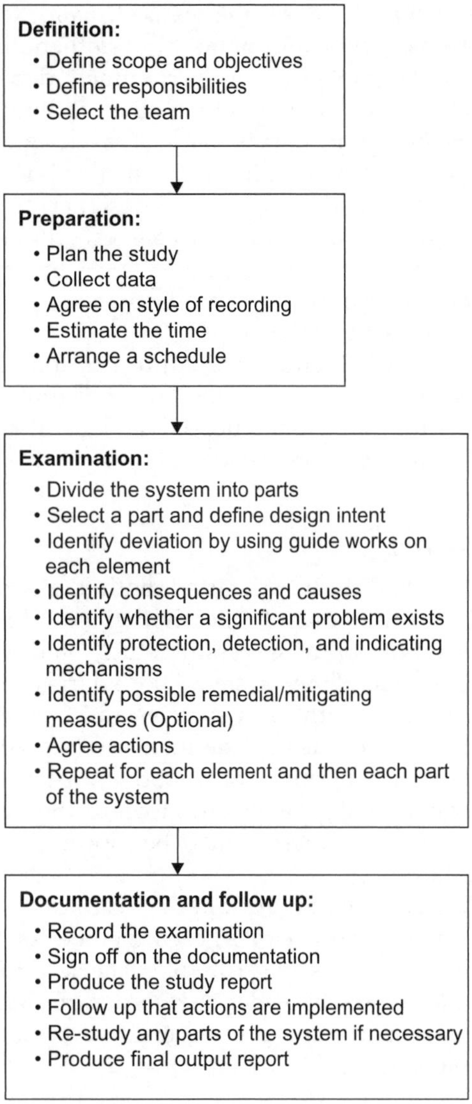

Definition:
- Define scope and objectives
- Define responsibilities
- Select the team

Preparation:
- Plan the study
- Collect data
- Agree on style of recording
- Estimate the time
- Arrange a schedule

Examination:
- Divide the system into parts
- Select a part and define design intent
- Identify deviation by using guide works on each element
- Identify consequences and causes
- Identify whether a significant problem exists
- Identify protection, detection, and indicating mechanisms
- Identify possible remedial/mitigating measures (Optional)
- Agree actions
- Repeat for each element and then each part of the system

Documentation and follow up:
- Record the examination
- Sign off on the documentation
- Produce the study report
- Follow up that actions are implemented
- Re-study any parts of the system if necessary
- Produce final output report

FIGURE 6.6 The HAZOP study procedure. *Source: British Standards Institution, 2001, p. 9.*

Key to the preparatory work is the development of a study plan which will include (BSI, 2001, p. 17):

- Objectives and scope of the study
- A list of participating members

- Technical details
  - A design representation divided into parts and elements (sometimes called nodes) with defined design intent, and for each element a list of components, materials, and activities and their characteristics
  - A list of proposed guide words to be used and the interpretation of the guide word- element/characteristic combinations
- A list of appropriate references (this may include data sheets, relevant design standards, etc.)
- Administrative arrangements, schedule of meetings including their dates, times, and locations
- Form of recording required (usually recorded using a word-processing package on a PC)
- Templates that may be used in the study

DESIGN DESCRIPTION

In clinical engineering the design description may include functional block diagrams, electronic circuit diagrams, printed circuit board layouts, component data sheets, and mechanical engineering diagrams. For programmable systems this would include programming information, logic and timing diagrams, and state transition diagrams. The combination of these technical diagrams will provide a rich source of reference material to inform the examination process.

GUIDE WORDS AND DEVIATIONS

It is the study leader's responsibility to compile the initial list of guide words, and care must be taken to ensure that they are neither too specific that they may limit ideas or too broad so they are not well defined. The compilation of the list of guide words is therefore of paramount importance. Some examples of

TABLE 6.5 Guide Word Examples

Deviation Type	Guide Word	Example
Negative	No	No physiological, control, or data signal passed
Substitution	Reverse	Of no consequence
	Other than	The physiological, control, or data signals are not correct
Time	Early	Signal not synchronized to clock, arrives early
	Late	Signal not synchronized to clock, arrival delayed
Order or sequence	Before	Signals or events occurred in the wrong sequence, before intended
	After	Signals or events occurred in the wrong sequence, after intended

guide words suitable for clinical engineering are given in Table 6.5.

The guide word/element associations as shown in Table 6.5 can be thought of as a risk matrix. To produce comprehensive hazard identification, all possibilities must be considered. There may be gaps in the matrix where combinations are not credible or realistic. There are many variations on risk matrices, for example: probability—impact grid, risk map, and summary risk profile. Further guidance can be found in the OGC book entitled *Management of Risk: Guidance for Practitioners* (OGC, 2007).

Once a deviation is identified and its risk impact in terms of consequences and probability is determined, the next step is to determine what actions, if any, can be taken to mitigate or reduce the risk. Following the implementation of these actions there may still be a residual risk present which must be graded, and then a decision is made whether this risk is considered acceptable for the organization.

The examination process should be well documented and the study leader will have determined the requirements for documentation at the planning stage. There are several iterations of the examination process for each guide word applied to every element or component of the system.

Examination

The examination meetings are where the deviations to elements of the system are considered, and they must be structured according to the study plan to achieve the best outcomes. A particular element/node (characteristic) should be selected and analyzed following its (correct) sequence and then the next element should be considered. It is often helpful to sequence the examination logically following either a process from the start to its destination or tracking a signal from the input to the output.

Central to the examination is the use of guide words, which structure a specific search for deviations from the intent of the design. This demonstrates the importance placed on developing an adequate guide word list with suitable definitions.

IEC 61882 describes elements as: "...discrete steps or stages in a procedure, individual signals [in clinical engineering this may also be psychological waveforms] and equipment items in a control system, equipment or components in a process or electronic system" (BSI, 2001, p. 10). The size of an element may depend on the complexity of the system. For a simple system this may be an individual component (e.g., a sensor, transducer, or valve actuator); for a larger system this may be a functional block such as the ECG front end in an electrocardiograph.

It is useful to express the element in terms of its input signals, functionality, and the output to either the destination (such as an LCD display) or linked to the next processing stage of the system that will be the next element for investigation.

Documentation

Like many systematic logical processes, HAZOP is extremely powerful if used with diligence and thoroughly documented. The extent of the documentation and recording may be determined by the complexity of the study, legislation, and regulatory or contractual obligations. Essentially recording falls unto one of two types: full recording or recording by exception. Full recording is obvious, whereas exception reporting is only documenting the problems highlighted, whether they are hazards or operability. For robustness, full recording is the gold standard.

The use of templates or worksheets assist in providing clarity and consistency and one worksheet should be used for each element or node. The design of the worksheet and level of detail recorded will be dependent on individual needs. Proprietary word-processing packages or spreadsheets provide simple and convenient solutions to recording. The header should include the project title, design intent, element or part under examination, and names of the team members. The footer will include the file name, date, and page number.

A table can then be constructed using the following column titles: reference number, element or part, guide word, deviation, cause, consequences, control measures or barriers, severity, risk rating, action plan, priority, the actionee, and finally the status.

For completeness it is useful to record all of the reference documents referred to during the examination, such as circuit diagrams, exploded views, and regulatory documents and design standards. The culmination of all the examinations will lead to the study report. The report should comprise:

- An executive summary
- Conclusions from the study

- Agreed scope and objectives of the project
- Identified hazards and operability problems
- Recommendations for changes in particular aspects of the design, and actions required to mitigate uncertainties and problems

Appendices to the report will include:

- Completed worksheets
- Listings of diagrams and drawings
- Reference list of standards and regulations together with relevant information from previous studies

Probability Impact Grid

The probability impact grid is one of the techniques used to assess the likelihood of a threat or opportunity materializing and their potential impact. It is a qualitative technique used to rank previously identified risks. The probability (sometimes called consequences or impact scale) is a measure derived from percentages. The scale can be divided into any number of parts; typically a five-point scale is used together with a five-point impact scale, thus producing a 5×5 risk matrix.

The Australian Standard AS/NZS 4360:2004 and the International Standard ISO 14971:2007 both use a 5×5 risk matrix. The scales for each access do not need to be the same, and can equally be linear or logarithmic. Table 6.6 shows an impact scale where the bands are all less than 1. Typically this may describe cost or time; in clinical engineering we would be interested in patient safety and reliability (failure rates).

Table 6.6 shows qualitative words to describe the impact scale specific to patient safety.

Risk Map

Another form of risk estimation is the risk map (Table 6.7). In this case the term *likelihood*

TABLE 6.6 Probability Impact Grid[a]

PROBABILITY	1.0	Certain 81–100%	0.2	0.4	0.6	0.8	1.0
	0.8	Almost certain 61–80%	0.16	0.32	0.48	0.64	0.8
	0.6	May happen 41–60%	0.12	0.24	0.36	0.48	0.6
	0.4	Unlikely 21–40%	0.08	0.16	0.24	0.32	0.4
	0.2	Very unlikely 0–20%	0.04	0.08	0.12	0.16	0.2
			No harm				Death
			0.2	0.4	0.6	0.8	1.0
			IMPACT				

[a]*Values given are for illustrative purposes only, representing patient safety.*

TABLE 6.7 Example of a Risk Map

Likelihood	Very high				
	High				
	Medium				
	Low				
	Very low				
	Very low	Low	Medium	High	Very high
	Impact				

TABLE 6.8 Example of RAGB Grading

Color	Status
Red	No progress, remains high risk
Amber	Moderate progress bus still poses a significant risk
Green	Good progress made with evidence of deliverables
Blue	Action completed and deliverables achieved

is used instead of probability. The classifications of risk at maximum would be high, medium, and low, which produces a 3 × 3 risk matrix. However, more often this is extended to a 5 × 5 matrix. Examples of this application are a tool for assessing the situational and contextual risk factors when using medical devices (Brown, 2004, 2007) and also assessing the risks associated with undertaking clinical trials to highlight the patient risk, PR risk, and financial risk for the organization (Brown, 2011).

Clearly this taxonomy of grading risk has little meaning unless each of the terms is given an explicit meaning. Failure to provide a definition will expose the term to be interpreted differently by people dependent partly on their experience and knowledge in the particular area where the assessment is taking place.

RAGB Status

Both the probability impact grid and risk map can be further enhanced by the use of RAGB (red, amber, green, blue) status which can translate them into risk registers/profiles (see Table 6.8). The RAGB status can be simplified to RAG rating (red, amber, and green) as is commonly used across many aspects of the healthcare sector from pressure score assessments to preventative maintenance requirements.

References

Association for Project Management (APM), 2010. Project Risk Analysis and Management Guide. second ed. Association for Project Management (APM), Princes Risborough, Buckinghamshire.

British Standards Institution (BSI), 2001. Hazard and operability studies (HAZOP studies) — Application guide BS IEC 61882. British Standards Institution (BSI), London.

British Standards Institution (BSI), 2007. Medical Devices — Application of risk management to medical devices BS EN ISO 14971. British Standards Institution (BSI), London.

Brown, A.S., 2004. Finding the hidden risks with medical devices: a risk profiling tool. J. Qual. Prim. Care. 12, 135–138.

Brown, A.S., 2007. Identifying risks using a new assessment tool: the missing piece of the jigsaw in medical device risk assessment. Clin. Risk. 13 (2), 56–59.

Brown, A.S., 2011. Clinical trials: a new risk assessment tool. Clin. Governance. 16 (2), 103–110.

Brown, S, Robbins, P (Eds.), 2007. Risk Management and its Application to Medical Device Management — Report. Institute of Physics and Engineering in Medicine, York.

Chartered Institute of Environmental Health (CIEH), 2005. Risk Assessment: Principles and Practice — Level 2 Course.

Cummins, E., Butler, F., Gormley, R., Brunton, N., 2009. A Monte Carlo Risk Assessment Model for Acrylamide Formation in French Fries. Risk Anal. 29 (10), 1410–1426.

European Parliament and the Council of the European Union, 2007. Directive 2007/47/EC. Official J. Eur. Union L247.

London, A.J., 2006. Reasonable risks in clinical research: a critique and proposal for the integrative approach. Stat. Med. 25 (17), 2869–2885.

Medicines and Healthcare Products Regulatory Agency (MHRA), 2003. Adverse Incidents — Our Role. MHRA, London.

Moubray, J., 1992. Reliability Centred Maintenance. second ed. Industrial Press Inc, New York.

Office of Government Commerce (OGC), 2007. Management of Risk Guidance for Practitioners. The Stationery Office (OGC), 2007, Norwich.

Office of Government Commerce (OGC), 2009. Managing successful projects with Prince 2. The Stationery Office (OGC), 2009, Norwich.

Reason, J., 2000. Human error: models and management. BMJ. 320, 768–770.

Rosa, E.A., 2003. The logical structure of the social amplification of risk framework (SARF): Metatheoretical foundations and policy implications. In: Pedgeon, N.F., Kasperson, R.E., Slovic, P. (Eds.), The Social Amplification of Risk. Cambridge University Press, Cambridge.

Royal Society, 1992. Risk: Analysis, perception and management. Report of a Royal Society study group. The Royal Society, London.

Sowden, R., 2011. Best Management Practice: Managing Successful Programmes. The Stationery Office, Norwich.

Further Reading

Brown, S, Robbins, P (Eds.), 2007. Risk Management and its Application to Medical Device Management Report. Institute of Physics and Engineering in Medicine, New York.

ISO 13485:2003 Medical devices _ Quality Management Systems — Requirements for Regulatory purposes

Glossary

Competent authority An agency or body that holds the power to regulate and conduct inspections. This is also sometimes referred to as the regulatory authority.

Mean time to failure (MTTF) The mean or average time to failure of a component or piece of of equipment.

Medicines and Healthcare Products Regulatory Agency (MHRA) The regulatory body and competent authority in the United Kingdom that oversees the safety of medical devices and equipment.

Quality management system (QMS) A system of documents that describes the planned and systematic work practices to ensure a consistent quality. It usually includes standard operating procedures (SOPs) and work instructions (WIs).

Risk A situation involving exposure to danger, or exposure (someone or something valued) to danger, harm, or loss.

Risk appetite This describes the level or amount of risk that an organization or individual considers to be acceptable and to which they are prepared to be exposed.

Statutory instrument The formal method by which a directive is enacted into British law.

The Role of Clinical Engineers in Hospitals

Fran J. Hegarty, John Amoore†, Richard Scott**,
Paul Blackett††, and Justin P. McCarthy****

*Medical Physics & Bioengineering Department, St James's Hospital, Dublin, Ireland, †Department of
Medical Physics, NHS Ayrshire and Arran, Scotland, U.K., **Sherwood Forest Hospitals NHS
Foundation Trust, Nottinghamshire, U.K., ††Lancashire Teaching Hospitals NHS Foundation Trust,
Lancashire, U.K., ***Clin Eng Consulting Ltd, Cardiff, U.K.

INTRODUCTION

Technology has been used from the early days of medicine. Records of early Greek medicine describe the use of simple tools, splints, and crude surgical tools (Milne, 1907). The development of healthcare technology can be illustrated by considering the evolution of a simple cutting blade, through carefully designed surgical knives and by way of sophisticated electrosurgical tools, to precise robotic controlled surgical arms, enhancing, but not replacing, the clinical staff who provide the healthcare. As medicine and technology have advanced over the millennia, healthcare technology has become more sophisticated and more complex, but its essential role of extending the ability of people to deliver healthcare remains. Over the past hundred years, healthcare has become increasingly reliant on medical technology with clinicians dependent on it for diagnosis and treatment. The appropriate deployment of technology contributes to the improvement in the quality of healthcare, the containment of cost, and to increased access to services (David and Jahnke, 2004). Thus an essential activity within hospitals is managing the medical technology and its use. Healthcare technology and its role and applications continue to evolve, as do hospitals and healthcare institutions. Consequently healthcare must continuously evolve, investigate, and, where justified, adopt new technologies to be able to deliver quality healthcare in a cost-effective way. An important development is the increasing trend for healthcare in the community to meet increasing demands for healthcare in response to demographic trends, to meet demands for patient-centered care, and to contain rising costs (Smith et al., 2012). The management and care of the healthcare technology in the community will continue to bring new challenges.

Engineers create products and processes to improve the delivery of healthcare. In this book the products and processes being considered are electro-medical devices, systems, and the processes within which they are used. The term *electro-medical* covers devices such as MRI scanners, defibrillators, endoscopes, and ECG recorders. Medical dramas and science documentaries on TV have brought these devices to the attention of the general public. Devices are rightly portrayed as complex high-tech machines that are reliable, safe and used expertly by clinicians. As with technology like the aircraft used in the travel industry, the public are optimistic in how they view medical technology and seem happy to put themselves and their safety in the hands of professionals who use these electro-medical devices. However, the ubiquitous electro-medical devices bring with them not only benefits but also risks. Electro-medical devices and systems, like aircraft, bridges and other pieces of technology infrastructure, need to be actively managed to ensure that the benefits they bestow far outweigh any risk associated with their use, and that these risks are managed, controlled, and have methods put in place to minimize them.

Today's electro-medical devices incorporate optics, electronics, mechanics, computing, digital signal processing, and sensors of all types. They often make intimate connection with the human body to deliver fluids, or energy in the form of electrical current, or ionizing and nonionizing radiation. These devices need to be carefully controlled to ensure their effectiveness, accuracy, and safety.

The complete range of electro-medical devices and systems bought by a hospital constitutes a valuable financial asset with capital and revenue resourcing requirements. Their selection, procurement, upkeep, and life cycle needs to be carefully managed to ensure they are cost effective, up to date, and continue to support the corporate objectives of the hospital. What technology is required to support the healthcare delivered by the hospital? What resources (clinical staff, space, infrastructure, consumables, and finance) are required for its application? When should technology be removed from service and if so should it be replaced? The requirement to actively manage these assets has resulted in the emergence of a specialist strand of engineering dedicated to the pursuit of excellence in the application of technology in the clinical setting. Clinical engineering is the name given to this discipline. Those who practice it are called Clinical Engineers. Clinical engineering is a particular specialization of the discipline of biomedical engineering. What differentiates clinical engineering from biomedical engineering is that the activity and those who deliver it are based in the hospitals at the point of care. Clinical engineering as a discipline is concerned with the application of engineering tools and theory to all aspects of the diagnosis, care and cure of disease and life support in general, all of which are embraced in the delivery of healthcare services (Geddes, 1977).

Clinical engineers are individuals employed in hospitals to advance care through actively supporting the application of technology at the point of care. They assure positive outcomes from clinical practices predicated on the use of electro-medical devices. On one level this requires measures to be put in place to assure the technical performance, maintenance, and quality control of the devices themselves, their usability, and support for their use, both in terms of staff training and in the supply of accessories and consumables. On another level it requires the provision of expert and independent advice on what technology to deploy and how to best use technology in the clinical setting. In some instances this may extend to the development of a particular or bespoke device, data processing, or other engineering solutions to solve a specific problem or to meet and advance a clinical need. Within the complex environment of the modern hospital, clinical engineering activity is concerned primarily with devices but recognizes that interactions between drugs, procedures, and devices commonly occur and must be understood and managed to ensure safe and effective patient care (Dyro, 2004). The clinical engineer is the expert in the engineering and system science underpinning the application of the electro-medical devices and systems. Clinical engineers operate at the point of care and it is this firsthand knowledge of the day-to-day needs within the patient environment and care process that allows the term *clinical* to be used (Aller, 1977).

Clinical engineering services delivered within healthcare delivery organizations are provided by a range of individuals each with specialist expertise, qualifications, and skill. In this chapter the term *clinical engineer* is used to describe all engineers and scientists who provide clinical engineering services regardless of their employment role or level of professional development.

CLINICAL ENGINEERING AS APPLIED BIOMEDICAL ENGINEERING

Clinicians working in modern hospitals use medical devices to diagnose, treat, and monitor patients. These technologies have been developed to improve clinical outcome, reduce trauma, and support staff in optimally delivering effective care.

Advances in the field of optics have resulted in the development of advanced laparoscopy and endoscopy with high-quality video imaging systems that have revolutionized practices in gastroenterology and surgery, where keyhole surgery has transformed the delivery of care with benefits for patients and healthcare institutions. Developments in keyhole and open surgical procedures have been made possible by sophisticated electrosurgical diathermy devices that allow surgeons to deliver radio frequency electrical currents to the body in precise ways allowing them to cut, dissect, coagulate, and ablate tissue in a highly controlled manner. Advances in modern electronics allow for feedback to be built into these devices so that the device controls and alters the cutting current and waveform in real time in response to changes in the tissue at the treatment site.

Modern anesthetic workstations are complex electromechanical-optical systems that deliver precise gas mixtures to patients during surgery. The development of small and reliable spectroscopy systems allows these gas mixtures to be measured and displayed in real time. Anesthetists use different optoelectronic sensors and electrophysiology measurement methods to monitor the patient's physiology during surgery. These devices are configured into systems that allow data from all these sources to be processed and analyzed to provide new information for anesthetists to help them guide the course of care. Often data from such devices are stored in databases in clinical information systems both for archiving and clinical audit purposes.

Similarly multiparameter physiological measurement devices are used extensively in intensive care units. Modern intensive care ventilators, like anesthetic machines, incorporate physiological measurement modules and software that allow the performance of the ventilator to be precisely set up and to act in harmony with the patient's own physiology and breathing patterns. Medical imaging systems use ionizing (CT scanners, X-ray machines, gamma cameras) and non-ionizing (MRI machines, ultrasound scanners, thermal imaging systems) to provide sophisticated images of anatomy and physiological function to support clinical decision making.

So doctors, nurses, and paramedics with a background in the life sciences find themselves using complex technology every day in their clinical practice. The technology relies on complex optical systems, lasers, and high frequency electrical current generators, electromechanical and optoelectronic sensors, bioamplifiers, and myriad associated software-based signal processing systems and information technology.

It is not surprising then that clinicians using such technology collaborate with clinical engineers at the point of care. With knowledge of the engineering, physics and system science underpinning electro-medical devices and systems, and working collaboratively in the clinical environment, clinical engineers are well placed to process, analyze and critically review the data produced by all of this technology. They play an important role in ensuring positive outcomes from the use of technology and reducing negative ones. Clinical engineers manage the devices, their integration into systems, and the clinical environment so that they perform accurately and reliably. Failure or misuse of a device may result in unintended harm to the patient or user. A failure or misuse that results in a data error, if it is not detected, may negatively influence the course of a patient's care. So in the day-to-day delivery of healthcare, which is predicated on the use of electro-medical devices and systems, there is a need for clinical engineers to be based at the point of care, to exercise due care, and conscientiously monitor the deployment and use of technology in the clinical setting. In doing so, clinical engineers should imaginatively foresee hazards and act to mitigate them. This can take the form of projects to redesign the processes in which devices are used, advice on how to better implement the device within clinical practice, or indeed advice on replacement of the device with new improved technology.

Clinical engineers also use their expertise and particular understanding of the use of devices at the point of care to contribute to international work concerned with the ongoing development of essential standards that new medical devices must meet. They also report back to industry and both national and international regulatory bodies on the performance of specific devices. Given the diversity, complexity, and ubiquity of medical devices in the modern hospital, their safe and effective use requires ongoing training. Clinical engineers contribute significantly to the development and delivery of training courses for all staff who use technology to deliver care. They also contribute to undergraduate and postgraduate education programmes and continuous professional development for doctors, nurses, paramedics, and biomedical and clinical engineers.

The practice of clinical engineering draws on all the other subdisciplines of biomedical engineering. Consequently, you can find clinical engineers who are experts in medical electronics, rehabilitation engineering, clinical measurement, and, increasingly, clinical informatics. In their practice they work directly with clinicians in supporting clinical practice and may work directly with patients as appropriate. What differentiates clinical engineers from other biomedical engineers working in academia, industry, or the regulatory environment is their direct involvement with the delivery of care in hospital and community settings.

CLINICAL ENGINEERING ACTIVITIES

The practice of clinical engineering in modern hospitals is difficult to define for two reasons. First the activity is interdisciplinary. Furthermore, the scope of the activity undertaken by a clinical engineering department varies from institution to institution depending on the jurisdiction, regulatory environment, corporate governance, maturity of the organization, and corporate priorities. Within this broad remit, the nature of the work depends on the experience and focus of the individuals, both within clinical engineering and within the healthcare institution. So it is impossible to describe here a single defined approach that would be applicable for all healthcare delivery organizations. The approach taken in this book is to develop a generic description, modeled on a large teaching hospital that will be used to explain, illustrate, and justify common approaches taken.

Clinical engineering activities will be discussed under the following two headings:

- Supporting and advancing care
- Healthcare technology management

Supporting and Advancing Care

The supporting and advancing care role includes a range of activities provided by clinical engineers to facilitate the hospital management and clinical staff to effectively integrate healthcare technology into clinical practice. The focus is on collaboration relating to the application of technology or using engineering skills to solve clinical, research, or process problems rather than specific technical issues with equipment.

Clinical Support

Clinical engineers based at the point of care are rightly regarded as a valuable resource to those who use advanced technology to deliver care. It is not uncommon for clinical engineers to contribute to the diagnosis and treatment of patients by facilitating the application of new devices or novel methods at the point of care. In some instances clinical engineers will work alongside clinicians applying complex or novel technology at the point of care, making measurements and analyzing data. Some

specialists, such as rehabilitation engineers or clinical measurement scientists, work directly with patients. Other clinical engineers based at the point of care assist with the application of technology at the bedside, but do not have a direct clinical role. With the increase in the reliance on technology comes a requirement for experts in engineering and science to support and advance the application of these technologies by clinicians.

Clinical Informatics

Clinical informatics is interdisciplinary, operating at the intersection of clinical care, the health system, and informatics and communications technology. Supporting this is the integration of medical devices with the eHealth systems (ECRI Institute, 2008). Increasingly, clinical engineers are taking a lead role in informatics projects. This is not surprising given their well-established role in supporting the application of technology at the point of care.

Managing the Clinical Environment

Clinical engineers can contribute to the design and control of the clinical environment itself. Where new facilities are being built or existing ones upgraded, clinical engineers can play a pivotal role in developing the design brief and acting as facilitators of a conversation between architects, civil and structural engineers, hospital capital planning teams, and the clinical staff. This includes understanding and discussing the interaction between and requirements for different medical devices within the clinical environment. For example, the development of a new CT scanning facility must consider the requirement for anesthetic procedures and hence the supply of anesthetic gases and the optimum position of the anesthetic machine and patient monitors. Similarly, the development of a renal dialysis unit must consider the space and utilities required for specialized chairs and dialysis equipment and the requirement for pure water and dialysis chemicals supply. Development of the equipment-intensive environments of critical care and theatre requires particular planning for the medical equipment that will support and provide critical clinical services.

Optimum Utilization of Electro-medical Resources

The optimum utilization of medical equipment requires that the type and quantity of medical equipment be correctly specified and that the equipment is appropriately tailored and configured to meet the needs. Modern electro-medical devices are complex, often software controlled, and offer a range of features, setting, and options. Similar to the typical general-purpose personal computer with its operating systems and general applications packages, the hardware and software features of medical equipment need to be tailored to the clinical requirements. Some of this is achieved before the procurement phase, with the specification of the required options. However, the process also requires that during the commissioning phase, the applicable features are selected and configured. In practice, the configuration is a complex task usually undertaken by a super user who understands the clinical requirements working with a clinical engineer who understands the technology.

Configuration includes mode of operation and selection of parameters and their details. The configuration required varies with the device. For a patient monitor it includes how the various vital signs are displayed and the setting of alarm limits and filters for signals such as the electrocardiogram. For an infusion device it includes infusion rate limits, operating mode, and alarms to help alert the user to problems with the device's operation. Configuration includes tailoring the user-operating environment and connections to networks and hospital information systems. The multidisciplinary expertise of clinical engineers, embracing both clinical and technical knowledge, leads them to be charged with managing the configuration of these devices.

Optimum utilization also requires that the extent to which equipment is in use is actively considered and managed. Planning the optimum utilization requires an understanding of the criticality of having devices available when required and ensuring their availability at the place and time of need. Equipment libraries have been developed to share commonly used devices such as infusion devices, with processes in place to ensure delivery of the devices to places of need. These libraries have traditionally been physical rooms and departments. An alternative is the concept of a virtual library that utilizes equipment-tracking systems (RFID systems) to know where equipment is and whether it is being used and to alert those requiring equipment where to find unused equipment within the organization.

Planning equipment numbers to ensure optimum utilization requires an understanding of the function of the equipment and the clinical demand. Planning the deployment of emergency equipment such as life-preserving/restoring defibrillators must consider the requirement for equipment to be available on demand at the point of need within a short period of time.

Teaching and Training

Optimal and safe use of medical technology requires more than the device to be properly commissioned, installed, and configured. It requires the user to have an understanding of the technology, its characteristics and limitations, and how it can be used to support healthcare. This knowledge and understanding will include an appreciation of the interplay between the patient, clinical and care staff, the technology, and the environment in which the equipment will be used. Teaching and training is another important activity undertaken by clinical engineers. They contribute to university undergraduate and postgraduate teaching and develop and deliver training programmes within the hospital on issues associated with the application of medical devices. Clinical engineers have unique knowledge, skills, and understanding of both the technology (its characteristics and limitations) and of its clinical applications and implementations. This gives the clinical engineer particular skills to support training. Some clinical engineering departments, recognizing the importance of user training, employ staff to provide this training to medical and nursing staff, working in co-operation with the medical and nursing leadership and training departments.

Risk Management

Problems can occur with medical devices arising from device failures, device usability problems, and failure to operate devices correctly (Amoore and Ingram, 2002). The failures may lead to or (but for the timely intervention of clinical staff) could have led to patient harm. Where incidents involving medical devices occur, clinical engineers have an important role in the investigation. When an incident occurs it is important to understand the underlying causes and identifying measures that will minimize the risk of recurrence. Clinical engineers also play a role in implementing the recommendations for improving hospital processes that emerge from such investigations. The need to learn from incidents and share the findings from incident investigations has led to national incident-reporting processes. Where an incident gives cause for concern regarding the function or design of the medical device the hospital's clinical engineer will need to report this to the local regulatory authority.

Innovating Care Processes and Quality Improvement

Many hospitals now have continuous quality improvement programmes in place focused on innovating how care is delivered with the aim of improving safety, controlling costs, and making the service more accessible and effective for patients. With both engineering and system science training, and a practical knowledge of the clinical workflow, clinical engineers can take a leadership role in quality improvement initiatives associated with the application of technology. This includes the implementation of health informatics systems and analysis of the data they produce. Quality improvement initiatives can be effectively delivered at the micro level, within individual units or wards. At this level, where the patient interacts with the healthcare provider, is where quality, safety, reliability, and efficiency are delivered and the patient's experience of care is created. Clinical engineers actively promote a culture of safety and quality improvement around the use of technology at both the healthcare institution-wide and individual ward levels.

Research and Development

Where the hospital has active research programs clinical engineers contribute scientific support. Research programmes may require computer programming, development of specific unique devices, or application of existing devices. Clinical engineers have the skill and knowledge to deliver solutions through the combination of engineering knowledge and skill and their knowledge and understanding of the clinical context in which a project will be delivered. Hospital-based clinical engineers, who deal every day with safety and quality assurance issues in the clinical environment, have a lot to offer when it comes to taking a research project out of the lab and into the clinical environment. Consequently they are also involved in the innovation phase of new device developments and can be key facilitators of getting prototype designs into clinical practice for evaluation. Clinical engineers' involvement with the development of international standards both draws on and contributes to their effectiveness in this regard.

Healthcare Technology Management

Healthcare technology management is concerned with the active management of the complete range of electro-medical devices and systems within the hospital. The American College of Clinical Engineering in 1992 proposed the following definition: "A clinical engineer is a professional who supports and advances patient care by applying engineering and managerial skills to healthcare technology." (Bauld, 1991.) The inclusion of managerial skills in this definition reflects the fact that within many large hospitals, the clinical engineering department takes direct responsibility for actively managing the electro-medical devices over their useful life.

An inventory of the electro-medical devices must be maintained and expenditure associated with their maintenance and upkeep reported. This activity, often referred to as equipment management, includes the technical maintenance and financial management of all the electro-medical equipment within the hospital. Equipment management activity covers the full life cycle of assets from procurement and their eventual write-off and disposal. Given the vast quantity and diversity of devices in the modern hospital, the financial resources required for their management, and the varied support mechanisms that need to be put in place to achieve this, effective equipment management requires clinical engineers to be skilled in management practice.

Clinical engineers who provide institution-wide equipment management usually work closely with the hospital's risk and finance departments to ensure that optimal support can be delivered within a managed cost control environment. The equipment management function can be considered support for the equipment at each stage in its life cycle. This includes management of healthcare technologies and computer systems that are highly integrated and interoperable. The role extends beyond managing the electro-medical devices themselves, but also associated and networked equipment, in particular where the devices are incorporated into information technology networks. Therefore the term *healthcare technology management* (AAMI, 2011) describes the role more completely. This activity embraces, but is not limited to, a focus on scientific and technical support of electro-medical devices and clinical information technologies, and includes financial stewardship.

The ongoing technical maintenance of healthcare technologies has three key components. These are *scheduled maintenance, performance verification*, and *unscheduled maintenance*. Scheduled maintenance consists of all proactive activities whose purpose is to prevent failure of the device. Performance verification includes all proactive processes that assure devices that appear to be working are working optimally. Unscheduled maintenance covers all reactive actions that are initiated as a result of a reported real or suspected fault or failure of a device or system. Clinical engineers use risk management methodologies to optimize the hospital resources to deliver a healthcare technology management programme that is focused on ensuring that the clinical work of the hospital can be delivered safely, cost-effectively, and optimally. These technology management functions can be delivered by in-house teams of clinical engineers, be contracted to the service engineers of the equipment manufacturers, or be contracted to third-party maintenance groups. In practice it is often a combination of these. One of the key roles of the clinical engineer is to determine the optimum mix of services to meet the organization's need.

Management of the devices takes place within a business model and so clinical engineers work closely with the organization's finance and procurement departments to ensure that not only is the technical equipment management effective but the support arrangements are also an efficient use of financial resources. Consequently, the clinical engineering role includes the development of strategies for the support of equipment, management and regular review of these strategies, and the management of both fiscal resources and personnel.

Clinical engineers are often the drivers of the healthcare technology management programmes within the hospital and develop processes by which the existing medical device inventory is assessed annually to ensure it is efficient and effective. This is a good example of how the advancing care and equipment management roles are complementary and integrated. To be able to advise effectively, clinical engineers should know both the current state of the existing technology deployed in the hospital and be able to provide an overview of new developments, and then be able to articulate how implementing a new technology will affect the hospital's ability to meet its corporate goals.

PATIENT-FOCUSED ENGINEERING

In this chapter we have looked at the specific activity of clinical engineering, based within the hospital. Clinical engineers have particular expertise and experience in the application of technology to the delivery of healthcare. Clinical engineers work in support of those delivering care by contributing to the management of healthcare technology, the clinical environment, and by providing direct support to those involved in clinical practice (Figure 7.1). In doing so, clinical engineers can support a wide range of activities, devices, and systems within the modern healthcare delivery organization.

Clinical engineering works at the point of care, and in many cases clinical engineers have direct patient contact. We have seen that by its nature, clinical engineering is interdisciplinary. Clinical engineers work collaboratively with doctors, nurses, paramedics and hospital finance, risk and quality departments. To function optimally clinical engineers need to remain connected to the wider biomedical engineering community, the medical device industry, and the standards and regulatory community.

The most visible activity of a clinical engineering department engaged in the full gamut of activity described in this chapter is often the maintenance processes, workshops, and test equipment used to deliver the maintenance function. Consequently, clinical engineers are often mistakenly viewed as being primarily electro-medical device maintenance managers. However, nothing could be further from the truth. As advances are made in material and manufacturing science, and improved standards drive the development of excellently designed and engineered devices, the challenge for hospitals has shifted from repairing and maintaining devices to prevent failure, toward optimizing the use and exploiting the potential of new devices and systems to meet both the clinical and corporate mission of the healthcare delivery organization.

Clinical engineering management is complex. Balancing the activities of supporting and advancing care and healthcare technology management is challenging. Medical equipment is sophisticated and complex. Financial controls, constraints, and decision making are complex and often emotional. Managing and balancing competing demands can sometimes blind clinical engineers away from the essential focus of the healthcare system: helping the patient. Whether advancing care through the provision of engineering and scientific support, improving safety and effectiveness of medical technology, or ensuring compliance

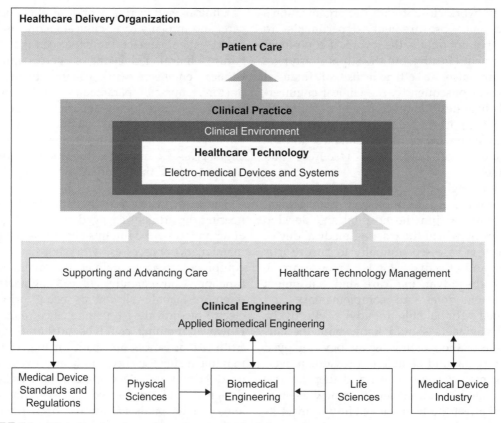

FIGURE 7.1   Clinical engineering and how it supports the delivery of patient care.

with regulations and guidelines, clinical engineering should always remain focused on the care of the patient. Clinical engineering should both support and advance patient care. Therefore patient care is the ultimate reason that clinical engineering exists.

Black and Amoore (2011) in their paper "Pushing the Boundaries of Device Replacement: Introducing the Keystone Model" proposed that during all stages of medical devices procurement, the needs of the patient should be kept in mind. The keystone metaphor introduced in that paper can be extended and used to illustrate the value of developing a patient-care-focused model of clinical engineering. The

supporting and advancing care and the healthcare technology management roles can be considered the pillars of any clinical engineering service. However, it is important to recognize and acknowledge at all times that these roles are complementary and are tightly integrated in practice. The archway shown in Figure 7.2 represents the integration of these two roles. It is only by fulfilling both roles that clinical engineering can completely support the delivery of patient care. The keystone at the apex of an arch is the final piece placed during construction and locks all the stones into position, allowing the arch as a whole to bear weight. We suggest that the keystone of any clinical

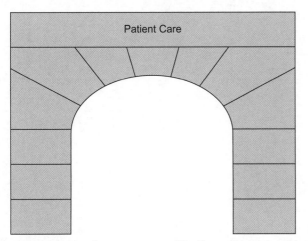

Patient Care

Supporting and Advancing Care   Healthcare Technology Management

FIGURE 7.2 The keystone model of clinical engineering supporting patient care.

engineering service should be that it is focused on delivering processes that improve the patient and carer experience of the application of technology to healthcare.

References

AAMI Press Release May 2011 <http://www.aami.org/news/2011/050611.future.forum.html>

Aller, J.C., 1977. Technology and people. In: Careers, C.A. (Ed.), The Practice of Clinical Engineering, 1977. Academic Press Inc., London, p. 34.

Amoore, J.N., Ingram, P., 2002. Learning from adverse incidents involving medical devices. Br. Med. J. 325 (7358), 272–275, August 2002.

Bauld, T.J., 1991. The definition of a clinical engineer. J. Clin. Eng. 403–405, Sept–Oct 16.

Black, P., Amoore, J.N., 2011. Pushing the Boundaries of Device Replacement: Introducing the Keystone Model. European Medical Physics and Engineering Conference, Trinity College Dublin, Ireland, September 2011.

David, Y., Jahnke, E.G., Planning Hospital Medical Technology. IEEE Engineering in Medicine and Biology Magazine, May–June 2004.

Dyro, J.F., 2004. Introduction to the Clinical Engineering Handbook. Elsevier Academic Press.

ECRI Institute, 2008. Coping with convergence: a road map for successfully combining medical and information technologies. Health Devices. 2008, 293–304.

Geddes, L.A., 1977. Historical background of interdisciplinary engineering. In: Careers, C.A. (Ed.), The Practice of Clinical Engineering. Academic Press Inc., London, p. 17, 1977.

Milne, J.S., 1907. Surgical Instruments in Greek and Roman Times. Clarendon Press, Oxford.

Smith, M., Saunders, R., Stuckhardt, L., McGinnis, J.M. (Eds.), 2012. Best Care at Lower Cost: the path to continuously learning healthcare in America. Institute of Medicine, September 2012. <www.iom.edu/Reports/2012/Best-Care-at-Lower-Cost-The-Path-to-Continuously-Learning-Health-Care-in-America.aspx> (accessed May 2013).

PART II

INFORMATION TECHNOLOGY AND SOFTWARE ENGINEERING

Paul S. Ganney, Sandhya Pisharody, Ed McDonagh, and Edwin Claridge

Overview

Probably the greatest leap forward in the field of medical technology in recent years has been the adoption and adaptation of computer hardware and software. Early uses centered around record-keeping (word processing of letters, then databases) but the design of the personal computer as a generic device (rather than a specific one) has enabled it to be used in more and more fields. It is therefore not a surprise that medical uses have been found and developed also.

Part II first introduces the regulatory issues that enable the effective use of computers in medicine, then deals with the requirements for the development and management of software, including databases. The part concludes with a chapter on web hosting and programming.

8

Information Communications Technology

Paul S. Ganney

University College London Hospitals NHS Trust, London, U.K.

THE REGULATION OF CLINICAL COMPUTING

The Use of a Computer as a Clinical Device—MHRA, FDA

Introduction

Putting a computer into clinical, as opposed to administrative, use brings with it additional requirements in the form of regulations. The requirements of IEC 601 Medical Electrical Equipment family of standards[1] and the Medical Devices Directive (MDD) (both of which are discussed in the next section, "Regulatory Standards") need to be addressed. It is worth noting that any electrical device that is within 1.5 m of a patient[2] is deemed to be in the "patient environment" and thus becomes subject to the safety requirements of IEC 60601, even if its primary purpose is not clinical. Thus equipment on trolleys may be

considered as likely to come into this environment and should be assessed in the same way as equipment permanently there.

It is allowable to bring nonmedical equipment into the patient environment provided that the following two points are satisfied:

- It meets the safety standards relevant to its own type (e.g., IEC 60950, safety of information technology equipment).
- It meets the single fault touch current requirements for medical equipment (i.e., 500 μA).

The U.S. Food and Drug Administration (FDA) and the United Kingdom's Medicines and Healthcare Products Regulatory Agency (MHRA) are national regulators setting standards of compliance that must be achieved before a product can be placed on the market in that country.

MHRA

The MHRA is an executive agency of the United Kingdom's Department of Health. It is responsible for the regulation of medicines

---

[1] Of which 60601 is the most pertinent.

[2] Some educators quote 3 m: a distance a person might reach when holding onto the patient.

---

and medical devices and equipment used in healthcare and the investigation of harmful incidents. It also oversees blood and blood products, working with U.K. blood services, healthcare providers, and other relevant organizations to improve blood quality and safety. The MHRA regulations implement the European Commission's Medical Device Directives into U.K. law. In 2012 there were four such sets of regulations,[3] the most recent of which came into force in 2010. For a device to be placed onto the EU market, a U.K. manufacturer must be registered with MHRA. Non-EU manufacturers may access the EU market through a registered authorized representative.

It is important to note that devices manufactured by healthcare establishments that are only used on their own patients are exempt from the requirements of the medical devices regulations. While this is unlikely to be true of information and communications technology (ICT) equipment per se, it is true of, for example, patient immobilization devices for radiotherapy and surgical implants, which will most likely have been designed using ICT. Custom-made devices (i.e., products manufactured specifically in accordance with a written prescription of a duly qualified medical practitioner or a professional user and intended for the sole use of a particular patient) are not required to follow the normal conformity assessment procedures. However, the manufacturers (and this includes small hospital departments and workshops) do need to register with MHRA. In both cases, while the end product is not subject to the regulations, the software producing them may be, as that is not single-use.

It follows that software can be developed for in-house use but must not be placed on the market. MHRA tries to provide

updates on this complex area and further information can be found on its website (*www.mhra.gov.uk*).

FDA

FDA regulations are generally regarded as being "tougher" than MHRA ones, although it is unclear whether this refers to the stringency of the regulations or the regulation process. The main difference is that MHRA is concerned with safety, while the FDA is also concerned about clinical effectiveness. In the United Kingdom clinical effectiveness is a separate role, dealt with by the National Institute for Health and Clinical Excellence (NICE). The FDA is the federal agency responsible for ensuring that foods are safe, wholesome, and sanitary; that human and veterinary drugs, biological products, and medical devices are safe and effective; cosmetics are safe; and electronic products that emit radiation are safe. The FDA also ensures that products are honestly, accurately and informatively represented to the public. It is the medical device remit that is of interest here.

Premarket approval (PMA) is the most stringent type of device marketing application required by the FDA and is based on a determination by the FDA that the PMA contains sufficient valid scientific evidence providing reasonable assurance that the device is safe and effective for its intended use or uses. PMA is therefore required before a product can be placed on the market. One interesting form of regulation the FDA offers is humanitarian device exemption. This is a form of PMA but is for humanitarian use devices (HUD), which are intended to benefit patients by treating or diagnosing a disease or condition that affects fewer than 4000 individuals in the United States per year. Thus the "effectiveness" section of a full PMA submission is not required.

---

[3]The Medical Devices Regulations 2002 and the amendments of 2003, 2007, and 2008.

Recalls

So far we have considered only the regulatory nature of the two agencies. Additionally, they both issue alerts and recalls. It is therefore important that a healthcare provider has someone tasked with receiving these and disseminating the information appropriately, although only from their own agency (i.e., the U.K.'s National Health Service need not receive FDA alerts).

From a medical physics/clinical engineering perspective, therefore, there are three main areas where interaction with MHRA/FDA is required:

- Ensuring that devices deployed have the correct accreditation/approval
- Determining when in-house developments require accreditation/approval (and seeking it when it is required)
- Receiving and disseminating alerts and recalls

In this section we have examined the overall organizational arrangements in the United States and the United Kingdom; the latter is also typical of Europe. The next section looks at the detail of IEC 60601, the MDD, and CE marking.

Regulatory Standards Including IEC 601, the Medical Devices Directive, and CE Marking as Applied to Software

IEC 601

IEC 601 refers to IEC 60601-1, Medical Electrical Equipment, Part 1: General Requirements for Safety, first published in 1977. It is now in its third edition and became effective in Europe in June 2012. It is the main standard for electromedical equipment safety and is first in a family of standards, with 10 collateral standards (numbered 60601-1-N) which define the requirements for certain aspects of safety and

performance (e.g., Electromagnetic Compatibility IEC 60601-1-2) and over 60 particular standards (numbered 60601-2-N) which define the standards for particular products (e.g., nerve and muscle stimulators IEC 60601-2-10). Two important changes from the previous versions are the removal of the phrase "under medical supervision" from the definition of medical electrical equipment and the inclusion of the phrase "or compensation or alleviation of disease, injury or disability" (previously only diagnosis or monitoring) meaning that many devices previously excluded are now included in the standard's coverage. Possibly the largest change is the requirement for manufacturers to have in place a formal risk management system that conforms to ISO 14971:2007 Medical devices – Application of risk management to medical devices (Sidebottom et al., 2006[4]).

A risk management system includes two key concepts:

- Acceptable levels of risk
- Residual risk

Once acceptable levels of risk have been established, all residual risks (as documented in the hazard log, a part of the risk management file) can be measured against them. That way, risks can be demonstrably determined to be acceptable prior to manufacture and certainly prior to deployment.

Furthermore, there are two compulsory ISBs (Information Standards Board—originally NHS Data Set Change Notices) that came into force on April 1, 2010. ISB 0129 (formerly DSCN 14; newly titled Clinical Risk Management: Its Application in the Manufacture of Health IT Systems) describes the risk management processes required to minimize risks to

---

[4]Sidebottom's paper references 14971:2000, which was the version in force at the time. Although the version has updated, the principle remains the same.

patient safety with respect to the manufacture of health software products either as new systems or as changes to existing systems. ISB 0160 (formerly DSCN 18; newly titled Clinical Risk Management: Its Application in the Deployment and Use of Health IT Systems) is similar, describing the deployment of such products. Thus all healthcare sites need to comply with ISB 0160 and many with ISB 0129. A fully populated hazard log may be used to demonstrate compliance, and compliance with ISB 0160 will encompass compliance with IEC 80001-1 (NPFIT, 2009).

IEC 60601-1 covers all aspects of the medical device, including classifying the parts that connect (directly or indirectly) to patients. Of interest here is its applicability to computer-based medical devices. It is almost certain that a standard PC power supply will not meet the regulation so this must either be replaced or an isolation transformer must be placed in line. As it is unlikely that any computer-based medical device will exist in isolation, the connections to other items of equipment must also be examined. One commonly overlooked connection is to a network; this can be isolated either via transformer or optically. However, some onboard network adaptors are unable to supply the power required to drive such an isolator; this may not pose a problem to a desktop machine (as a separate adaptor can be added), but would to a laptop, netbook, or tablet device (especially if it is operating on battery power). Wi-Fi may provide the best electrical isolation, but brings with it other problems leading to a risk-balancing exercise.

IEC 60601-1 applies to medical devices. For a definition as to what constitutes such a device, we must turn to the Medical Devices Directive.

The Medical Devices Directive

The Medical Devices Directive of 1993 (less well known as Directive 93/42/EEC)

contained within it a definition of a medical device. This definition lies at the heart of all subsequent directives, regulations, and guidance. The Medical Devices Directive defines a "medical device" as meaning "any instrument, apparatus, appliance, material or other article, whether used alone or in combination, including the software necessary for its proper application intended by the manufacturer to be used for human beings for the purpose of:

- Diagnosis, prevention, monitoring, treatment or alleviation of disease
- Diagnosis, monitoring, treatment, alleviation of or compensation for an injury or handicap
- Investigation, replacement or modification of the anatomy or of a physiological process
- Control of conception

and which does not achieve its principal intended action in or on the human body by pharmacological, immunological or metabolic means, but which may be assisted in its function by such means."[5]

Note that accessories (defined as "an article which while not being a device is intended specifically by its manufacturer to be used together with a device to enable it to be used in accordance with the use of the device intended by the manufacturer of the device"[6]) are treated as medical devices in their own right and must be classified, examined, and regulated as though they were independent.

There have been several additional directives (2003, 2005, and 2007; the year of adoption is not the same as the year of implementation) which have modified the original directive. Generally these modifications have been to increase the scope of the

[5]Eur-Lex 1, Council Directive 93/42/EEC of 14 June 1993 concerning medical devices. Available: *http://eur-lex.europa.eu/LexUriServ/LexUriServ.do?uri = CELEX:31993L0042:EN:HTML*
[6]Ibid.

Medical Devices Directive to cover more types of devices. At the time of writing, the most recent technical revision is Directive 2007/47/EC. The interest in this directive from the point of view of this chapter lies in items 6 and 20 and article 1. Article 1 redefines a medical device adding the word "software" to the list, becoming "any instrument, apparatus, appliance, software, material or other article." The effect of this is to mean that software alone may be defined as a medical device and not just when it is incorporated within hardware defined as a medical device. This is made clear by article 6 of this directive, which states:

> "It is necessary to clarify that software in its own right, when specifically intended by the manufacturer to be used for one or more of the medical purposes set out in the definition of a medical device, is a medical device. Software for general purposes when used in a healthcare setting is not a medical device."[7]

The ability to run software that is a medical device on a computer that is not originally designed as a medical device thus redesignates the hardware as a medical device and must be evaluated and controlled accordingly. The non-mandatory guidance document "Qualification and Classification of Stand Alone Software" (EC, 2012) provides decision flowcharts and definitions to assist the determination as to whether standalone software is a medical device (Figure 8.1) and, if so, the class to which it belongs. Without repeating the full content here, it is worth noting that it is recognized that software may consist of multiple modules, some of which are medical devices and some of which are not.

[7]Eur-Lex 2, Directive 2007/47/EC of the European Parliament and of the Council of 5 September 2007. *http://eur-lex.europa.eu/LexUriServ/LexUriServ.do?uri = OJ: L:2007:247:0021:0055:EN:PDF*

CE Marking

Directive 2007/47/EC embraces the concept of software as a medical device and therefore clearly links it to a discussion of CE marking. It is important to understand European legislation concerning the CE mark. "CE" is an abbreviation of the French phrase *Conformité Européenne* ("European Conformity"). While the original term was *EC mark*, it was officially replaced by *CE marking* in the Directive 93/68/EEC in 1993, which is now used in all EU official documents. The key points are:

- CE marking on a product is a manufacturer's declaration that the product complies with the essential requirements of the relevant European health, safety, and environmental protection legislation.
- CE marking on a product indicates to governmental officials that the product may be legally placed on the market in their country.
- CE marking on a product ensures the free movement of the product within the EFTA and European Union (EU) single market.
- CE marking on a product permits the withdrawal of the nonconforming products by customs and enforcement/vigilance authorities.

CE marking did not originally encompass medical devices, but they were brought into the scope of the general directive by a series of subsequent directives from 2000 onwards.

Note that a CE mark does not imply independent testing, although this will usually be done. It is a manufacturer's declaration of their belief that the device complies, and in simple cases they can prepare appropriate tests and documentation themselves.

Custom-made devices, devices undergoing clinical investigation, and in-vitro medical devices for clinical investigation do not require CE marks but must be marked "exclusively for clinical investigation."

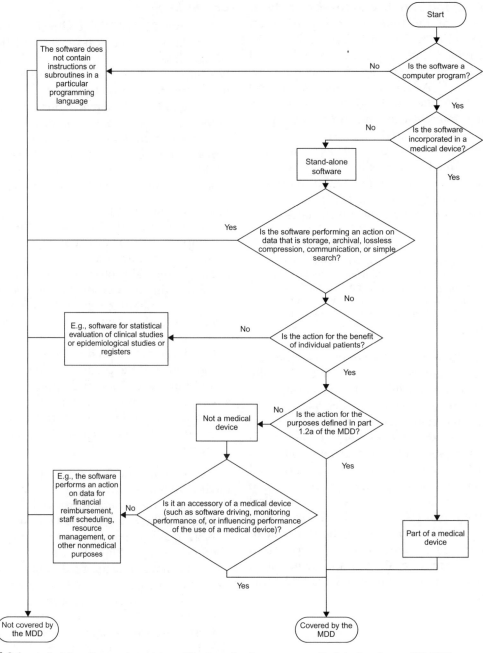

FIGURE 8.1   A decision diagram to assist qualification of software as a medical device. *Source: EC, 2012.*

The MHRA is the U.K. body that provides advice and guidance on matters concerning the relationship between the MDD and the CE requirements; its website contains helpful documents.

Three international standards offer valuable contributions to those working to provide and support medical devices:

- IEC 62304-2006 Medical device software – Software life cycle processes. This offers a risk-based approach and makes reference to the use of Software of Unknown Provenance (SOUP).
- IEC/ISO 90003-2004–Guidelines for the application of ISO 9000-2000 to computer software, which offers similar concepts to those embraced in the Tick-IT scheme (TickIT, 1992).
- ISO 13485-2003 Medical devices – Quality management systems – Requirements for regulatory purposes. This deals with the production and management of medical devices in a manner that parallels ISO 9000.

Of more recent interest is IEC 80001-1 (2010). This standard, titled "Application of risk management for IT-networks incorporating medical devices – Part 1: Roles, responsibilities and activities" contains many definitions. The main one for consideration here is that of the "medical IT network," which is defined as "an IT-network that incorporates at least one medical device." An IT-network is defined as "a system or systems composed of communicating nodes and transmission links to provide physically linked or wireless transmission between two or more specified communication nodes" and is adapted from IEC 61907:2009, definition 3.1.1. The medical device definition is from the MDD. Thus, a hospital that connects even one medical device into its standard network (or, indeed, loads medical device software onto a nonmedical device so connected) has thereby created a medical IT-network. The bounds of this network are that

of the responsible organization[8] but do bring different responsibilities into play, as detailed in the standard. In particular, the role of the medical IT-network risk manager, the person accountable for risk management of the medical IT-network, is specified.

This family of standards could stimulate cross-disciplinary teams in hospitals, involving IT departments, informatics departments, and clinical departments in establishing quality systems for clinical computing. This would include assurance of finance, planning of procurement and upgrades, and the monitoring of adequate support arrangements. Medical physicists and clinical engineers would have an important role to play in these groups, particularly with regard to day-to-day running and relationships with device suppliers.

Having considered the regulation of medical devices we now consider the regulation of data.

The Data Protection Act, the Freedom of Information Act, and the Caldicott Committee

In the United Kingdom there are two main pieces of general legislation covering the use of information/data and the right to disclosure. Personal data are covered by the Data Protection Act and official information by the Freedom of Information Act. Both of these are described below. A third item, the Environmental Information Regulations (U.K., with a separate version for Scotland), covers more public bodies but is only likely to impact on the NHS in terms of buildings and waste disposal. The particular sensitivities surrounding medical data were considered by the Caldicott committee in 1997. The legislation will first be discussed, followed by consideration of

---

[8]Thus a connection to the Internet does not render a network a medical IT-network.

required practice within the medical environment.

The Data Protection Act

The Data Protection Act 1998 (DPA) is a U.K. law which came into force in 2000. It is designed to prevent the misuse of personal data. It encapsulates eight main principles, which are set out in Schedule 1, as follows (ICO, 2011):

1. Personal data shall be processed fairly and lawfully and, in particular, shall not be processed unless
    a. at least one of the conditions in Schedule 2[9] is met, and
    b. in the case of sensitive personal data, at least one of the conditions in Schedule 3[10] is also met.
2. Personal data shall be obtained only for one or more specified and lawful purposes, and shall not be further processed in any manner incompatible with that purpose or those purposes.
3. Personal data shall be adequate, relevant and not excessive in relation to the purpose or purposes for which they are processed.
4. Personal data shall be accurate and, where necessary, kept up to date.
5. Personal data processed for any purpose or purposes shall not be kept for longer than is necessary for that purpose or those purposes.
6. Personal data shall be processed in accordance with the rights of data subjects under this Act.

[9]Briefly, that the subject has given consent or that the processing is necessary for legal compliance, either of which may apply to the NHS.

[10]Clause 8 of this schedule says "The processing is necessary for medical purposes and is undertaken by (a) a health professional, or (b) a person who in the circumstances owes a duty of confidentiality which is equivalent to that which would arise if that person were a health professional."

7. Appropriate technical and organisational measures shall be taken against unauthorised or unlawful processing of personal data and against accidental loss or destruction of, or damage to, personal data.
8. Personal data shall not be transferred to a country or territory outside the European Economic Area unless that country or territory ensures an adequate level of protection for the rights and freedoms of data subjects in relation to the processing of personal data.

There are exemptions to these principles, provided for by the Act, as follows (ICO, 2011):

Exemption from the requirement:

- To notify the Information Commissioner
- To grant subject access to personal data
- To give privacy notices
- Not to disclose personal data to third parties

Entitlement to an exemption depends in part on the purpose for processing the personal data in question; for example, there is an exemption from some of the Act's requirements about disclosure and nondisclosure that applies to processing personal data for purposes relating to criminal justice and taxation. However, each exemption must be considered on a case-by-case basis because the exemptions only permit departure from the Act's general requirements to the minimum extent necessary to protect the particular functions or activities the exemptions concern. It is therefore often simpler to assume that there are no exemptions. Most of the time this assumption is true.

The implications of the Act when applied to patient data are as follows:

- Data must only be used for the purpose for which it was collected. It is therefore imperative to structure consent forms appropriately and to take into account potential future uses of the data. Retrospective research without the appropriate consent in place is therefore not possible, unless the data was collected before the Act came into force.

- Only adequate and relevant data may be collected: each item requested must therefore be justifiable.
- Data must be kept accurate. Therefore old databases must be decommissioned unless part of a research project that acknowledges the time frame of the data.
- Data cannot be kept forever, only for as long as their purpose is still current. Many medical records have to be kept for fixed periods of time, which may depend on the age of the patient, the severity of the condition, and whether the data are to be used as a test or reference case.
- The data subjects (patients) have rights concerning this data. There are exemptions concerning personal data relating to an individual's physical or mental health. In general, this applies only in certain circumstances and only if granting access would be likely to cause serious harm to the physical or mental health of the individual or someone else. Such data should therefore be clearly marked as "do not disclose." Another exemption covers personal data relating to human fertilization and embryology.
- The data must be technologically secured against unauthorized access and accidental damage.
- The data may not be transferred outside of the European Economic Area (EEA) without appropriate safeguards. This has implications for obtaining a second opinion from the United States or using a nighttime reporting service in Australia. These safeguards are not just to do with the integrity or security of the data, but also the rights and freedoms of patients regarding their data.

It should be noted that the DPA does not just cover electronic records, but all data, be it paper, knots in string, or smoke signals. However, it does expect the data to have an organizational structure to it so that it can be easily retrieved. String is unlikely to have this and smoke signals are debatable (as is paper, judging by most engineers' desks).

Finally, the DPA also applies to the disposal of electronic records such as on hard drives. This must be done in such a way that the data are completely destroyed; clicking "delete" is not sufficient as they can be recovered too easily (even after the recycle bin has been emptied).

The Freedom of Information Act

The Freedom of Information Act 2000 (FOI) is a U.K. law (although there are many similar laws elsewhere, especially in Europe) that describes the rights of groups and individuals to request information and the obligations of the organization to respond to such requests. All public authorities in England, Northern Ireland, and Wales and those that are U.K.-wide have a legal obligation to provide information through an approved publication scheme and in response to requests. Scotland has a similar Act, the Freedom of Information (Scotland) Act 2002 which is regulated by the Scottish Information Commissioner's Office.

Under the FOI Act, a member of the public has the right to ask for any information they think a public authority may hold. The right only covers recorded information, which includes information held on computers, in emails, and in printed or handwritten documents as well as images, video, and audio recordings.

A request is made directly to the relevant authority, in writing. It may be in the form of a question and must contain the requester's real name and a contact address for the reply. The request must be responded to within

20 working days. This response can take one of the following forms:

- Provide the information requested.
- Respond that the authority doesn't have the information.
- Respond that another authority holds the information. The authority may (but does not have to) transfer the request.
- Respond that the information is available and can be provided for a fee (but there are rules surrounding this).
- Refuse to provide the information, and explain why.
- Respond that more time is required to consider the public interest in disclosing or withholding the information, and state when a response should be expected. This should not be later than 40 working days after the date of the request. The time limit can only be extended in certain circumstances, and an explanation as to why the information may be exempt must be provided.

The request may be refused for the following reasons:

- It will cost too much to provide. The limit is currently set at £600 for any government department, Houses of Parliament, Northern Ireland Assembly, National Assembly for Wales, Welsh Assembly Government, and armed forces, and £450 for any other public bodies (ICO, 2011). This is interpreted as being 24 and 18 hours' effort, respectively (i.e., about two and a half days' work for the NHS).
- The request is vexatious or repeated.
- The information is exempt from disclosure under one of the exemptions in the Act. There are 23 exemptions in the Act, broadly divided into three categories:
  - Those that apply to a whole category (or class) of information, for example, information about investigations and proceedings conducted by public authorities, court records, and trade secrets.
  - Those that are subject to a "prejudice" test, where disclosure would, or would be likely to, prejudice, for example, the interests of the United Kingdom abroad, the prevention or detection of crime, or the activity or interest described in the exemption.
  - Requests that should be dealt with under the DPA either as a subject access request or by releasing information that would contravene the DPA (e.g., where the FOI request is about a third party).

Note that a request may not be refused because part of a document that would have to be provided is exempt. In this case, a redacted version of the document with the exempt information removed must be provided.

This leads to the question of anonymized data, where all person-identifiable information has been removed. For example, a request to provide an anonymized MRI image. This would seem to satisfy the requirements of an FOI request, but is not one of the purposes for which the data was collected, so is exempt via the DPA.[11]

While we are considering anonymized data, it is worth mentioning synonymization, that is, the process of removing all patient-identifiable data (for example, from a DICOM header) and replacing it with another identifier, such as a research number, so that all data relating to that study can be kept together, without having to reference the patient.

The Caldicott Committee

The Caldicott committee was established in 1997 to review the flow of patient-identifiable information around and out of the NHS. Its

---

[11]Unless, of course, it was included in the consent form.

work resulted in an instruction to all NHS Chief Executives to appoint Caldicott Guardians by the end of March 1999.

A Caldicott Guardian is a senior member of staff whose role is to perform the following tasks:

- Establish the highest practical standards for handling patient information
- Produce a year-on-year improvement plan for ensuring patient confidentiality
- Monitor yearly improvement against the improvement plan
- Agree and review protocols governing the protection and use of patient identifiable information
- Agree and review protocols governing the disclosure of patient information
- Develop the trust security and confidentiality policy

The implication of this is that approval of the Caldicott Guardian must be gained for all new flows of patient data except data to be used for research purposes. The decision of the Caldicott Guardian is final and may not be challenged.

On the international scene information is protected through the ISO 27000 family of standards. In particular, ISO 27799: 2008 Health informatics — Information security management in health using IEC/ISO 20002 covers the principles described in this section, together with the provision of accredited information security management systems.

In summarizing the principles surrounding the proper protection of data, the NHS information strategy adopted the acronym CIA (confidentiality, integrity, and access). While data must be kept confidential and its integrity is paramount, it must also be readily accessible to those who need it.

This section has dealt with the legislative aspects; the next section covers the engineering aspects used to help achieve the objectives. Overriding clinical governance issues will be discussed after both aspects have been described, in a separate section.

DATA SECURITY REQUIREMENTS

Information Communications Technology Security: Firewalls, Virus Protection, Encryption, Server Access, and Data Security

Introduction

There are great advantages in connecting together information communications technology (ICT) equipment to enable data sharing, together with the enhanced safety from a reduction in transcription errors and the increased availability and speed of access to information. However, this connectedness brings with it additional system security issues: a failure in one part may be swiftly replicated across the IT estate. There are many ways to tackle these issues and this section details some of these. It should be noted that best practice will utilize a range of security methods.

Firewalls

The first method is one of segregation, using a firewall. A firewall is, in the simplest sense, a pair of network cards and a set of rules. A network package arrives at one card, is tested against the rules, and (if it passes) is passed to the other card for transmission. In this way, a part (or the whole of) a network can be protected from activity on the rest of the network by restricting the messages that can pass through it to a predefined and preapproved set. The rules controlling this may be as simple as only allowing a predefined set of IP addresses through. Refinements include port numbers, the direction the message is travelling in, whether the incoming message is a response to an outgoing one (e.g., a web page), and specific exceptions to general rules. This is all achieved via packet filtering, where the header of the packet is examined to extract the information required for the rules.

The above description is of a hardware firewall. Software firewalls run on the machine after the network traffic has been received. They can, therefore, be more sophisticated in their rules with additional information such as the program that made the request. Software firewalls can also include privacy controls and content filtering. As the software firewall runs on the device, if the device becomes compromised, the firewall may also be compromised. The Windows 7 firewall only blocks incoming traffic, so will not prevent a compromised device from sending malicious network packets.

Malware

Probably the most common security issue faced by NHS IT systems is that of viruses. The term *virus* covers many different types of malicious software (or "malware") such as trojans, worms, spyware, and rootkits. There are two important features of such malware: it does something malicious and it can replicate itself, thereby passing from one device to another. The trojan is probably the simplest of these. This is a piece of software that purports to perform some useful function (such as a system scan), which it may do. However, it also contains within it another program that performs the malicious action (such as seeking out passwords and emailing them). It may also copy itself to all other programs on the device, so that deleting the original program does not remove the trojan from it.

A worm is a program that seeks to exploit vulnerabilities in a device (or network), usually by probing certain access points or by appearing to be a "trusted" service. Having installed itself on the device ("infected" it), it is then able to activate its malicious element (the "payload").

Viruses may be resident or nonresident. Both types attach themselves to a legitimate executable program to be executed by the device. They initiate (possibly complete) and then pass control to the host program. A nonresident virus searches for other files to infect and does so by copying its code into the new host (thereby replicating itself). A resident virus does not perform this action immediately, but will instead install itself into memory and attach itself to an operating system function, running each time this function is called. A virus scanner that fails to recognize such a virus may itself become the initiator, thereby causing each scanned file to become infected (via replication) as soon as the scan completes.

Virus technology is not new—John von Neumann described them in 1949—but it is the high level of connection in networks (and especially the Internet) that has seen the most widespread proliferation.

An antivirus program will work in many ways, such as trapping unpermitted behavior (see "The Human Factor" section), but the most common is the file scan. In this method, a file (usually an executable program, but remember that many spreadsheets and word processing documents contain executable code in the form of macros) is examined byte by byte. The virus scanner has a set of patterns known as "virus definitions" that it is looking for. These are unique to the virus in question and are therefore not simply a few bytes long. In this way, trojans and spyware can also be trapped. Once found, the scanner will perform some corrective action, the most common of which are quarantine (moving the infected file somewhere else for further examination) or deletion. "Cleansing" the file is the removal of the virus, but as viruses become more sophisticated this is not always successful.

For an antivirus program to be successful, it must scan the files on the system. There are two main ways to do this: scheduled scanning and on-access scanning (frequently the two are combined).

On-access scanning scans files as and when they are accessed. In the case of an executable, this is just prior to its execution. Scheduled scanning scans the full system at a predefined time. On-access scanning is clearly the most secure, yet is not always applicable due to the processing overhead and time delay it introduces. This may not have any critical effect on entering figures into a spreadsheet, but is likely to have an effect in real-time control and acquisition. For this reason most medical devices employ scheduled scanning, but it is therefore imperative that the scan is scheduled for a time when the device is operational.[12]

Encryption

Assuming that the devices are secure and the connection also, the next issue to address is that of interception. Data may be deliberately intercepted (via packet logging) or simply mislaid. In either case, the next level of security is encryption. Successful encryption means that only the authorized receiver can read the message.

Encryption is, of course, a very old science. It has gone from simple substitution ciphers,[13] through the complexity of the Enigma machine, to today's prime-number based techniques. The most common encryption is RSA developed by Rivest, Shamir, and Adelman in 1978, which relies on the difficulty of factoring into primes. It works as follows:

- Let p and q be large prime numbers and let $N = pq$. Let e be a positive integer which has no factor in common with $(p - 1)(q - 1)$. Let d be a positive integer such that $ed - 1$ is divisible by $(p - 1)(q - 1)$.
- Let $f(x) = x^e$ mod N, where a mod N means "divide N into a and take the remainder."
- Let $g(x) = x^d$ mod N.

- Use $f(x)$ for encryption, $g(x)$ for decryption.[14]

Therefore, to transmit a secure message only the numbers e and N are required. To decrypt the message, d is also required. This can be found by factoring N into p and q then solving the equation to find d. However, this factorization would take millions of years using current knowledge and technology.[15]

A simpler method is PGP ("pretty good privacy") developed by Phil Zimmerman in 1991 and subsequently revised multiple times. In this there is a single public key (published) which is used for encryption and a single private key used for decryption. In PGP a random key is first generated and is encrypted using the public key. The message is then encrypted using the generated (or "session") key. Both the encrypted key and the encrypted message are sent to the recipient. The recipient then decrypts the session key using their private key, with which they decrypt the message.

So far we have only considered data transmissions. Encryption can also be used on data "at rest," that is, on a storage device. RSA encryption is therefore usable in this context, although PGP is not. There are two forms of encryption: hardware and software. Both use similar algorithms but the use of hardware encryption means that the resultant storage device is portable as it requires no software to be loaded to be used. A device may be fully encrypted (i.e., the entire storage, sometimes including the master boot record) or filesystem-level, which just encrypts the storage being used. Devices may use multiple keys for

---

[12]I once found a hemodynamic system that was set to scan at 3 A.M., but was switched off every night when the lab closed for the day.

[13]Where each letter of the alphabet is exchanged for another. Decryption is a simple matter of reversing the substitution.

[14]Clay Mathematics Institute, The RSA algorithm. *www.claymath.org/posters/primes/rsa.php*

[15]It is not worth beginning to speculate on the effect of a mathematician discovering a fast factorization method.

different partitions, thereby not being fully compromised if one key is discovered.

Three final concepts must be considered before we move on from encryption: steganography, checksums, and digital signatures. Steganography is a process of hiding files within other files, often at bit level; image files are therefore very suitable for this.

Checksums were originally developed due to the unreliability of electronic transmission. In the simplest form, the binary bits of each part of the message (which could be as small as a byte) were summed. If the result was odd, a bit with the value 1 would be added to the end of the message. If even, then the bit would be 0. Thus, by summing the entire message's bits, the result should always be even. Extensions to this were developed to detect the corruption of multiple bits and also to correct simple errors. Developed to ensure the integrity of the message due to electronic failure, these techniques can also be used to detect tampering. (See the discussion of RAID in the next section for further information.)

Digital signatures verify where the information received is from. It uses a similar technique to PGP, in that a message is signed using a public key and verified (decrypted) using a private key. A more complex version also uses the message, thereby demonstrating (in a similar fashion to checksums) that the message has not been altered.

Access Control

Having secure messaging and secure storage, we can now consider further access controls on a server. There are two methods to consider here: direct (shared filesystem) and remote (or terminal). Both require a system of authentication.

For filesystem access (such as to a shared folder) the server administrator will grant permission for the folder[16] (and subfolders) to be accessed by specific users. This may be done individually, or by groups. The latter is simpler to maintain, but takes longer to set up. A mixture of methods is possible. Permissions on a folder include permission to read, to write, to delete, and to create/delete subfolders, all of which may be granted or denied individually. It is also possible to grant/deny these permissions on individual files. Where this is not done, files inherit the permissions of the folder they are contained in. Direct access will usually be achieved by mounting the server's shared folder as a remote folder on the accessing device. If there are insufficient privileges, then mounting will fail.

Remote access is achieved through a separate program, such as Telnet, SSL (Secure Socket Layer), or RDP (Remote Desktop Protocol). For all of these, the server[17] must be configured to accept such connections and will need to be verified via a username/password combination. Telnet is the most trusting of the three and SSL the most secure, but once they are set up the experience of using them is very similar.

The Human Factor

Finally, we consider the security of the data itself. Data loss prevention (DLP) is of particular interest to NHS organizations as the movement and transfer of patient-identifiable data is tightly controlled (see the previous "The Caldicott Committee" section). DLP may take many forms: systemwide policies can be applied in which no data can be written to removable devices (or restricted to encrypted ones); emails may be examined to determine if they contain patient-identifiable information (usually via keyword searches); all emails to non-NHS addresses may be encrypted on departure; firewall rules may permit data to be sent from certain devices only at specified times. Sadly, the most common cause of data

---

[16]Or directory, on UNIX and Linux servers.

[17]Which may be an individual PC.

loss in the NHS is human error. Technology merely reduces this.

The implicit assumption, so far, has been that we have one copy of the data. This is not the case in reality because strategies are in place to maintain copies, in various forms, so seeking to avoid catastrophic loss. This is the subject of the next section.

Server/Database Replication, Backup, Archiving, RAID, Bandwidth, and Infrastructure

A reliable data center, be it for a web presence, a clinical database, or document repository requires resilience to be built into the design. As with security, there are many ways of achieving this, and the best designs will incorporate a mixture of these. In this section we consider four examples.

Replication

Replication, either of a server or a database (or both), is as the name implies: a copy exists of the server/data enabling it to be switched to in the event of a failure in the primary system. There are two main ways of achieving this, and the required uptime of the system determines which is the most appropriate. The simplest to achieve is a copy, taken at a specified time, to the secondary system. The two systems are therefore only ever briefly synchronized and in the event of a system failure bringing the secondary system online means that the data will be at worst out-of-date by the time interval between copying. Replication of this form is usually overnight (thereby utilizing less busy periods for systems), meaning that the copy is at worst nearly 24 hours out-of-date. If changes to the system have been logged, then these may be run against the secondary system before bringing it online, but this will lengthen the time taken to do so.

The most common use of this form of replication is in data repositories (see the section "Links to Hospital Information Systems" for a description), thereby removing some of the workload from the primary system and improving its reliability (and response time).

The second method of replication is synchronized: that is, both copies are exactly the same at all times. There are two primary ways of achieving this. The simplest is a data splitter: all changes sent to the primary system are also sent to the secondary (see the section "HL7" for details on messaging). This is clearly the best method when the system receives changes via an interface (which may be from an input screen or from a medical device) as the primary system only has to process the incoming message and the work of replication is external to it (e.g., in an interface engine).[18] The second way is where every change to the primary system is transmitted to the secondary system for it to implement. This introduces a processing overhead on the primary system, especially if it has to also receive acknowledgment that the change has been applied to the secondary system. As this method introduces a messaging system, the system could instead be designed to use the first method described.

The most basic (and therefore most common) method of replication is the backup. This is simply a copy of the system (or a part of it) taken at a specified time, usually onto removable media (for high volumes, usually tape). As such backups are generally out-of-hours and unattended, systems that exceed the capacity of the media are

---

[18]It is often postulated that giving the primary and secondary systems the same IP address will also achieve this. It is left as an exercise to determine whether or not this is a good idea.

backed up in portions, a different portion each night.

While a backup enables a quick restore of lost or corrupted data (and simplifies system rebuilds in the case of major failure), data errors are usually not so swiftly noticed and may therefore also exist on the backup. The grandfather-father-son (GFS) backup rotation system was developed to reduce the effect of such errors. In this, three tapes are deployed: on day one, tape 1 is used; on day 2, tape 2 is used; on day 3 tape 3 is used; on day 4 tape 1 is reused, and so on. Using this method there are always three generations (hence the name) of backup. Most backup regimes are variants on the GFS scheme and may include a different tape for each day of the week, 52 tapes (e.g., every Wednesday) also rotated, or 12 tapes (e.g., every first Wednesday). In this way data errors tracing back as far as a year may be corrected.

Archiving

Despite the massive increases in storage capacity in recent years, medical imaging has also advanced and thus produces even larger data sets. It has been estimated that 80% of picture archiving and communication system (PACS) images are never viewed again. However, as a reliable method for identifying those 80% has not yet been achieved, all images must be kept. Keeping them online (on expensive storage) is not a sensible option, thus old images are generally archived onto removable media (again, tape is usual) or onto a slower, less expensive system.[19] There are several algorithms for identifying data suitable for archiving, but the most common is based on age: not the age of the data, but the time since it was last accessed. To implement such a system it is therefore imperative that each access updates the record, either in the database (for single items) or by the operating system (in the case of files).

Resilience Using RAID

Probably the most common type of resilience, especially on a server, is RAID, or redundant array of inexpensive disks.[20] There are several forms of RAID and we consider two (levels 1 and 5) here. RAID level 1 is a simple disk image, as per replication above, only, of course, in real time. The replication is handled by the RAID controller[21] which writes any information to both disks simultaneously. If one disk fails, the other can be used to keep the system operational. The failed disk may then be replaced (often without halting the system, known as "hot swapping") and the RAID controller builds the new disk into a copy of the current primary over a period of time, depending on the amount of data held and the processing load.

Other forms of RAID do not replicate the data directly, but spread it across several disks instead, adding in some error correction as well. The form of spreading (known as "striping") and the type of error correction are different for each level of RAID.

RAID 5 uses block-level striping and parity data, spread across all disks in the array. In all disk storage, the disk is divided up into a set of blocks, being the smallest unit of addressable storage. A file will therefore occupy at least one block, even if the file itself is only one byte in size. A read or write operation on a disk will read or write a set of blocks. In RAID 5 these blocks are spread across several disks, with parity data stored on another one (see Figure 8.2). Thus RAID 5 always requires a minimum of three disks to implement.

Parity is a computer science technique for reducing data corruption. It was originally

---

[19]It is debatable whether or not a slower system is actually an archive and not, in fact, a replication.

[20]"I" is sometimes rendered "independent."

[21]A disk controller with additional functionality.

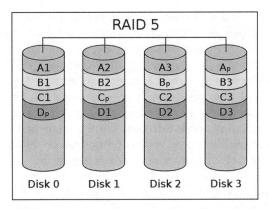

FIGURE 8.2 Diagram of a RAID 5 setup. Each letter represents the group of blocks in the respective parity block (a stripe). *Source: Wikipedia.*

designed for data transmission and consisted of adding an extra bit to the data.[22] This extra bit forced the sum of the bits to be odd (odd parity) or even (even parity). It was added at transmission and checked at reception (although this would only detect one error).[23] More complex forms used more bits which enabled the data not just to be better checked, but also to be corrected. Such codes are known as Hamming codes. The parity used within RAID 5 not only checks that the data are correct, but enables data to be rebuilt, should a disk fail and have to be replaced.[24]

In Figure 8.2, the distribution of the blocks and the parity can clearly be seen. Distributing the parity blocks distributes the load on these across all the disks, as this is where bottlenecks may appear (to read blocks B3 to C1, for example, also requires two parity blocks to be read—in this case each disk only has one read operation to perform). RAID 5 has found favor as it is viewed as the best cost-effective option providing both good performance and good redundancy. As write operations can be slow, RAID 5 is a good choice for a database that is heavily read-oriented, such as image review or clinical look-up.

Traditionally, computer systems and servers have stored operating systems and data on their own dedicated disk drives. With the data requirements expanding it has now become more common to have separate large data stores using network-attached storage (NAS) or storage area network (SAN) technology. These differ in their network connectivity but both rely on RAID for resilience. NAS uses TCP/IP connections and SANs use fiber channel connections.

Bandwidth

Finally, for this section, we consider bandwidth. Bandwidth in a computer network sense is its transmission capacity, which (as it is a function of the speed of transmission) is usually expressed in bps (bits per second). The most common wired bandwidths are 1 Gbps (often called gigabit Ethernet), 10 Mbps (standard Ethernet), and 100 Mbps (fast Ethernet). Wireless is generally slower; 802.11 g supports up to 54 Mbps, for example. Note that these are maximums and a wired network stands a better chance of providing the full bandwidth due to less interference. As bandwidth is actually the capacity, binding together several cables can increase the total bandwidth while not increasing the speed. Although this would not normally be done in a departmental network, the point at which a hospital meets the national N3 network may be implemented this way, provided both sides of the connection can handle it; it is achieved by routing predefined packets to specified lines (e.g., by IP address range).

[22]An alternative was to use one of the 8 bits in each byte for parity. Hence ASCII only uses 7 bits and simple integers often only have a range of 0 to 127.

[23]For a 7-digit binary number, 3 bits are required to check for all possible errors.

[24]Therefore, also, a failed disk can be ignored as the data on it can be computed in real time by the RAID controller.

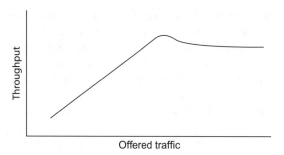

FIGURE 8.3 A generalized throughput versus offered traffic curve showing a deterioration in performance when optimal levels are breached.

It is never a good idea to reach 100% bandwidth utilization. The average to avoid this may be as low as 30%, although 50% would be more common. The amount of "spare" capacity is often termed *headroom*. At 75% the throughput versus offered traffic curve starts to depart from a linear proportional increase of throughput for increase of offered traffic (Figure 8.3). At 80% the channel could be approaching overload. Much is dependent on the traffic type: data traffic can cope with higher utilization levels than voice as delay and jitter have more effect on the user experience for voice traffic than data traffic. Optimization techniques such as QOS[25] can be used to prioritize voice traffic (or any other traffic that is time-critical).

The above utilization levels are generally for noncollision-based channels. In the case of Ethernet, which uses CSMA[26] with collision detection as the access mechanism, utilization should be much lower. An overdriven CSMA channel can result in throughput reduction rather than an increase with increasing offered traffic. Retrys as a result of a collision lead to more retrys and more collisions, and so on.

[25]Quality-of-Service, a Cisco product.

[26]Carrier Sense Multiple Access, a protocol in which a node verifies the absence of other traffic before transmitting on a shared transmission medium.

Collision detection with a limitation on the number of retrys and backoff between the retrys is aimed at keeping the channel stable, but throughput will tail off. Kleinrock (1975, 1976) provides good further reading.

All the resilience methods outlined here require a level of redundancy, be it a copy, checksums, or headroom. Thus a resilient system will always be over-engineered. In the case of bandwidth, over-engineering can remove the need for optimization systems such as QOS, thus making the design (and therefore the support) simpler.

Infrastructure

The core of modern hospital infrastructure is the supervised, air-conditioned server room, with redundant power supply provision. The users link to this facility by Ethernet and off-site electronic replication is likely. The bulk of the front-line data storage will use SAN or NAS and racks of servers will often be running virtual systems. Clinical scientist/engineer user specialists will have remote administrator access to their systems. Critical clinical systems, such as radiotherapy facilities, will be likely to have some form of segregations, such as virtual local area networks (VLANs) or firewalls. Virtual systems are particularly useful for enabling the local testing of commercial software upgrades, as it is relatively simple to destroy the server and recreate it.

Information Governance

Information governance (IG) is the function of corporate governance that ensures the confidentiality, integrity, and availability of an organization's information assets. There is a range of complex legal and professional obligations that limit, prohibit, or set conditions in respect of the management, use, and disclosure of information and, similarly, a range of statutes that permit or require information to be used or disclosed (see the previous section "The Data Protection Act, the Freedom of Information Act, and the Caldicott Committee"[27]).

IG is achieved through a mix of policy and protocol. It is not unusual for IG policies to form part of an employee's contract.

The major tension in IG for healthcare (and arguably any organization) is between openness and confidentiality. Too confidential (restrictive) and the delivery of healthcare suffers; too open and patient confidentiality suffers (so might the financial integrity of the organization which, in turn, would impact on healthcare delivery: an imbalance results in reduced healthcare, whichever way the imbalance lies). There is also the tension of the adoption of leading-edge technology without introducing unacceptable levels of risk. The protocols must be open enough to permit this advance, while being restrictive enough to prevent breaches.

It is important for IG protocols to recognize the need to share information between health organizations and other agencies in a controlled manner consistent with the interests of the patient (and in some circumstances, the public interest). Underpinning this is the need for electronic and paper information to be accurate, relevant, available when required, and processed appropriately (otherwise the data does not form information; "information = data + structure," as noted in the next chapter).

It is impossible to create policies that achieve complete restriction and complete openness. Therefore a large element of IG is in risk reduction: a pragmatic approach to

[27]For a full set of legislation, standards, codes of conduct, and guidelines refer to "NHS Information Governance – Guidance on Legal and Professional Obligations." The September 2007 edition lists 44 pieces of legislation, 14 standards and guidelines, and 8 codes of conduct.

ensuring the organization can function and develop, without leaving it open to damage.

IG policies cover all *information assets*, which include all information and data held by the organization (whether held electronically or in manual, paper-based systems) and all information processing/computer systems and networks owned or operated by the organization, including all systems operated on behalf of it by third parties and those entrusted to it by third parties (e.g., N3).

All organizations have an IG manager/officer/director who has top-level (e.g., executive board) influence and authority. It is that person's responsibility to interpret legislation, devise the organization's implementation of this, and to advise and rule on risk.

Up to 2008, NHS Trusts had to make an Information Governance Statement of Compliance (IGSoC), which was part of the Connecting for Health IG Toolkit. This IGSoC declaration has now been superseded by the requirement in the IG Toolkit to accept the IG Assurance Statement. This is reconfirmed each year by the submission of the annual IG Toolkit Assessment. The IGSoC process needs only be completed outside of the annual return of the IG Toolkit Assessment if the legal status of the organization changes or if there is a change of signatory.

Two systems are required to implement IG. Firstly, to monitor compliance, an organization should have a reporting mechanism for incidents; this will most likely be as a part of an existing incident reporting system. Secondly, an information asset's inability to comply fully with policies and procedures does not necessarily mean it cannot be deployed. However, it must be fully risk assessed and entered onto a risk register to do so.

Information governance may include computer forensic readiness—the ability of an organization to maximize its potential to use digital evidence while minimizing the costs of an investigation. This is normally utilized after an incident (criminal or otherwise) and may involve log files, emails, backups, and so on, which therefore need to be retrievable in a suitable interpretable format.

Part of the restriction IG implements comes under the heading "data security;" the NHS has adopted information security standards ISO 20007/1 and 20007/2 which address this. Data security addresses the physical security for data centers and communication/hub rooms, technical measures/standards to achieve appropriate access control (e.g., password complexity and expiry), standards for the development of software, standards for the procurement of software, disaster recovery/business continuity standards, technical standards, processes to prevent malicious code attack (e.g., intrusion detection systems and antivirus), standards for technical documentation, and so on.

Though IG can at first appear to be a restrictive mechanism and the role of the IG manager one of bottleneck or killjoy, if implemented properly with all aspects of the organization's activities considered, it can assist in releasing information, especially for research purposes. As such, it should be regarded more as an enabler (by doing things properly and efficiently) than a disabler.

DATA EXCHANGE PROTOCOLS

Data Exchange Standards: Digital Imaging and Communications in Medicine and Health Level Seven

Introduction

We have now considered legislation and data protection. To make clinical computing work we also need to deal with protocols and standards surrounding data communication and exchange. For computer systems to be able to process data produced by other systems, a data standard must be agreed on. Even in proprietary systems from the same

manufacturer, the format of the data has to be consistent so that different modules can access that data and understand its meaning.

The simplest form of this is *positional meaning*. In this standard, the data will always consist of the same items in the same order (and each item may be of a fixed length). For example, the data in Figure 8.4(a) may be converted for transmission to an external system via the process shown in Figures 8.4(a) through (c).

While the advantages of this system for data sharing are obvious, so are the limitations.[28] To overcome these, we might introduce a header to the data stream that describes the data that is to follow. Such a header might, for this example, be:

```
4^Surname^9^Forename^7^Address^22^e-
mail^22^
```

This header first states the number of items per record (and thus also the number of items in the header), then the name and length of each field in turn, all separated by a special character. Extensions to this to describe the type of data (numeric, currency, textual, Boolean, etc.) are left to the reader to envisage,[29] along with a solution to the problems posed by the control character appearing in the data.

DICOM

The addition of the header therefore solves the data length problem and the "new field" problem and can be very flexible. However, there are still proprietary elements to this data: specifically the construction of the header and the control character selected. It was to overcome such problems (and more besides) that ACR and NEMA[30] proposed a standard.

---

[28]If they are not immediately obvious, consider a longer piece of data or the addition of a new field.

[29]Or you could look up the dBase III file format.

[30]American College of Radiologists and National Electrical Manufacturers Association.

(a)

Surname	Forename	Address	e-mail
Snail	Brian	Magic roundabout	brian@roundabout.com
Flowerpot	Bill	Garden	bill@weed.co.uk
Cat	Bagpuss	Shop window	bagpuss@catworld.net
Miller	Windy	The Windmill, Trumpton	wmiller@chigley.ac.uk

(b)

Surname	9
Forename	7
Address	22
Email	22

(c)

Snail^^^^Brian^^Magic Roundabout^^^^^^brian@roundabout.com^^FlowerpotBill^^^Garden^^
^^^^^^^^^^^^^bill@weed.co.uk^^^^^^^^^^Cat^^^^^^BagpussShopWindow^^^^^^^^^^^bagpus
s@catworld.net^^Miller^^^Windy^^TheWindmill, Trumptonwmiller@chigley.ac.uk

FIGURE 8.4 (a) A set of data, arranged in a table. The items in the first row are field names (but may also be thought of as column headings). (b) The maximum lengths of the data in each field. (c) The data converted to one long data stream. Note that unused characters are rendered with ^ and not as spaces, as these appear in the data itself.

(a)

4^Surname^9^Forename^7^Address^22^e-
mail^22^Snail^^^^Brian^^Magic Roundabout^^^^^^brian@roundabout.com^^

(b)

Surname^9^Snail^^^^Forename^7^Brian^^Address^22^Magic Roundabout^^^^^^e-
mail^22^ brian@roundabout.com^^EOF

FIGURE 8.5 (a) The first record of the data stream from Figure 8.4(c), complete with header. (b) The same data stream, in a tagged format. Note the "EOF" indicating the end of the stream. Also that the order of the fields is now unimportant as they are prefixed by their tag in all cases.

ACR/NEMA 300 was published in 1985 and (after several revisions and additions) was renamed DICOM (Digital Imaging and Communications in Medicine) in 1993. DICOM specifies not only the file format but also a networks communication protocol (based on TCP/IP) and a set of services. DICOM, as the name implies, is used extensively in digital imaging (although it is also used for textual information exchange, such as worklists, and in specialized information, such as a radiotherapy treatment plan[31]). Standardization also addresses one additional problem: longevity. Data format standards ensure the data can still be read in many years' time, a key consideration for medical data.

The DICOM file format is based on data sets and embeds data such as the patient identifier into this, ensuring that the image cannot be separated from the patient to whom it belongs. A DICOM data object consists of multiple data elements, each of which is tagged to describe the data it contains, such as name, date of examination, and so on.[32] A tagged version of the example shown is in Figures 8.5(a) and (b). It can be seen from this example that a tagged

[31]Not, as you may suppose, an image, but a series of instructions to a linac for patient treatment.

[32]Two very common tagged file formats in current use which may assist in the understanding of DICOM are JPG and MP3.

format is not suitable for data sets that are comprised of multiple rows.

A DICOM data object contains only one image element; this may in turn comprise several frames. A DICOM data element consists of the tag, an optional value representation (the values for which are defined as part of the standard), the length of the data, and the data itself. When a DICOM data object is exported as a DICOM file, several of the key elements are formed into the header (though they do also exist within the object—they are purely copies), along with details of the generating application. This simplifies the import of objects as the entire object need not be read prior to storage.

DICOM *information objects* are definitions of the information to be exchanged. These are effectively templates into which a new image is placed. Each image type, and therefore information object, has specific characteristics; an MRI image requires different descriptors to an ultrasound image, for example. These information objects are identified by unique identifiers, which are registered by NEMA. An Information Object Definition (IOD) is an object-oriented abstract data model used to specify information about real-world objects. An IOD provides a common view of the information to be exchanged.

An IOD does not represent a specific instance of a real-world object, but rather a class of real-world objects that share the same properties. An IOD used to generally represent a single class of real-world objects is called a *normalized information object*. An IOD that includes information about related real-world objects is called a *composite information object*.[33]

As mentioned, DICOM also defines services. These are again defined in the standard, and include the following:

- Query/retrieve
- Storage commit
- Worklist management
- Print
- Verification

As there have been multiple versions of the DICOM standard and no application is required to implement the full standard (for example, a service may not be applicable to it) the *DICOM conformance statement* is an essential part of any system, describing the parts of DICOM that it implements (and to which version). Just as "runs on electricity" does not fully explain how to connect up a device, neither does "DICOM compliant."

DICOM is very useful for the exchange of a large quantity of data, such as a worklist or an image. It is not so useful, however, for exchanging incremental changes within a database. This case is very common in healthcare, where many clinical systems take a "feed" from the Patient Administration System (PAS) and return results to it. Here the PAS exchanges small changes such as date of appointment, time of arrival at reception, and so on, as well as large ones such as a new patient being registered (although in a database of over a million patients this may also be considered "small").

HL7

Such incremental changes are achieved through a messaging interface, and the most common standard adopted for these is Health Level Seven (HL7). (Note that this is not "Health Language Seven" as some translate it.) "Level 7" refers to the position it occupies in the Open Systems Interconnection (OSI) 7-layer model—the application layer. HL7 is administered by Health Level Seven International, a not-for-profit ANSI-accredited organization.[34] Although HL7 develops conceptual standards, document standards, and

---

[33]NEMA. The DICOM standard. *http://medical.nema.org/standard.html*

[34]It was announced in 2012 that the HL7 standard will become an open standard in 2013.

(a)

```
PID|||555-44-4444||EVERYWOMAN^EVE^E^^^^L|JONES|19620320|F|||153
FERNWOOD DR.^  ^STATESVILLE^OH^35292||(206)3345232|(206)752-
121||||AC555444444||67-A4335^OH^20030520<cr>
```

(b)

```
<recordTarget>
  <patientClinical>
  <id root="2.16.840.1.113883.19.1122.5" extension="444-22-2222"
    assigningAuthorityName="GHH Lab Patient IDs"/>
  <statusCode code="active"/>
   <patientPerson>
    <name use="L">
      <given>Eve</given>
      <given>E</given>
      <family>Everywoman</family>
    </name>
    <asOtherIDs>
      <id extension="AC555444444" assigningAuthorityName="SSN"
        root="2.16.840.1.113883.4.1"/>
    </asOtherIDs>
   </patientPerson>
  </patientClinical>
</recordTarget>
```

FIGURE 8.6 (a) A section of an HL7 v2.4 message, detailing the patient a test is for. (b) A similar section to the message in (a), using HL7 v3. *Source: Ringholme.*

application standards, it is only the messaging standard that we will consider here.

Version 2 of HL7 was established in 1987 and went through various revisions (up to 2.7). Version 2 is backward-compatible, in that a message that adheres to 2.3 is readable in a 2.6-compliant system.

Version 3 appeared in 2005 and, unlike v2, is based on a formal methodology (the HL7 Development Framework, or HDF) and object-oriented principles. The HDF "documents the processes, tools, actors, rules, and artifacts relevant to development of all HL7 standard specifications, not just messaging."[35] As such, it is largely UML (Universal Modeling Language) compliant, although there are currently exceptions.

The cornerstone of the HL7 v3 development process is the Reference Information Model (RIM). It is an object model created as part of the Version 3 methodology, consisting of a large pictorial representation of the clinical data. It explicitly represents the connections that exist between the information carried in the fields of HL7 messages.

An HL7 v3 message is based on an XML encoding syntax. As such, it is far more versatile than v2, but with an attendant overhead. Due to the installed userbase of v2 and the difference in message structure, v3 is not yet in widespread use. A very good comparison of the two formats is available on Ringholme's website,[36] of which a small part is reproduced in Figure 8.6.

A further layer of data exchange is Cross-Enterprise Document Sharing (XDS) which allows structured data documents of any type to be shared across platforms. Key elements in this are the document source (e.g., an EHR), the document repository (a shared store), and the document registry (essentially an index of the documents). Because the data are structured, the registry is able to index not only the title or metadata from within a header, but the data contained within the document itself. This makes searching the registry

[35]*www.hl7.org*

[36]*www.ringholm.de/docs/04300_en.htm*

more powerful. XDS is of particular interest in a healthcare setting where the source material may be produced from a large range of systems and devices (e.g., for a PACS or EHR).

Links to Hospital Administration Systems

The major hospital administration system in use is the Patient Administration System (PAS), sometimes called an *electronic health record* (EHR).[37] This system handles all patient registrations, demographics, clinic appointments, and admissions (planned and emergency). It is the gold standard for data, especially the demographics and doctor contact information.

An EHR has a heavy processing load. For this reason, it is common practice to have a second data repository that is a copy of the data, created via an overnight scheduled task. Reports may thereby be run against this data without incurring a performance hit on the main system. The data are at worst 24 hours out of date, but for most administrative reporting (as opposed to clinical reporting) this is acceptable. A notable exception is bed state reporting, which is therefore normally a feature of the main system.

The functionality of the EHR is greatly enhanced via interfaces to clinical systems. This is usually via HL7 messages (see the previous section "HL7") and may be outbound (where the data flow from the EHR to the clinical system), inbound (the other direction), or two-way. Demographics are usually outbound, test results are usually inbound, and bed state information may be two-way.

As the EHR interfaces to many other systems, the use of an interface engine is usual.

This may be thought of as a sophisticated router, in that not only will it pass on the relevant messages to the appropriate systems, it can also process these messages, so, for example, codes used by the EHR may be converted into ones used by the downstream system. An interface engine normally includes a large cache, so that downstream systems that go offline (e.g., for upgrade work) are able to collect the relevant messages when they come back online. It can thus be seen that acknowledgment messages form an important part of the interface engine protocol.

Finally, it is worth mentioning the e-Government Interoperability Framework (e-GIF), which "sets out the government's technical policies and specifications for achieving interoperability and information systems coherence across the public sector."[38] The e-GIF applies to NHS Trusts as well as to the Department of Health and NHS Health Authorities. It outlines a set of standards for systems that interact with citizens (defined as "user or recipient of NHS services, including patients and their relatives, formal and informal carers, and staff"), other government or NHS agencies, or private sector organizations. The main points are that such interfaces should be web browser interfaces and XML/XSL. Note that HL7 is built on XML, meaning that interfaces using HL7 are automatically compliant.

References

European Commission (EC), 2012. Medical Devices: Qualification and classification of stand alone software (MEDDEV 2.1/6) [online]. Available: <http://ec.europa.eu/health/medical-devices/files/meddev/2_1_6_ol_en.pdf> (accessed 18.04.12).

Kleinrock, L., 1975. Queuing Systems. Volume 1: Theory. John Wiley & Sons.

---

[37]Technically there is a difference: A PAS consists of demographics and appointments, whereas an EHR contains the full medical record including correspondence. As PAS functionality increases, the distinction becomes blurred.

[38]DoH, The e-GIF and the NHS — a policy statement: *www.dh.gov.uk/en/Publicationsandstatistics/Publications/PublicationsPolicyAndGuidance/DH_4130864*

Kleinrock, L., 1976. Queuing Systems. Volume 2: Computer Applications. John Wiley & Sons.

NPFIT, 2009. Nine steps for safer implementation [online]. Available: <www.isb.nhs.uk/documents/isb-0160/dscn-18-2009/0160182009guidance.pdf> (accessed 13.06.13).

Sidebottom, C., Rudolph, H., Schmidt, M., Eisner, L., 2006. IEC 60601-1 — the third edition, Journal of Medical Device Regulation, May 2006 pp 8–17 [online] Available: <http://www.eisnersafety.com/downloads/IEC60601-1_JMDRMay2006.pdf> (accessed 09.02.12).

TickIT making a better job of software, Guide to software quality management system construction and certification using EN29001, DTI, 1992.

Software Engineering

Paul S. Ganney*, Sandhya Pisharody[†],
and Edwin Claridge**

*University College London Hospitals NHS Trust, London, U.K.,
[†]Varian Medical Systems, U.K.,
**University Hospitals Birmingham NHS Trust, Birmingham, U.K.

SOFTWARE DEVELOPMENT AND MANAGEMENT

Operating Systems

In the previous chapter we looked at governance and general hardware issues; now we turn our attention to software. We first give a brief review of the most common software operating systems. An operating system (OS) is a very complex piece of software that interfaces directly with the hardware on which it is running: all user software runs under the operating system and will call routines within the OS to achieve hardware effects, such as displaying information (on screen or printer), receiving input (from the keyboard or mouse), and reading from or writing to storage (hard disks or RAM drives). The three outlined in this section are all multitasking operating systems which is why they are found on servers as well as end-user machines.

Microsoft Windows

There are two families of the Windows operating system: end-user (such as Windows XP, Windows 7, and Windows 95) and server (normally dated, e.g., 2012, with a release number, e.g., 2008 R2). Both may also be referred to by their service pack status, for example, XP SP3 indicates Windows XP where service packs 1 to 3 have been applied.

Windows is a graphical user interface, employing the WIMP (windows, icons, mouse, pointer[1]) paradigm pioneered by Xerox and brought into popularity by Apple and Atari. It is an event-driven operating system, in that events are generated (e.g., by a mouse click, a keyboard press, a timer, or a USB device insertion) and these are offered by the operating system to the programs and processes that are

---

[1]Some variants interpret "M" as "menus" and "P" as "pull-down menus."

currently running (including itself) for processing. For this reason a program may not necessarily cancel just because a user has clicked on a button labeled "Cancel"—the event merely sits in a queue, awaiting processing.[2]

Windows has gained great popularity, partly because of its relative openness for developers (compared to, say, Apple) but mostly due to its common user interface: similarly displayed buttons and icons perform similar functions across programs (e.g., the floppy disk icon for saving, even though the program is unlikely to be saving to a floppy disk) and almost-universal keyboard shortcuts (e.g., Ctrl-C for "copy"). This makes the learning of new programs easier and more intuitive.

Windows is most likely to be found running desktop end-user machines, departmental servers, and (in its embedded form) medical devices.

UNIX

UNIX is also a family of operating systems[3] and may also run on end-user machines as well as on servers. However, its most prevalent use is for servers, due to its robust stability. While there are graphical user interfaces, UNIX is most commonly accessed through a command-line interface, a "shell" (such as Bourne or C) that accepts certain commands (usually programs in their own right rather than embedded into the OS), and logic flow. UNIX was originally developed for dumb-terminal access (where all the user has is a keyboard and VDU, the processing taking place on the server) so it can be accessed easily via a terminal emulator on any end-user machine (e.g., a Windows PC or an Android phone).

A key concept in UNIX is that of the pipe. In this, the output of one program can be "piped" as the input to the next in a chain, thereby producing complex processing from a set of relatively simple commands.

UNIX is generally seen as an expensive operating system, which is one of the reasons it is less likely to be found on end-user machines. Its more common use is for running departmental and critical enterprise servers.

Linux

Linux was developed as an open-source variant of UNIX by Linus Torvalds. It gained rapidly in popularity due to its open-source nature: developers are able to access the source code to the OS and produce new versions of the programs within it, thereby expanding functionality and correcting errors. The caveat is that this new version must also be freely available, with source code, to anyone who wishes to use it. There are therefore various different versions (referred to as "flavors") of Linux, such as Ubuntu and openSUSE.

As with UNIX, Linux is mostly command-line, yet graphical user interfaces do exist and (as the cost pressure does not exist) are more likely to be found on end-user machines. Linux is most likely used to run desktop end-user research machines, departmental servers, and especially web servers.

Language Selection

There are many factors that influence the selection of a programming language with which to tackle a specific problem. The following lists nine criteria for consideration (see also Table 9.1 for a different take on this). The key consideration (repeated often in this list) is of maintainability of the produced code. It is tempting to think that code is written once and that the stable system never requires reworking. While true of some projects, it is not true of the majority.

---

[2]Therefore well-written code will frequently "peek" the event queue for such an occurrence.

[3]Such as UNIX 98, POSIX, SunOS.

TABLE 9.1 Ratings of SASEA Language Characteristics

Language		4GL or 5GL	3GL	3GL	3GL	2GL
Language Characteristic	WEIGHT					
			C++	Fortran	Java	Assembly
Clarity of source code		5	6	5	8	1
Maintainability		5	7	2	9	0
Object-oriented programming support		0	10	0	10	0
Portability		1	7	3	9	1
Reliability		3	5	1	8	0
Reusability		1	8	3	8	1
Safety		0	3	0	4	0
Standardization		1	5	5	8	0
Overall Language Rating						

The specific problem provides the weights, from which the rest of the table is completed.
Extracted from Lawlis (1997). A full table with explanations is available at the referred website.

1. **Ease of learning:** Does the project timescale provide sufficient time for the programmer to attain the required level of competence?
2. **Ease of understanding:** While a language will be learned once, the code produced will be read many times (especially during debugging). If the code statements are very complex then code maintenance becomes problematic. This can be mitigated to an extent by documentation, especially in-code comments.
3. **Enforcement of correct code:** Development is frequently time-pressured and there is always a temptation to write "quick and dirty" code and "tidy it up later." It is possible to write good, well-structured code in every programming language. One that forces good practice (e.g., variable typing) produces code that is easier to maintain.
4. **Productivity of the language:** All development systems contain tools to increase the programmer's productivity. These include the integrated development environment (IDE), the debugger, and libraries (for example, to provide TCP/IP support).
5. The availability of peer support, both inside and outside (e.g., Internet forums) the organization: While programming can be a very solitary activity, someone who knows how to implement a poorly documented feature is worth a hundred textbooks.
6. Applicability to the problem domain (e.g., writing a device driver in a scripting language is a poor choice).
7. Performance of the compiled code: While computing power is always increasing, some projects still require more: hardware acceleration (e.g., in a graphics card) or the use of a low-level language may provide better speed.
8. Platforms: If the solution is only ever required for one platform, then the language should be chosen accordingly. Likewise, if multiple platforms are required, then a language such as Java is required.

9. Portability: If the base platform changes, how portable is the code? How portable does it have to be?

There are formalized methods for selecting a language, by listing the attributes of each and weighting them in importance.[4] Each weight is multiplied by the score for that attribute for each language and summed. The final score then indicates the language that should be used. A sample of such an approach is in Table 9.1.

Finally, we consider mixed-language projects. These are often found in large software projects, where layers (or components) of the project are written in different languages to take advantage of the features (thereby implementing point 4 in the previous list). Another reason for mixed language is when components for the project already exist and are being reused (either from in-house development or purchased modules or components).

In this case, the primary language chosen for the project will be the interface between these components, ensuring that the system operates safely and efficiently. Once this language has been selected, it may be appropriate to develop new code in it also.

Mixing languages is never as straightforward as using just one language and it may prove simpler to redevelop the paradigms of one component into another language (thereby enforcing point 9 in the previous list).

Software Coding and Coding Management

There isn't enough room in this section to completely describe the art of computer programming; indeed, Donald Knuth devoted several large volumes to it.[5] What we will consider here are instead those elements of software coding that are to be found in most high-level (and in some cases low-level) languages. Illustrations are provided in several languages. Additional examples can be found in the section, "Procedural, Object-Oriented, and Functional Programming" and in the Chapter 10 section "Web Programming."

The first thing any program requires is somewhere to store values. The base types are usually numbers (integers, text, Booleans) but from these bases more complex types can be built (see the section "Procedural, Object-Oriented, and Functional Programming" for examples). Some languages require this storage (often termed *variables*) to be declared prior to use (e.g., C), whereas others (e.g., shell script) allow declaration at use. The naming of variables may be prescribed by the language (e.g., in some forms of BASIC a variable ending in "$" is of type text; in Fortran a variable that starts with a letter between I and N is an integer; in M, formerly known as MUMPS, variables prefixed with "^" indicate disk storage, i.e., permanent). The most common form of variable naming is what is known as "Hungarian notation"[6] where the variable name (starting with a capital letter) is prefixed with a lowercase letter(s) that denotes its type; for example, nTotal is numeric, sName is string (text), and bCapsLockOn is Boolean.

Assigning values to variables is generally accomplished using the "$=$" operator, although Algol 68 used "$:=$" to distinguish assignment from comparison (see later). More complex variable types may use a "set"

---

[4]Weights are from 1 to 10, attributes from 0 (no support) to 10 (full support).

[5]Knuth's *The Art of Computer Programming* (2011) was originally planned as a 7-volume series; the first 3 were published from 1968–1973 with Volume 4 in 2005.

[6]Named after Charles Simonyi, a Hungarian employee at Microsoft who developed it. Hungarian names are, unusually for Europe, rendered surname first.

operator so that multiple values can be assigned at once to a single variable.[7]

All languages contain key words that are part of it and form instructions. These are called "reserved words" and cannot be used as variable names (although they can be used as part of a name; e.g., "If" is a reserved word so cannot be used as a variable name, whereas "IfPrinting" is an acceptable name, though "bIfPrinting" would be better).

The "intelligence" within software, however, is provided through the flow-of-control mechanisms. Through these mechanisms decisions are made and different algorithms invoked according to the current state of the variables. Simple flow control is achieved through the *if-then-else* mechanism. An example in Visual Basic (VB) would be:

```
If nMonth = 9 or nMonth = 4 or nMonth = 6 or
nMonth = 11 Then
    nDays = 30
ElseIf nMonth <> 2 Then
    nDays = 31
Else
    nDays = 28⁸
End If
```

Note the use of "End If" to terminate the clause. The reserved word "Then" is optional, but does improve the readability of the code.

The other control mechanism we will consider is *switch*. This takes an input value and executes a clause corresponding to it. The previous example might thus be rendered in C as:

```
switch(nMonth)
{
    case 9:
    case 4:
    case 6:
    case 11:
```

```
    nDays = 30;
    break;
    case 2:
    nDays = 28;
    break;
    default:
    nDays = 31;
    break;
}
```

Note the use of "break" to indicate that processing should continue after the switch clause. It is therefore superfluous in the final "default" clause but doesn't aid readability, however it is good practice as other "case" clauses can be added after it and it may thus prevent erroneous processing.

The other mechanism we consider here is repetition. The simplest of these is the *for-next* loop, such as this one in JavaScript, which generates the first 10 powers of 2:

```
n = 1;
for (i = 0; i < 10; i++)
{
    n = n*2;
    document.write("2^" + i + " = " + n + ",
                   ");
}
```

The *for* statement has three clauses: the starting condition (i = 0), the continuing condition (i < 10), and the iterative condition (i++; i.e., i is increased by 1 on each circuit around the loop).

An alternative repetition is *while*. Whereas *for* executes the given code a set number of times (although there are alternatives in languages such as C), *while* executes the code until a condition is no longer true. Thus, to output powers of 2 that are less than 1 million, a C program might be:[9]

```
n = 1;
i = 0;
```

[7]See the date examples in the section "Procedural, Object-Oriented, and Functional Programming."

[8]The special processing for leap years is left as an exercise, in this and all subsequent examples.

[9]Actually this code also outputs the first power of 2 over a million. It is left to the reader to correct this.

```
while (n < 1000000)
{
    n* = 2;
    i++;
    cout << "2^" << i << " = " << n << ", ";
}
```

The final item a language requires is a mechanism for input and output. In VB this is achieved through statements such as:

```
FileOpen(1, "PATIENTLIST", OpenMode.Input)
Input(1, sPatName)
Debug.WriteLine(sPatName)
```

Some languages use the same statements to input from the user as to input from a file, and to output to a file as to output to screen or printer. While this aids the speed of learning the language, it does not assist code readability.[10]

While all these things aid the development of computer programs, one common feature (sadly probably the most poorly used) greatly aids the maintenance of code: the comment. This is a piece of text that is ignored by the compiler/interpreter and plays no part whatsoever in the execution of the program. It does, however, carry information. There have been many attempts to codify the use of comments, from the simple "one comment every 10 lines of code,"[11] to a more complex rigidly defined comment block at the head of every subroutine describing variables used, the execution path, and all revisions to the code since it was originally authored. The most prevalent use is probably as an aide-mémoire to the programmer; a line of code that required the author to have to think carefully about when writing certainly requires a comment so that

the logic behind it need not be recreated every time the code is altered. This, however, is insufficient when multiple programmers are working on a system at the same time and a solid commenting methodology should form part of the design.[12]

As with any document, version control is vital in software development. There are two elements to this:

- Ensuring developers are all working on the latest version of the code.
- Ensuring users are all executing the latest version of the code.

There are multiple ways (and multiple products) to achieve this, so we will examine only one example of each.

A common code repository enables multi-programmer projects to be successfully developed. This repository allows all developers access to all the code (although some allow version locking). When a developer wishes to work on a piece of code, this is "checked out" and no other developer may then access the code for alteration until it is "checked in" again. Dependency analysis is then utilized to alert all developers to other code modules that depend on the altered code and may therefore require revision or revalidation.

In a database project (or any software that accesses a database, regardless of whether that is core functionality—and it is worth remembering that a password list is a simple database) it is possible to ensure that all users are using the latest version of the software by hard-coding the version into the source code and checking it against the version in the database. If the database has a later version number, the program can alert the user, advising them to upgrade or even cease to progress (depending on the nature of the upgrade).

---

[10]See discussion of *overloading* in the section "Procedural, Object-Oriented, and Functional Programming."

[11]Resulting in useless comments such as "this is a comment" and failing to document more complex coding.

[12]There are two useful rules of thumb: "you cannot have too many comments" and "there are not enough comments yet."

Software Life Cycle

The life cycle of a software system runs from the identification of a requirement until it is phased out, perhaps to be replaced by another (Figure 9.1). It defines the stages of software development and the order in which these stages are executed. The basic building blocks of all software life cycle models include:

1. Gathering and analyzing user requirements
2. Software design
3. Coding
4. Testing
5. Installation and maintenance

Each phase produces deliverables required by the next stage in the life cycle. Requirements are translated into design and code is produced driven by the design. Testing verifies the deliverable of the coding stage against requirements; tested software is installed and maintained for its lifespan. Maintenance requests may involve addition or revision to requirements and the cycle repeats.

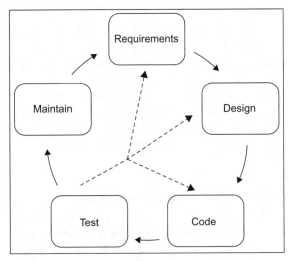

FIGURE 9.1 The software life cycle. If testing reveals errors, the process may return to any one of the three preceding stages. Release would normally come between "Test" and "Maintain."

To see how software progresses through its life cycle, we consider an example. A need has been identified for software for recording training and competencies of staff within a department. In the following subsections, we look at what is involved at each stage of this software's life cycle.

Requirements Specification

This phase of the life cycle identifies the problem to be solved and maps out in detail what the user requires. The problem may be to automate a user task, to improve efficiency or productivity within a user group, to correct shortcomings of existing software, to control a device, and many more. Requirements on software are usually complex combinations of requirements from a variety of users associated with a problem.

For our example, this would be staff as users who would enter their details, supervisors who might use the system to verify competencies, and managers who would use it to ensure staff are competent for the tasks assigned.

Various techniques are available for requirements gathering and analysis:

1. Observation of existing workflows, pathways, and processes and any related documentation
2. Interviews with potential users, individually or in groups
3. Prototyping the concept, not necessarily as software but as mock screenshots or storyboards
4. Use cases from different perspectives: user, programmer, maintainer

Requirements should be analyzed for consistency and feasibility. It is essential for software requirements to be verifiable using available resources during the software acceptance stage. A change management process for requirements (see the section "Software

Quality Assurance") is also helpful depending on the expected lifespan of the software.

The final step is to record these requirements including functional ones (e.g., what training data needs to be recorded), nonfunctional ones (e.g., how to ensure data confidentiality), design (what tools are required to build the software), and so on, in a software requirements specification (SRS). A prototype may also be built during this stage to help the user visualize the proposed software solution.

Software Design

Software design typically involves two levels of design: architectural and detailed design. The architectural design specifies the basic components of the software system such as user interface, database, reporting module, and so on, often using tools such as data flow diagrams (DFD) and entity-relation diagrams (ERD; see the section "Databases: The Use of Spreadsheets, Flat-File, and Structured Databases"). Detailed design elaborates on each of these components in terms of tables and fields in the database, layout and data to be displayed on the graphical user interface, and often pseudocode for any data analysis modules (Bennett, McRobb, and Farmer, 2010, Chapter 18).

Common design strategies adopted include:

1. **Procedural:** Software is broken down into components based on functionality following a top-down approach and the process continues with each component until sufficient detail has been achieved. For our example, this could mean different components dealing with staff login, data management, user-specific functionalities, etc. Data management could then be further split into data entry, data storage, data retrieval, and so on.
2. **Object-oriented:** Software is described in terms of objects and their interactions with each other. This enables a component-based approach to be followed enabling modular deployment and improving reuse. In developing object-oriented software the design can be greatly assisted through the use of Unified Modeling Language (UML) tools (Miles and Hamilton, 2006; Bennett, McRobb, and Farmer, 2010). For our training records system, the objects could be staff member (which could then be used as a base class for specialized objects for supervisor and manager[13]), data record, competency, database handler, etc. Interactions are defined between objects. For example, "staff member" adds "data record" which includes one or more "competencies."
3. **Data-oriented:** In this case, the input and output data structures are defined and functions and procedures are simply used as a tool to enable data transformation. Here, you could perhaps have data structures defined for a competency record and defined functions for its insertion, updating, verification, and deletion and build the software around this basic framework.

A requirements traceability matrix is used to map design components to requirements and can be used to ensure that all the key requirements have been captured.

Risk analysis and management options are commonly carried out at this stage. Identifying potential problems or causes of failure here can influence the development stage of the software.

Software Development and Coding

This phase of the software development life cycle converts the design into a complete software package. It brings together the hardware, software, and communications elements for

---

[13]See the subsequent section "Object-Oriented Programming."

the system. It is often driven by the detailed design phase and has to take into consideration practical issues in terms of resource availability, feasibility, and technological constraints.[14] Choice of development platform is often constrained by availability of skills, software, and hardware. A compromise must be found between resources available and ability to meet software requirements. A good programmer rarely blames the platform for problems with the software.

Installing the required development environment, development of databases, writing programs, refining them, and so on, are some of the main activities at this stage. More time spent on detailed design can often cut development time, however, technical stumbling blocks can sometimes cause delays. Software costing should take this into account at the project initiation stage.

For the training and competency software example, a database will need to be created to hold the staff details and records. User interfaces will need to be built for inputting data, signing off competencies, and other user interactions. Database handling routines for insertion, updating and validation of data records, logic for supervisor and management roles, and so on, would be some of the functional modules required. The back end chosen should take into account the number of records expected to be held, simultaneous multiuser accessibility and, of course, software availability.[15] As far as possible it is a good idea for departments to select a family of tools and to use them for several projects so that local expertise can be efficiently developed.

Software coding should adhere to established standards and be well documented. The basic principles of programming are simplicity, clarity, and generality (Kernighan and Pike, 1999). The code should be kept simple, modular, and easy to understand for both machines and humans. There are several books (e.g., Bennett, McRobb, and Farmer, 2010; Knuth, 2011) that describe best practices in programming both in terms of developing algorithms as well as coding itself. Code written should be generalized and reusable as far as possible and adaptable to changing requirements and scenarios. Automation and reduced manual intervention will minimize human errors.

Unit testing is often included in this phase of the software life cycle as it is an integral part of software development. It is an iterative process during and at the end of development. This includes testing error handling, exception handling, memory usage and leaks, connections management, and so on, for each of the modules independently.

Software Testing[16]

Software testing is an ongoing process along the development to maintenance path. There are three main levels of testing:

1. Unit testing, where individual modules are tested against set criteria.
2. Integration testing, where the relationships between modules are tested.
3. System testing, which tests the workings of the software system as a whole.

Test criteria should include functionality, usability, performance, and adherence to standards. Test cases are usually generated during the design stage for each level of testing. These may be added to or modified along the pathway but should, in the least, cover the basic criteria. Testing objectives also influence the set of test cases. For example, the test cases for

[14]There almost always is something that was not picked up during design.

[15]Sadly, this is often the major deciding factor within medical physics/clinical engineering.

[16]See also the "Software Validation and Verification" section.

acceptance testing might differ from those for a beta test or even a usability test.

Software Maintenance

Software maintenance is defined in the IEEE Standard for Software Maintenance (IEEE 1219) as the modification of a software product after delivery to correct faults, to improve performance or other attributes, or to adapt the product to a modified environment. It lasts for the lifespan of the software and requires careful logging and tracking of change requests as per the guidelines for change management set out at the end of the requirements phase.

A maintenance request often goes through a life cycle similar to software development. The request is analyzed, their impact on the system as a whole determined, any required modifications are designed, coded, tested, and finally implemented. Training and day-to-day support are also core components of the software maintenance phase. It is therefore essential for the maintainer to be able to understand the existing code. Good documentation and clear and simple coding at the development stage is most helpful at this point especially if the developer is not available or if there has been a long time gap since development.

There are many tools that provide help with the discipline of software development, such as in UNIX/Linux, SCCS (Source Code Control System), and MAKE, which codifies instructions for compiling and linking. In addition, OSS wiki document management systems can help departments log their activities, including the queries that are raised during the complete software life cycle.

Software Life Cycle Models

Software life cycle models help manage the software development process from conception through to implementation within time and cost constraints. We consider three such here.

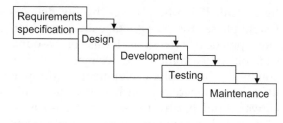

FIGURE 9.2   The waterfall model.

WATERFALL MODEL

The waterfall model is the most common and classic of life cycle models, also referred to as a linear-sequential life cycle model (Figure 9.2). It is very simple to understand and use and follows a structured sequential path from requirements to maintenance. It sets out milestones at each stage, which must be accomplished before the next stage can begin.

The rigidity of the waterfall model aids project management with well-defined milestones and deliverables. It does, however, restrict flexibility and does not provide much scope for user feedback until software development has been completed. It is only suitable for small-scale projects where user requirements are clearly defined and unlikely to change over the software lifespan.

INCREMENTAL MODEL/PROTOTYPING MODEL

The incremental model is an intuitive approach to the waterfall model (Figure 9.3). There are multiple iterations of smaller cycles involving requirements, design, development, and testing, each producing a prototype of the software. Subsequent iterations improve or build on the previous prototype. In situations where there is no manual process or existing system to help determine the requirements, the prototype can let the user plan its use and help determine the requirements for the system. It is also an effective method to demonstrate the feasibility of a certain approach for

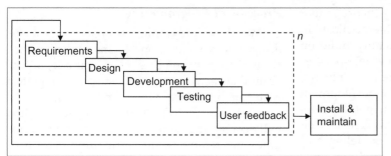

FIGURE 9.3 The incremental water-fall model.

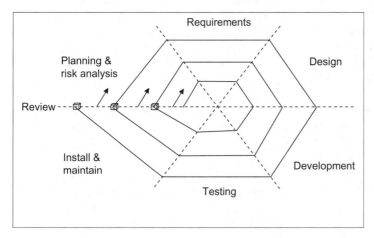

FIGURE 9.4 The spiral model.

novel systems where it is not clear whether constraints can be met or whether algorithms can be developed to implement the requirements. Testing and managing is easier with smaller cycles and errors can be detected early. It provides a better system to users, as users have a tendency to change their mind in specifying requirements and this method of developing systems supports this.[17]

A drawback, however, is the "scope creep" that can often result from frequent user feedback and the associated risk of getting stuck in a never-ending development loop. Also, since the requirements for the entire system are not gathered at the start of the project, the system

architecture might be affected at later iterations. Another issue is the risk of a build-and-patch approach through development, leading to poor code design.

SPIRAL MODEL

The spiral model was designed to include the best features from the waterfall and prototyping models (Figure 9.4). It is similar to the incremental model, but each iteration (called a *spiral* in this model) produces a robust working version of the software which is released for user evaluation. The development of each version of the system is carefully designed using the steps involved in the waterfall model. The first version is called the *baseline spiral* and each subsequent spiral builds on the baseline spiral, each producing a new version

[17]This model is very similar to agile programming in this respect.

with increased functionality. The theory is that the set of requirements is hierarchical in nature, with additional functionality building on the first efforts. The spiral model specifies risk analysis and management explicitly, which helps to keep the software development process under control. This is a good model for systems where the entire problem is well defined from the start, such as modeling and simulating software, but not so much for database projects where most functions are essentially independent.

The models discussed above are some of the basic models; individual software projects may sometimes combine techniques from different models to suit their specific needs. Medical physics/clinical engineering projects often follow an iterative approach since the requirements are not always clearly defined and may undergo frequent revisions. It is essential in such cases to maintain a robust software change control process (see the subsequent section "Software Quality Assurance").

Procedural, Object-Oriented, and Functional Programming

In the early days of computing, when memory was small and processing power slow (compared to today), many programs were short and simple. They followed a linear path from start to finish. As computers grew in complexity and memory (both internal and external) increased, along with processing power,[18] more complex programs were possible. However, to better utilize what were still limited resources, procedural programming was developed.

[18]Moore's law states that the number of transistors on a microprocessor (and thereby the processing power) will double every two years. This was stated prophetically rather than empirically, but does seem to have held.

Procedural Programming

Procedural programming breaks a problem down into several subproblems and identifies similarities between these. For example, a program that displays a set of test results, together with the reference range and a note as to whether the result is inside or outside of this, will format the same type of output several times. By removing this formatting code to a separate routine (often called a subroutine) and parameterizing it, the code becomes easier to write and more efficient to store (see Figure 9.5). It also has two added advantages: firstly that a change to the way in which these results are to be displayed (e.g., wanting results outside of the reference range to be displayed in red) only has to be made in one place. The second advantage is that if a subsequent program is written that requires the same formatting, it is easy to extract the relevant code and reuse it elsewhere.

The previous description also provides two examples of code reuse: internal and external. Code that is to be reused externally often requires a more stringent design (as it is impossible to predict exactly all the uses the code may be put to) and provides the basis for a code library. A code library may be precompiled so that other programs may use the routines, but not alter the source code, thereby ensuring continued integrity.

Procedures are used by calling them. The values in the brackets following the procedure name (in its definition) are the parameters. They are given the values, in order, that are in the brackets when the procedure is called. These may be explicit values (3.14, "Pi", etc.), the values stored in variables (e.g., nPi), or pointers to variables (e.g., *nPi). The last of these three forms means that the procedure cannot only use the value passed to it, but—as it is only a pointer—it can alter the value also. In this way a procedure can return more than a single result. In the example in Figure 9.5(b), the procedure

(a)
```
sText.Format("Potassium: %.2f. Reference range: 0-60. This value is
%s", nPotassium,(nPotassium>0 && nPotassium<60?"Normal":"Suspect");
pDC->TextOut(100,100,sText);

sText.Format("Sodium: %.2f. Reference range: 0-60. This value is
%s",nSodium,(nSodium>0 && nSodium<60?"Normal":"Suspect");
pDC->TextOut(100,200,sText);

sText.Format("Urea: %.2f. Reference range: 0-60. This value is %s",nUrea,(nUrea>0
&& nUrea<60?"Normal":"Suspect");
pDC->TextOut(100,300,sText);
```

(b)
```
DisplayResult("Potassium",nPotassium,0,60,pDC,100,100);
DisplayResult("Sodium",nSodium,0,60,pDC,100,100);
DisplayResult("Urea",nUrea,0,60,pDC,100,100);

void DisplayResult(CString sName, float nValue, float nLower, float nUpper, DC
*pDC, int x, int y)
{
   CString sText;
   sText.Format("%s: %.2f. Reference range: %.2f-%.2f. This value is
%s",sName,nValue,nLower,nUpper,(nValue>nLower          &&
nValue<nUpper?"Normal":"Suspect");
   pDC->TextOut(x,y,sText);
}
```

(c)
```
Potassium: 30. Reference range: 0-60. This value is Normal

Sodium: 30. Reference range: 0-60. This value is Normal

Urea: 30. Reference range: 0-60. This value is Normal
```

FIGURE 9.5 (a) Nonprocedural code in C++ with MFC.[19] Text is a CString (a text string) and pDC a device context on which the results will be displayed. nPotassium, etc., are numbers calculated elsewhere. (b) The code from (a) written in a procedural style. (c) The output from both pieces of code.

does not return anything (hence it is declared as having type void) but if it did not contain the display code, it might have been of type CString and returned the value of sText which it had assembled, for the calling program (which itself may have been a procedure) to display.

sText in Figure 9.5(b) is worthy of further investigation. As it is declared inside a procedure, it only exists there. As soon as the procedure completes, sText is destroyed and the memory it occupies is freed up for use by other procedures. sText is therefore known as a *local variable*. A *global variable* is one that exists throughout the entire program and is

declared differently. In aiming for code reuse, it is unwise to rely on global variables, as they may not exist in the next use of the code. It is therefore always better to pass all variables required by the procedure into it. The parameters of the procedure are also local to it. Therefore a procedure may freely alter the value that these variables are initially given, because the variable that the procedure was called with will remain untouched (unless it is passed as a pointer, as previously discussed). As sText is local to the procedure, it does not matter if a variable outside of the procedure also has the name sText: the procedure will use local variables first and only use global variables for those it cannot find locally. If

[19]Microsoft Foundation Classes.

there exists a global variable called sText then the procedure will still use the local variable. There are ways to get a procedure to use a global rather than a local variable, but it is generally better to use different names as this avoids confusion and makes the code more readable and therefore more maintainable—another key consideration when writing code for reuse. Where a variable exists is known as its *scope*; a procedure will therefore always use the ones that are in scope. Ones outside of a routine's scope are inaccessible.

The program keeps track of the local variables and which ones are currently in scope by use of a stack (see the section "Artificial Intelligence and Expert Systems" for a description of this).

Object-Oriented Programming

Object-oriented programming builds on procedural programming and introduces the concept of an *object*. There are two main concepts to cover here: *encapsulation* and *polymorphism*. As an introduction to these, we first consider a *structure*.

A structure is a collection of data items into one. It can therefore be manipulated as a single item, passed around the program as a single item, and stored and retrieved as a single item. The data items that comprise the structure do not need to be of the same type (indeed, it would be unusual if they were). As an example, let us consider a structure that holds information about an event. For the sake of simplicity we will restrict an event to having a name, a location, and a date on which it will occur. Two structures to represent this might be:

```
    struct date {int nDay; int nMonth; int
nYear};
    struct event {CString sName; CString
    sLocation; date dDate};
```

The individual members of the structure are addressed using dot notation. For example, if we have a variable evWedding, then the location is evWedding.sLocation and the year evWedding.dDate.nYear.

It can be seen that, through the use of multiple structures, the tables of a relational database can be simply represented in code, which makes the manipulation of the data simpler.

Operations on structures such as this can then be written, for example, to determine the day of week that the date represents or to move an event forward or backward in time (e.g., for implementing a to-do list, a recurrent action might be moved to a week later on completion rather than deleted). However, these are external to the structure and must also be moved to another program should the structure be reused (which, in the case of simple structures such as the ones described, is very likely). This is the issue that *encapsulation* addresses. A *class* is a combination of one or more data items that describe that class (i.e., a structure) together with one or more methods that act on or for it. An instance of a class (i.e., a variable defined as having this type) is called an *object* and leads to object-oriented programming. The methods are addressed in the same way as the data items, using dot notation, for example, evWedding.dDate. GetDayOfWeek() or evWedding.SetLocation ("Westminster Abbey"). These are actually called *methods* to avoid confusion with *operators*.

Operators are similar to mathematical operators and can also be defined for classes. Thus, it is possible to create a method dDate. AddDays(7) and also an operator + which adds the number of days to the date and delivers a new date, so that you can write dDate2 = dDate1 + 7. It can be seen that dDate. AddDays(1) and dDate ++ will yield the same result in this case. The operators are defined by the programmer for the class so need not bear any similarity to a mathematical one. Hence evWedding ++ could reverse the letters in the name of the location, although this is not helpful for maintainability and would be better accomplished through a method ReverseLocation().

```
dDate2 = dDate1 + 7;

dDate3 = dDate1 + dDate2;

dDate1.SetDay(5);

dDate1.SetDay("Wednesday");

sText = dDate1.GetDay();

nDay = dDate1.GetDay();
```

FIGURE 9.6 Some overloaded methods and operators. The one used depends on the operands, parameters, or assignment.

Object-orientation encourages code reuse in that classes developed for one application can easily be used in another. As they encapsulate the methods as well as the data, all the functionality is in one place, making it easily transportable.

Polymorphism, the other prime concept in object-orientation, means "many forms." There are three forms of polymorphism: method polymorphism, operator polymorphism, and class polymorphism. Method and operator polymorphism are similar in nature and implement *overloading*. An overloaded operator (or method) is one where there are more than one defined with the same name, the difference being either the type of the result, the type of the operand, or type (or number) of the parameters. Figure 9.6 gives some examples.

Class polymorphism relies on another concept in object-orientation: *inheritance*. Consider an event planner application. There may be several different types of event, a wedding, a birthday, a party. All require slightly different data members and may require different operators or methods. However, they also have a lot in common. Inheritance allows a class to be derived from another, *base class*. A derived class automatically inherits all the member data and methods of the base. These can be overridden (in much the same way as a local variable replaces a global) if different

functionality is required or added to. Class polymorphism allows a descendant class to be used where the base class is defined; for example, if the subclass dateofbirth is created which inherits from the class date (with the addition of a place name and registering office, for example) then the operator = which is defined for type date will also accept an operand of type dateofbirth.

Functional Programming

Functional programming is a paradigm that evaluates mathematical functions, avoiding state and mutable data. It contrasts with the imperative style (as used elsewhere in the previous two subsections) in that it emphasizes the application of functions rather than changes in state (in other words, it evaluates expressions, whereas imperative programming evaluates statements). Its roots are in lambda calculus which was developed in the 1930s to investigate function definitions, function application, and recursion (see later). As functional programming cannot alter the state of the program, it cannot be affected by it either. Therefore a function will always return the same output for a given input, which may not be true of imperative programming. It is therefore much easier to validate a functional program than an imperative one.

The most likely languages a scientist will meet which implement functional programming are Mathematica, Lisp, and (implemented to a lesser extent) SQL. C# and Perl, although strictly imperative languages, have constructs that allow the functional style to be implemented. Many functional concepts can be implemented in other languages.

The most important concepts are first-class and higher-order functions; pure functions; recursion; and lazy evaluation.

Functional programming requires that all functions are first-class, that is, they can be treated as any other values and can thus be used as arguments to other functions or return values from them. Such a function that takes another as an argument is a *higher-order function*. A good example of such a function is *map*, which takes a function and a list as its arguments. When executed, *map* applies the function to all the elements of the list. For example, to subtract a given number from a list lNum:

```
subtractFromList lNum nNum = map (\x -> x -
nNum) lNum
```

Some functional programming languages allow actions to be yielded as well as return values. These actions are termed *side effects*, indicating that the return value is the most important part. Languages that prohibit side effects are known as *pure*. Thus a *pure function* is one with no side effects.

Recursion is where a procedure calls itself. The classic example of this is in the evaluation of the mathematical function factorial (denoted !), where $n!$ is defined as 1 where $n = 1$ and $n \times (n-1)$ where n is greater than 1. It is only defined for positive integers. Thus a simple implementation would be:

```
factorial(int n);
{
    if(n<1) return 0; // error value
    if(n==1) return 1;
    return n*factorial(n-1);
}
```

Note the use of zero as an error value (see the section "Software Validation and Verification" for another error condition). Many procedures return predefined values to indicate that an error has occurred (usually in the input values, but sometimes in terms of failing to retrieve data from an external source, such as a networked device). The only occasions on which this error condition need not be tested is when the error conditions have been preevaluated, for example, when the input could never be less than 1 due to previous calculations.

Lazy evaluation is the deferment of the computation of values until they are needed. For example, in Python 2:

```
r = range(10)
print r
    [0, 1, 2, 3, 4, 5, 6, 7, 8, 9]
print r[3]
    3
```

All members of the list are calculated at declaration, whereas in Python 3:

```
r = range(10)
print(r)
    range(0, 10)
print(r[3])
    3
```

The value of r[3] is not computed until it is referenced. This can have a great saving in execution time when not all of a range might be referenced.

The documentation for Haskell, a functional programming language, purports the following:

> Functional programming is known to provide better support for structured programming than imperative programming. To make a program structured it is necessary to develop abstractions and split it into components which interface each other with those abstractions. Functional languages aid this by making it easy to create clean and simple abstractions. It is easy, for instance, to abstract out a recurring piece of code by creating a higher-order function, which will make the resulting code more declarative and comprehensible.
>
> Functional programs are often shorter and easier to understand than their imperative counterparts. Since various studies have shown that the average programmer's productivity in terms of lines of code is more or less the same for any programming language, this translates also to higher productivity.[20]

Real-Time System Programming

Real-time programming relates to programming where no delays are acceptable in the data processing. For example, a cardiac

[20]*www.haskell.org/haskellwiki/Haskell*

monitoring system must alarm the instant an event is detected, not five minutes later when other processing has completed. Thus the correctness of the system depends not just on the logical result, but also on the time frame in which it was delivered. A failure to respond is as critical an error as an incorrect result. Probably the most common use of real-time programming in a medical physics/clinical engineering sense is that of signal processing, and most real-time systems are part of embedded systems (see the next section).

Nothing is instantaneous and therefore real-time programming works according to a "real-time constraint," that is, the acceptable limit between an event and the system response. It is also worth noting that a system may have real-time and non-real-time elements, probably as independent threads. In this scenario a real-time thread will receive and process the incoming data/event within the required time frame, thereby being ready to process the next one. The other thread will take the result of this action and process it further, for example, by displaying only certain events or by implementing a localized trendline.

There are three forms of real-time, as defined by Burns and Wellings (2009):

- Hard real-time, where it is imperative that the system delivers a response within the required deadline, e.g., a flight control system.
- Soft real-time, where the deadlines are important but the system still functions correctly if occasional deadlines are missed, e.g., data acquisition.
- Firm real-time, where a missed deadline is tolerable but there is no benefit from the late delivery of service.

Systems may be time-triggered (e.g., every 25 ms or at 18:00) or event-triggered (where an external or internal event triggers activity).

Important characteristics of a real-time system are guaranteed response times, concurrency (rather than parallel processing), an

(a)
```
procedure guitar-solo is
begin
  left-hand-fingers-on-fretboard;
  right-hand-pluck-strings;
  face-screwed-up;
end
```

(b)
```
procedure guitar-solo is
  task right-hand;
  task body right-hand is
  begin
    right-hand-pluck-strings;
  end right-hand;
  task left-hand;
  task body left-hand is
  begin
    left-hand-fingers-on-fretboard;
  end left-hand;
  begin
  face-screwed-up;
end
```

FIGURE 9.7 (a) Sequential programming, which does not achieve the desired effect as each procedure (defined elsewhere) completes before the next commences. (b) Concurrent and parallel programming. The "begin" just before "face-screwed-up" activates the two previously defined tasks concurrently, after which face-screwed-up will run in parallel. Thus the desired effect is achieved.

emphasis on numerical computing (and algorithms), hardware interaction, and extreme reliability and safety.

The difference between concurrent, parallel, and sequential processing can be illustrated in Ada (Figure 9.7).

Examples of real-time programming languages are Ada, C, Java (although these two require extensions), RTL/2, Modula-2, and Mesa. There is insufficient space to reproduce sample code here, but good examples can be found on Andy Welling's web pages.[21]

Embedded System Programming

An embedded system is one that is contained ("embedded") into the electronics of the

[21]*www.cs.york.ac.uk/rts/books/CRTJbook.html*

device itself. There are several reasons for wishing to do this:

- Safety: An embedded system is very difficult to tamper with.
- Speed of execution: An embedded system is often preloaded into ROM[22] and so starts "instantly."
- Optimization: Code written for and developed around electronics to serve a specific purpose will always be more optimal (and hence faster to execute) than code written for a generalist machine (e.g., a standard PC).

The final point is a major characteristic of embedded systems: they are designed for specific purposes rather than general. Thus a more complex device may actually be composed of several modules,[23] each with its own embedded code.

The simplest form of embedded code is machine code (the lowest language of all) but this is by no means the simplest to create and maintain.[24] For many years embedded systems and languages were reflections of the electronics and especially of the microprocessors at the heart of them. Such electronic systems are often termed *microcontrollers* and the program that controls them *firmware*. This firmware is traditionally loaded into a ROM but is more likely to be an EPROM[25] now. The widespread use of the Internet means that such firmware can be easily updated, whereas older systems remained as they were once deployed.

While traditional embedded systems used very specialized programming languages and operating systems (the more specialized being both), three more general ones have risen in recent years, which we will consider further: Android (from Google), Windows Embedded (from Microsoft), and Java (from Sun). They can be found in devices as diverse as mobile phones and coffee machines.

Java is built on a tiered model. The key to this is the intermediate stage, or Java virtual machine (JVM). There are JVMs available for all the most common hardware and software platforms. Thus a Java program does not need to interact directly with the local operating system or with the hardware; the JVM does that. The Java program thus instructs the JVM, which instructs the OS or hardware. A Java program can therefore run on multiple platforms, just by using different JVMs.

Windows Embedded (originally Windows CE[26]) originally appeared in PDAs and was developed to have a small footprint, thereby fitting into a low-RAM device. It has the advantage that developers used to working with Visual Studio and Silverlight can quickly produce embedded systems.

Android is a relative newcomer to the field. Originally developed for mobile phones, it has also appeared in tablet computers (thus blurring the division between specialized and general). It is close to being open source, enabling developers to work closer to the heart of the OS. The distribution model for apps so produced is simple (certainly compared to Apple's), meaning that developers can release them swiftly. However, the openness of Android is also seen as its greatest weakness, in that it is viewed as being less secure for critical operations and data.

None of the three more general systems are as efficient as a microcontroller-specific

---

[22]Read-only memory.

[23]For example, an input module, a computation module, a data logging module, and an output module.

[24]Machine code is, however, the fastest. The author learned to write C6502 code by implementing a version of the game "Breakout." He then learned how to implement delay loops as it was too fast to play.

[25]Electrically programmable read-only memory.

[26]Compact Edition.

system, but with increased storage and faster processing that is rarely an issue anymore.

Software Validation and Verification

The primary purpose of validation and verification (when applied to medical software) is safety. Functionality is a secondary purpose, although without functionality the code is pointless. The way to reconcile this is to enquire as to the failure of the code: a system that is 75% functional but 100% safe is still usable (albeit annoying) but if the figures are reversed, it is not.

In validating and verifying a system as safe, one starts from the premise that all software contains "bugs." These "bugs" may be classified as faults, errors, or failures. A fault is a mistake in the design or code, which may lead to an error (but equally may not), such as declaring an array to be the wrong size. An error is unspecified behavior in execution, which may lead to a failure, such as messages starting with nonnumeric codes being discarded as they evaluate to zero. A failure is the crossing of a safety threshold due to an uncontained error.

There are two main approaches to testing, often referred to as "black box" and "white box." Applying this to software testing, the "box" is the program, or module, that is to be tested.

In black box testing, the contents of the box are unknown.[27] Therefore, tests comprise of a known set of inputs and the predetermined output that this should provide. This is very useful when the software has been commissioned using an output-based specification (OBS) or for end-user testing. It also removes any effect that may be caused by the application of the debugger environment itself.

In white box testing[28] the contents of the box are known and are exposed. In software terms, this may mean that the source code is available or even that the code is being tested in the

development environment via single-stepping. It is therefore usually applied to structures or elements of a software system, rather than to its whole. It is also not unusual for a black box failure to be investigated using white box testing.

In generic terms, therefore, black box testing is functional testing whereas white box testing is structural or unit testing. Thus a large system comprising multiple components will often have each component white box tested and the overall system black box tested to test the integration and interfacing of the components.

Testing should normally be undertaken by someone different than the software author. A draft British Computer Society standard (BCS, 2001) lists the following increasing degrees of independence:

a. The test cases are designed by the person(s) who writes the component under test.
b. The test cases are designed by another person(s).
c. The test cases are designed by a person(s) from a different section.
d. The test cases are designed by a person(s) from a different organization.
e. The test cases are not chosen by a person.

There are multiple test case design techniques with corresponding test measurement techniques. The British Computer Society (BCS, 2001) lists the following:

- Equivalence partitioning
- Boundary value analysis
- State transition testing
- Cause—effect graphing
- Syntax testing
- Statement testing
- Branch/decision testing
- Data flow testing

---

[27]But not in a Schrödinger sense.

[28]Also known as clear box, glass box, or transparent box testing, which may be a better descriptor of the process.

- Branch condition testing
- Branch condition combination testing
- Modified condition decision testing
- LCSAJ (linear code sequence and jump) testing
- Random testing

It is instructive to examine one of these, together with its corresponding measurement technique. The one we will select is boundary value analysis. This takes the specification of the component's behavior and collates a set of input and output values (both valid and invalid). These input and output values are then partitioned into a number of ordered sets with identifiable boundaries. This is done by grouping together the input and output values, which are expected to be treated by the component in the same way; thus they are considered equivalent due to the equivalence of the component's behavior. The boundaries of each partition are normally the values of the boundaries between partitions, but where partitions are disjoint the minimum and maximum values within the partition are used. The boundaries of both valid and invalid partitions are used.

The rationale behind this method of testing is the premise that the inputs and outputs of a component can be partitioned into classes that will be treated similarly by the component and, secondly, that developers are prone to making errors at the boundaries of these classes.

For example, a program to calculate factorials[29] would have the partitions (assuming integer input only):

- Partition a: $-\infty$ to 0
- Partition b: 1 to n (where $n!$ is the largest integer the component can handle, so for a standard C integer with a maximum value of 2147483647, n would be 12)
- Partition c: $n + 1$ to $+\infty$

The boundary values are therefore 0, 1, n, and $n + 1$. The test cases that are used are three per boundary: the boundary values and ones an incremental distance to either side. Duplicates are then removed, giving a test set in our example of $\{-1, 0, 1, 2, n - 1, n, n + 1, n + 2\}$. Each value produces a test case comprising the input value, the boundary tested, and the expected outcome. Additional test cases may be designed to ensure invalid output values cannot be induced.

It can clearly be seen that this technique is only applicable for black box testing.

The corresponding measurement technique (boundary value coverage) defines the coverage items as the boundaries of the partitions. Some partitions may not have an identified boundary, as in our example where Partition a has no lower bound and Partition c no upper bound. Coverage is calculated as follows:

Boundary Value Coverage
$$= \frac{number\ of\ distinct\ boundary\ values\ executed}{total\ number\ of\ boundary\ values} \cdot 100\%$$

In our example, the coverage is 100% as all identified boundaries are exercised by at least one test case.[30] Lower levels of coverage would be achieved if all the boundaries we had identified were not all exercised, or could not be (for example, if 2147483647 was required to be tested, where 2147483648 is too large to be stored). If all the boundaries are not identified, any coverage measure based on this incomplete set of boundaries would be misleading. (A fuller worked example can be found in BCS, 2001.)

Let us now consider a further level of complexity.

In theory, a software system is deterministic. That is, all of its possible states can be determined and therefore tested and the resultant system

---

[29]See the section "Functional Programming" for a definition.

[30]Although $+\infty$ and $-\infty$ were listed as the limits of Partitions c and a, they are not boundaries as they indicate the partitions are unbounded.

verified. However, although the states and the transitions between them may be finite, the use of multithreaded code and of multicore processors means that the number of test cases becomes unfeasibly large to process. This resultant complexity means that it is more practical to treat the system as being nondeterministic in nature and test/validate accordingly.

Testing therefore becomes a statistical activity in which it is recognized that the same code, with the same input conditions, may not yield the same result every time. To demonstrate this, consider the code in Figure 9.8.

On simple inspection, this code would be expected to produce a final value of x of between 10 and 20. However, it can produce values as low as 2 in 90 steps. (As an aside on complexity, this simple piece of code has in excess of 77,000 states; Hobbs, 2012.)

Hobbs (2012) defines "dependability" as "A system's [...] ability to respond correctly to events in a timely manner, for as long as required. That is, it is a combination of the system's availability (how often the system responds to requests in a timely manner) and its reliability (how often these responses are correct)." He goes on to argue that as dependability is inseparable from safety and dependability results in increased development cost, systems only need to be "sufficiently dependable" where the minimum level is specified and evidenced.

In conclusion, validation and verification may be fully possible. However, it may only be statistically demonstrable.

Software Quality Assurance

The terms *software quality assurance* (SQA) and *software quality control* (SQC) are often mistakenly used interchangeably. The definition offered by sqa.net is:

> Software Quality Assurance [is] the function of software quality that assures that the standards, processes and procedures are appropriate for the project and are correctly implemented.
>
> Software Quality Control [is] the function of software quality that checks that the project follows its standards processes, and procedures, and that the project produces the required internal and external (deliverable) products.

The two components can thus be seen as one (SQA) setting the standards that are to be followed with the other (SQC) ensuring that they have been. The process for SQC is one that should be specified as part of SQA so that the method for measurement and thus the criteria for compliance are known up-front. For example, the SQA may specify that ISO 14915 be used to define the multimedia user interfaces. SQC will therefore test to ensure that all multimedia user interfaces comply. It is thus clear that SQC will be undertaking a level of testing and it is not unusual for the full test suite to be part of the SQC process, testing not just the standards employed, but the functionality and safety of the software also.

It is common to break the software down into attributes that can be measured[31] and in this there is a similarity with software testing (see the previous "Software Validation and Verification" section). There are several

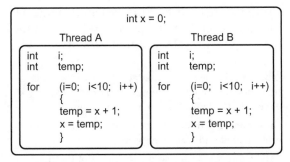

FIGURE 9.8 A simple two-threaded program. *Source: Hobbs, 2012.*

[31]If it can't be measured, then there's not a lot of point in specifying it.

definitions of software quality attributes: McCall (1977) and Boehm (1978) described in SQA,[32] Robert Grady's FURPS +,[33] and Microsoft's common quality attributes.

Some such attributes are:

- Accuracy: The ability of the system to produce accurate results (and to what level of accuracy this is required).
- Availability: The proportion of time that the system is functional and working.
- Compatibility: The ability of the system to work with, for example, different input devices.
- Functionality: What the system is actually supposed to do.
- Manageability: The ease with which system administrators can manage the system, through tuning, debugging, and monitoring.
- Performance: The responsiveness of a system to execute a required action with a set time frame.
- Security: The ability of a system to resist malicious interference (see Chapter 8, "Information Communications Technology Security: Firewalls, Virus Protection, Encryption, Server Access, and Data Security").
- Supportability: The ability of the system to provide information to assist in rectifying a performance failure.
- Usability: How well the system meets the users' requirements by being intuitive[34] and accessible.

Three further areas fall into the remit of SQA, all of which assume that the system (and especially the software at the heart of it) will not remain constant over time:[35] configuration management, change control, and documentation, which we will now examine.

Software configuration management (SCM) is the tracking and controlling of changes in the software. It therefore requires a robust change control (CC) method. The TickIT guide (1992) has two quality control elements covering this: "maintain and enhance" and "support," listing as the control mechanisms:

- Change control approval points: authority to proceed with change
- Full release approval
- Regression tests
- System, integration, and acceptance test
- Reviews and audits
- Configuration and change management
- Quality plans and projects plans

It can therefore be seen that changes to a live system should not be undertaken lightly. They require possibly more consideration than the initial system to ensure that a fix doesn't cause more problems.[36] A risk assessment, covering both the risk of making the change as well as the risk of not doing so, must be undertaken. Requiring authorization ensures that SQA is undertaken.

Like the software, the documentation must be kept up to date. It must form an accurate reflection of the software in use and should thus undertake the same version control as the software at the same time (see the section "Software Coding and Coding Management"). It is therefore clear that the system documentation (in several parts: user guide, programmers' guide, system administrators' guide,

---

[32]An independent website that presents information about SQA, SQC, and software development: www.sqa.net

[33]Functionality, Usability, Reliability, Performance, Supportability, constraints (such as design, implementation, interface, and physical).

[34]Standards may seem limiting and anticreative, but Microsoft's success is built on them, especially the common Windows user interface (left-click to select, right-click for a context menu, commonality of icons so "save" always looks the same, for example).

[35]It is often said that software is never finished, only implemented and passed to users.

[36]An IT director once told me that his aim for his support staff was that they fixed more problems than they caused. Sadly this was aspirational at the time.

etc.) must have a clear structure to it. Without such structure, the documentation becomes difficult to maintain. To enable programmers to work across multiple projects employing several languages, it is vital that the documentation has a consistent format to it so that information can be swiftly found and easily updated. In other words, the documentation also requires a standard, which should be part of the SQA process that it also enables.

Databases: The Use of Spreadsheets, Flat-File, and Structured Databases

While data can be stored in many ways and in many formats, all of the useful ones have this in common: structure. It has been noted that "Data + Structure = Information" and it is to turn data into information that structure is required.

The simplest data structure is the list: a single column of items. Arranging this list alphabetically (for example) brings structure to the data and enables it to be used for rapid lookups, using simple algorithms such as binary chop.[37] It is a simple step to expand this structure into multiple columns, allowing the overall structure to then be sorted by any one of the columns to yield different information. A basic spreadsheet (i.e., one without any formulae or computed cells[38]) forms one such database. This information can equally be represented as a set of text with one line per row in the spreadsheet. This, however, introduces two new problems: how to separate the data in each row and how to uniquely identify each row. Both of these are addressed by the spreadsheet structure, but not by the extracted data, or *flat file*.

A flat file can be given some structure by adding two things: a unique identifier on each line and a unique separator between each item in the lines. Taking the second solution first, the most common separator is the comma (hence "comma separated values," or CSV, a common file format used for data exchange), however this can produce problems of its own when recording data containing commas, for example addresses. There are two common solutions to this: encapsulating the data within double inverted commas and using an "escape character." The first of these brings problems where double inverted commas are present (e.g., addresses again). The second, although more robust, requires slightly more processing.

An escape character is a character that changes the interpretation of the character that follows it. A common escape character is "\". Hence, a comma that separates data would stand alone:, whereas a comma that should be taken to be part of the data would be preceded: \,. This notation does not suffer from the problems mentioned above, in that a backslash becomes \\.

Returning to the problem of adding structure to a flat file, we examine the unique identifier. A unique identifier may not need to be added, of course, if a part of the data is already unique. Such an item (whether added or already present) is called the *primary key* and its importance will become clearer as more structure and abstraction are added to our data. A spreadsheet already contains this identifier, in that each item of data can be uniquely referenced using its row and column identifiers.

At this point it is worth introducing some terminology (Figure 9.9):

- *Relational database*: A data structure through which data are stored in tables that are related to one another in some way. The way the tables are related is described through a relationship (see later definitions).

---

[37] A search mechanism that works on ordered data, repeatedly dividing the data into two halves and then proceeding to repeat the algorithm on the half containing the data searched for, until only the result (found or not) remains.

[38] This section is concerned with data storage—the use of spreadsheets in numerical processing is thus not covered here.

(a)

Surname	Forename	Address	e-mail
Snail	Brian	Magic roundabout	brian@roundabout.com
Flowerpot	Bill	Garden	bill@weed.co.uk
Cat	Bagpuss	Shop window	bagpuss@catworld.net
Miller	Windy	The windmill, trumpton	wmiller@chigley.ac.uk

(b)

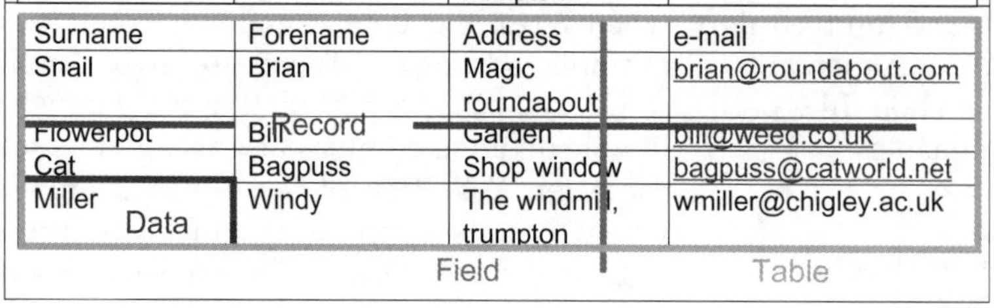

Surname	Forename	Address	e-mail
Snail	Brian	Magic roundabout	brian@roundabout.com
Flowerpot	Bill Record	Garden	bill@weed.co.uk
Cat	Bagpuss	Shop window	bagpuss@catworld.net
Miller Data	Windy	The windmill, trumpton	wmiller@chigley.ac.uk

Field   Table

FIGURE 9.9 (a) A set of data, arranged in a table. The items in the first row are field names (but may also be thought of as column headings). (b) The table in (a) labeled to show the terminology described above. An entity in this case is equivalent to a record.

- *Data*: Values stored in a database.
- *Entity*: A person, place, thing, or event about which we want to record information; an object we're interested in.
- *Field*: The smallest structure in a relational database, used to store the individual pieces of data about the object; stores a single fact about an object that we're interested in; represents an attribute.
- *Record*: A single "row" in a table; represents the collection of information for a single occurrence of the entity that the table represents.

- *Table*: The chief structure in a relational database, composed of fields and records, whose order is unimportant. A single table collects together all of the information we are tracking for a single entity; represents an object or an entity.

Let us now examine a specific example. Sam and Dakota decide to merge their music collections. Being computer scientists, they decide to create a database to catalogue the full collection. Their first attempt is a single-table database, such as described in the previous terminology list.

(a) Fields:

Media	CD/DVD/LP/MD/Tape/MP3 etc.
Artist	
Title	
Track1	
Track2	
(etc.)	
Price	In pence
Copies	
Total value	In pence

(b) Sample row (with just 3 tracks for simplicity):

CD	Example	Playing in the shadows	Skies don't lie	Stay awake	Changed the way you kiss me	999	2	1998

It is clear that this design is inefficient[39] and an inefficient design is difficult to maintain. At this point we need to introduce some more terminology:

- *Entity-relationship diagram (ERD)*: Identifies the data/information required by the business by displaying the relevant entities and the relationships between them.
- *Key*: A field in the database (or an attribute in an ERD) that is used to uniquely identify records and establish relationships between tables or entities; used for the retrieval of data in the table.
- *Primary key*: Uniquely identifies each record in a table, the key is part of the table for which it operates. (Note that this is normally a single field but may be a combination of fields—a *composite key*).
- *Foreign key*: A key from another table that is used to define a relationship to another record in another table. It has the same name and properties as the primary key from which it is copied.
- Rules for foreign keys:
  - *1-1*: Primary key from the main table is inserted into the second table.
  - *1-Many*: Primary key from the "1" table gets inserted into the "many" table.
  - *Many-many*: Primary key from each side gets placed into a third intermediate linking table that (usually) includes nothing but both keys.
- *Non-key*: A "regular" field; describes a characteristic of the table's subject.
- *Relationship*: Establishes a connection or correspondence or link between a pair of tables in a database, or between a pair of entities in an ERD.

- *One-to-one relationship*:[40] A single record in table A is related to only one record in table B, and vice versa.
- *One-to-many relationship*: A single record in table A can be related to one or more records in table B, but a single record in table B can be related to only one record in table A.
- *Many-to-many relationship*: A single record in table A can be related to one or more records in table B, and vice versa. There are problems with many-to-many relationships in that one of the tables will contain a large amount of redundant data, both tables will contain some duplicate data, and it will be difficult to add/update/delete records because of the duplication of fields between tables.

Returning to our example, we will work through a process to make the design more efficient. First we eliminate columns from the same table.

Change Track 1, Track 2, and so on into Tracks.

Media	Artist	Title	Tracks	Price	Copies	Value
CD	Example	Playing in the Shadows	Skies Don't Lie	999	2	1998
CD	Example	Playing in the Shadows	Stay Awake	999	2	1998
CD	Example	Playing in the Shadows	Changed the Way You Kiss Me	999	2	1998

We then create separate tables for each group of related data[41] and identify each row with a unique column (the primary key).

---

[39]For an indication of inefficiency, consider how many tracks you can fit on a box set of CDs, and then consider how many times you will fill all those fields.

[40]This and the following two relationships are intentionally similar to the rules for foreign keys.

[41]There is only one in this example.

Item_ID	Media	Artist	Title	Tracks	Track_No	Price	Copies	Value
1	CD	Example	Playing in the Shadows	Skies Don't Lie	1	999	2	1998
2	CD	Example	Playing in the Shadows	Stay Awake	2	999	2	1998
3	CD	Example	Playing in the Shadows	Changed the Way You Kiss Me	3	999	2	1998

Item_ID	Media	Artist_ID	Title_ID	Track_ID	Track_No	Price	Copies	Value
1	CD	1	1	1	1	999	2	1998
2	CD	1	1	2	2	999	2	1998
3	CD	1	1	3	3	999	2	1998

Artist_ID	Artist
1	Example

Title_ID	Title
1	Playing in the Shadows

Track_ID	Track
1	Skies Don't Lie
2	Stay Awake
3	Changed the Way You Kiss Me

Artist_ID	Artist
1	Example

Track_ID	Track
1	Skies Don't Lie
2	Stay Awake
3	Changed the Way You Kiss Me

them in separate rows and create relationships between these new tables and their predecessors through the use of foreign keys:

Finally, to move to *third normal form* or *3NF*, we must remove columns that are not fully dependent on the primary key:

Item_ID	Artist_ID	Title_ID	Track_ID	Track_No
1	1	1	1	1
2	1	1	2	2
3	1	1	3	3

Title_ID	Title	Media	Price	Copies
1	Playing in the Shadows	CD	999	2

This is known as *first normal form* or *1NF*. Note that field names do not contain spaces or punctuation. While some systems allow this, delimiters are then required. To move to *second normal form* or *2NF*, we must remove subsets of data that apply to multiple rows of a table and place

We can see that this final structure has the following relationships (naming the tables Item, Artist, Title, and Track, respectively):

- Artist to Item is one-to-many.
- Title to Item is one-to-many.
- Track to Item is one-to-one.

In this simple example there is little efficiency gained from this abstraction, but by the addition of just one extra field (track length) it can be seen how this structure is more adaptable than the one we started with.

The table Item has a primary key (Item_ID) and many foreign keys (Artist_ID, etc.) which are the primary keys in their own tables. An analogy from programming would be the use of pointers: not the data itself, but a link to where the data may be found (an analogy with the World Wide Web is similar).

A structure such as this is dependent on and representative of the relationships between the data and is known as a *relational database*. 4NF and 5NF also exist but are beyond the scope of this text.

One issue does arise through the use of normalizing a database, however. The relationships are achieved through the use of indexes. These are rapidly-changing files which therefore have a risk of corruption. If a corrupt index is used to retrieve records matching a certain key, the records returned may not all match that key. It is therefore imperative (depending on the criticality of the data retrieved) that these data are checked prior to use. This can be as simple as ensuring that each returned record does contain the key searched for. This will ensure that no erroneous results are used, but does not ensure that all results have been returned. To achieve this, a redundant data item, such as a child record counter, must be used. In most cases, this is not an issue but it does need to be considered when the criticality of the results is high (e.g., a pharmacy system).

Operations on relational databases are often carried out using Structured Query Language (SQL). As with any language there are various dialects, but the underlying principles are the same. The four most common commands are *SELECT*, *INSERT*, *UPDATE*, and *DELETE*. Commands do not need to be in uppercase, but are written this way here for clarity. The basic structure of a SELECT command is:

```
SELECT {fields} FROM {table} WHERE
{condition}
```

In the example above, "SELECT Track FROM Track WHERE Track_ID = 2" will return one result: "Stay Awake." The WHERE clause is a logical statement (using Boolean logic and can therefore include Boolean operators) which will evaluate to either TRUE or FALSE. The statement thus finds all rows for which the WHERE clause is TRUE and then returns the fields listed in the SELECT clause.

To make use of the relational structure, however, we need to return data from more than one table. There are two main ways of doing this:

```
SELECT {fields} FROM {tables} WHERE
{condition, including the relationship}
```

And

```
SELECT {fields} FROM {table} JOIN {table} ON
{relationship} WHERE {condition}
```

Examples of these might be:

```
SELECT a.Artist, t.Track FROM Artist a, Track
t, Item i WHERE a.Artist_ID = i.Artist_ID AND
t.Track_ID = i.Track_ID and i.Title_ID = 1
```

Which returns

Artist	Track
Example	Skies Don't Lie
Example	Stay Awake
Example	Changed the Way You Kiss Me

Note the use of the notation "Artist a" (in the FROM clause) which provides an alias for the table name (without an alias a.Artist—in the SELECT clause—could be referred to as

Artist.Artist). The importance of this can be seen from the table structures: several field names appear in multiple tables (as can be seen from the WHERE clause); SQL requires that no ambiguity exists in the statement and the use of aliases enables this.

The second form, to achieve the same result, might be written as:

```
SELECT a.Artist, t.Track FROM Item i JOIN
Artist a ON a.Artist_ID = i.Artist_ID JOIN
Track t ON t.Track_ID = i.Track_ID WHERE i.
Title_ID = 1
```

An INSERT statement adds a record to a single table. To add the track "The Way" to the example above would require the following statements:

```
INSERT INTO Item {Item_ID, Artist_ID,
Title_ID, Track_ID, Track_No} VALUES {4, 1,
1, 4, 3}
INSERT INTO Track {Track_ID, Track} VALUES
{4, "The Way"}
```

The field list is not always necessary: If the list is omitted it is assumed that the values are in the same order as the fields. If the primary key is an autonumber field (i.e., one that the system increments and assigns) then this should not be specified, meaning that the field list is required.[42]

An UPDATE statement has the form:

```
UPDATE {table} SET {field1 = value1,
field2 = value2, ...} WHERE {condition}
```

Examples (to correct the error introduced by the INSERT example) might be:

```
UPDATE Item SET Track_No = Track_no + 1 WHERE
Title_ID = 1 AND Track_ID > 3
UPDATE Item SET Track_No = 4 WHERE Item_ID = 4
```

---

[42]If the primary key is an autonumber field, adding data may require several steps, as the value assigned will have to be retrieved so it can be provided to the other INSERT statements.

Note that the first form may update multiple records, whereas the second will update only one as it uses the primary key to uniquely identify a single record.

Finally the DELETE statement has the form:

```
DELETE FROM {table} WHERE {condition}
```

If the *condition* is omitted, then all records from the specified table will be deleted. An example might be:

```
DELETE FROM Track WHERE Track_ID = 4
```

This statement alone would create a referential integrity error, in that Item now refers to a record in Track that no longer exists. To correct this, either a new Track with Track_ID of 4 must be created, or the following must be done:

```
DELETE FROM Item WHERE Track_ID = 4
```

This book's website describes two other common data structures: the linked list and the tree. See also the section "Artificial Intelligence and Expert Systems" for a description of another common data structure, the stack.

Relational databases are very common in hospital informatics and clinical computing. Examples include oncology management systems, equipment management systems (see later), the electronic patient record, and cardiology patient monitoring systems.

TYPICAL APPLICATIONS

Image Processing Software

Image processing is the application of a set of techniques and algorithms to a digital image to analyze, enhance, or optimize image characteristics such as sharpness and contrast. Most image processing techniques involve treating the image as either a signal or a matrix and applying standard signal-processing or matrix manipulation techniques, respectively, to it.

Terminology

A *pixel* or "picture element" is the smallest sample of a two-dimensional image that can be programmatically controlled. The number of pixels in an image controls the resolution of the image. The pixel value typically represents its intensity in terms of shades of gray (value 0–255) in a grayscale image or RGB (red, green, blue, each 0–255) values in a color image.

A *voxel* or "volumetric pixel" is the three-dimensional counterpart of the 2D pixel. It represents a single sample on a three-dimensional image grid. Similar to pixels, the number of voxels in a 3D representation of an image controls its resolution. The spacing between voxels depends on the type of data and its intended use. In a 3D rendering of medical images such as CT scans and MRI scans, the size of a voxel is defined by the pixel size in each image slice and the slice thickness. The value stored in a voxel may represent multiple values. In CT scans, it is often the Hounsfield unit which can then be used to identify the type of tissue represented. In MRI volumes, this may be the weighting factor (T1, T2, T2*, etc.).

Image arithmetic is usually performed at pixel level and includes arithmetic as well as logical operations applied to corresponding points on two or more images of equal size.

Geometric transformations can be applied to digital images for translation, rotation, scaling, and shearing, as required. Matrix transformation algorithms are typically employed in this case.

For binary and grayscale images, various *morphological operations* such as image opening and closing, skeletonization, dilation, erosion, and so on, may also be employed for pattern matching or feature extraction.

An *image histogram* represents the distribution of image intensity values for an input digital image. Histogram manipulation is often used to modify image contrast or for image segmentation when the range of values for the desired feature is clearly definable.

Some common image processing applications are introduced as follows.

Feature extraction is an area of image processing where specific characteristics within an input image are isolated using a set of algorithms. Some commonly used methods for this include contour tracing, thresholding, and template matching. Image segmentation is a common application of feature extraction which is often used with medical imaging to identify anatomical structures.

Pattern and template matching is useful in applications ranging from feature extraction to image substitution. It is also used with face and character recognition and is one of the most commonly used image processing applications.

There are several image processing software packages available, from freely distributed ones such as ImageJ to expensive suites such as MATLAB and Avizo which range in functionality and targeted applications. We'll discuss only a few of the commonly used ones within medical physics/clinical engineering here.

The image format most commonly used in medical applications is DICOM, providing a standardized structure for medical image management and exchange between different medical applications (see Chapter 8, "DICOM").

Image Processing Software Packages

ImageJ is an open source, Java-based image processing program developed at the National Institute of Health. It provides various built-in image acquisition, analysis, and processing plugins as well as the ability to build your own using ImageJ's built-in editor and a Java compiler. User-written plugins make it possible to solve many bespoke image processing and analysis problems.

ImageJ can display, edit, analyze, process, save, and print 8-bit color and grayscale, 16-bit integer and 32-bit floating point images.[43] It

[43]*http://rsbweb.nih.gov/ij/docs/intro.html*

can read many standard image formats as well as raw formats. It is multithreaded, so time-consuming operations can be performed in parallel on multi-CPU hardware. It has built-in routines for most common image manipulation operations in the medical field including processing of DICOM images and image stacks such as those from CT and MRI.

Mimics[44] is an image processing software for 3D design and modeling, developed by Materialise NV. It is used to create 3D surface models from stacks of 2D image data. These 3D models can then be used for a variety of engineering applications.

Mimics calculates surface 3D models from stacked image data such as CT, micro-CT, CBCT, MRI, confocal microscopy, and ultrasound, through image segmentation. The region of interest (ROI) selected in the segmentation process is converted to a 3D surface model using an adapted marching cubes algorithm that takes the partial volume effect into account, leading to very accurate 3D models. The 3D files are represented in the STL format.[45]

The most common input format is DICOM, but other image formats such as TIFF, JPEG, BMP, and raw are also supported. Output file formats differ, depending on the subsequent application, but common 3D output formats include STL, VRML, PLY, and DXF.

Mimics provides a platform to bridge stacked image data to a variety of different medical engineering applications such as finite element analysis (FEA; see the next section), computer-aided design (CAD), rapid prototyping, and so on.

MATLAB

MATLAB is a programming environment for algorithm development, data analysis, visualization, and numerical computation.[46] It has a wide range of applications, including signal and image processing, communications, control design, test and measurement, financial modeling and analysis, and computational biology. The MATLAB Image Processing Toolbox™ provides a comprehensive set of reference-standard algorithms and graphical tools for image processing, analysis, visualization, and algorithm development. It also has built-in support for DICOM images and provides various functions to manipulate DICOM data sets. This makes it a widely used tool in various medical physics/clinical engineering research groups and related academia.

IDL

IDL is a cross-platform vectorized programming language used for interactive processing of large amounts of data including image processing.[47] IDL also includes support for medical imaging via the IDL DICOM Toolkit add-on module.

The image processing software packages mentioned here are but a few of the commonly used ones within a medical physics/clinical engineering environment partly due to their extensive libraries for medical image processing and partly for historical reasons.[48] There are many more free-for-use as well as commercial software packages available providing varying degrees of functionality for different applications and, if there is a choice available, it is advisable to explore the options available for a particular task.

Finite Element Analysis

The finite element method is an engineering technique used for analyzing complex

---

[44]Materialise Interactive Medical Image Control System.

[45]http://biomedical.materialise.com/mimics

[46]www.mathworks.co.uk/products/matlab/

[47]www.exelisvis.com/language/en-US/ProductsServices/IDL.aspx

[48]These two factors feed each other.

structural problems. Finite element analysis (FEA) is an effective and widely used computer-based simulation technique for modeling mechanical loading of various engineering structures, providing predictions of displacement and induced stress distribution due to the applied load (Pao, 1986).

The basic idea of the finite element method is to break up a continuous model into a geometrically similar model made up of disjoint linked components of simple geometry called *finite elements*. The behavior of an individual element is described with a relatively simple set of equations. These elements are made up of a collection of *nodes* which define both the element's shape as well as its degrees of freedom.

The geometry of the element is defined by the placement of the geometric nodal points. Most elements used in practice have fairly simple geometries. In one dimension, elements are usually straight lines or curved segments. In two dimensions they are of triangular or quadrilateral shape. In three dimensions the most common shapes are tetrahedra, pentahedra, and hexahedra.

Elements are joined together at nodes along *edges*. When adjacent elements share nodes, the response field is shared across boundaries. Material and structural properties such as Young's modulus are assigned to each element that describes how it would respond to applied loading conditions. Nodes are assigned at a certain density throughout the material depending on the anticipated stress levels of a particular area. Regions that receive large amounts of stress (for example, areas of fracture-risk) usually have a higher node density than those that experience little or no stress. The response of each element is then defined in terms of the nodal degrees of freedom. The equations describing the responses of the individual elements are joined into an extremely large set of simultaneous equations that describe the response of the whole structure. Analysis of the model provides nodal displacements and induced stress distribution due to the applied load. (See Figure 9.10.)

FEA has become a solution to the task of predicting failure due to unknown stresses by showing problem areas in a material and allowing designers to see all of the theoretical stresses within. This method of product design and testing is far superior to the manufacturing costs that would accrue if each sample was actually built and tested.

In practice, a finite element analysis usually consists of three principal steps (Roylance, 2001):

1. Preprocessing: The discretized model of the structure is divided up into nodes and elements. Constraints and loads are then applied as required. Several commercial finite element analysis packages such as ANSYS, FEMtools, and QuickField have graphical user interfaces for this stage. Some of these packages also offer an option to import CAD models and overlay a mesh on these to generate the FE model.
2. Analysis: The data set prepared by the preprocessor is used as input to the finite element code itself, which constructs and solves a system of linear or nonlinear algebraic equations that compute displacements corresponding to externally applied forces at the nodal points.
3. Postprocessing: Most FEA packages present FE analysis results graphically with color coding or contours depicting hotspots and the range of deformation.

Artificial Intelligence and Expert Systems

The creation of artificial intelligence (AI) has been a human goal for a very long time. One of the most famous, the Grand Turk (a chess-playing automaton), was a hoax, the perpetration of which was aided by the desire for such a device to exist.

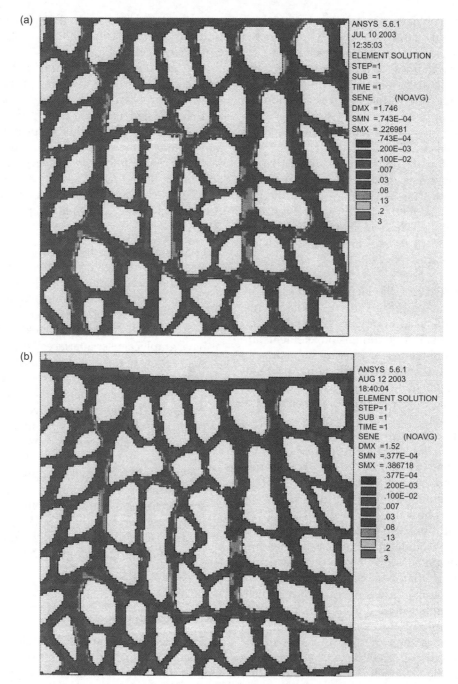

FIGURE 9.10 (a) A finite element model (FEM) of cancellous bone with uniform load applied across the top. (b) The same model showing the deformation due to this loading, after 16 iterations.

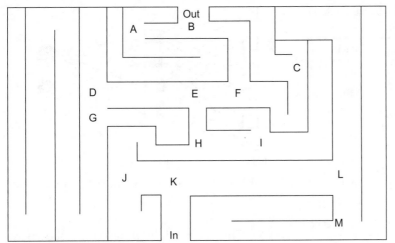

FIGURE 9.11 A simple maze. The "decision points" (i.e., the points at which a decision must be made in order to progress) are labeled as uppercase letters.

In computer science terms, AI is popularly perceived to be something that can pass the "Turing test" (Turing, 1950) in which an interrogator puts questions to a "thinking machine" and a human and is unable to tell the difference between the two. Currently AI is defined more as "the study and design of intelligent agents" (Poole et al., 1998, pp. 7–21).

AI is predominantly concerned with the solution of problems and can be best illustrated using a maze problem (Figure 9.11).

In Figure 9.11, the quickest route through the maze can be seen to be In-K-J-H-E-F-B-Out. Artificial intelligence is concerned with how the decision is made at each point, using a set of rules, or heuristics.[49] A simple heuristic might be "turn left" (to avoid confusion, "left" is taken to be "not right" where "right" is clear but left may not be—there are no cases in this example where neither left nor right is clear). This yields the path In-K-J-Fail. "Turn right" yields In-K-L-M-Fail, whereas "alternate left and right" gives either In-K-J-H-E-F-B-A-Fail (the best result so far) or In-K-L-Fail (an equal worst result). Additional complexity is

therefore required and may be introduced as follows:

1. If the decision point has not been encountered before, turn left.
2. If a dead end is reached, return to the most recently encountered decision point and turn right.
3. If both routes from a decision point reach dead ends, return to the most recently encountered decision point before this and turn right.

This yields the result In-K-J-Fail-J-H-E-D-G-Fail-G-Fail-D-Fail-E-F-B-A-Fail-A-Fail-B-Out, which is successful but long-winded. It does, however, demonstrate the use of heuristics to solve a problem. It also introduces a new requirement not previously present: that of recording the decision points encountered and the order in which they were encountered. A more complex maze would also require the decision made at that point to be recorded.

Assuming that we have some knowledge of the final target (for example, its coordinate position) we may postulate a further enhancement where the direction of turn (left or right) is determined by taking a small step in each direction and determining the absolute distance to the target from each and taking the

---

[49] A heuristic is "a way of directing your attention fruitfully." (Wikipedia)

TABLE 9.2 A Stack Implemented to Record the Maze Tracing

Step	Action	Stack
1	Turn left at K	Kr
2	Turn left at J	KrJr
3	Dead end—retrace; take top element from stack and follow this (turn right at J)	Kr
4	Turn left at H	KrHr
5	Turn left at E	KrHrEr
6	Turn left at D	KrHrErDr
7	Turn left at G	KrHrErDrGr
8	Dead end—retrace; turn right at G	KrHrErDr
9	Dead end—retrace; turn right at D	KrHrEr
10	Dead end—retrace; turn right at E	KrHr
11	Turn left at F	KrHrFr
12	Turn left at B	KrHrFrBr
13	Turn left at A	KrHrFrBrAr
14	Dead end—retrace; turn right at A	KrHrFrBr
15	Dead end—retrace; turn right at B	KrHrFr
16	Success	

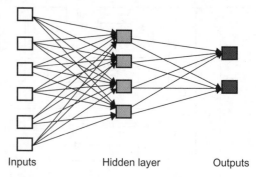

FIGURE 9.12 A simple feedforward[53] network. *Source: Stergiou and Siganos.*[54]

direction yielding the minimum value. An absolute rule (e.g., "turn left") is still required in the event of a tie. Such a heuristic adds the requirement to record the decision taken at each point encountered and would yield the path   In-L-Fail-L-M-Fail-M-Fail-K-J-H-E-F-B-Out which (aside from a poor start) is very efficient.

The best way to record this process is by using a *stack*. This is also known as a *heap*[50] and consists (as the name implies) of a last-in-first-out list[51] (just like a pile of paper on your desk). The stack might progress as shown in Table 9.2 for the heuristic outlined in points (1) through (3) (using "r" and "l" to represent "right" and "left"— note that this implementation records the action(s) left to be taken should you return, not the one(s) taken, which is thus more scalable).

It can be seen that such rigid heuristics may occasionally give rapid success, but may equally give poor results. While this example will always produce a result (as the number of options is finite) other problems may not do so or may lead into a cyclic path from which there is no escape.[52] One vital component of intelligence is missing from this heuristic-based system: the ability to learn. Artificial intelligence therefore seeks to develop algorithms that are self-modifying, based on the results that have been achieved. To illustrate this, we investigate the artificial neural network (ANN).

An ANN is an information-processing paradigm whose intention is to take a large number of inputs (for example, parameters from an ECG trace) and produce an output

---

[50]Specific implementations in programming languages such as C++ and Java give these different meanings (primarily to do with scope), but they are essentially the same structure.

[51]A similar structure, the *queue*, is a first-in-first-out structure, often referred to as FIFO.

[52]Minotaurs are an end to the problem but not necessarily a solution. The reader may select their own analogy as to what a Minotaur may represent.

representing a decision (for example, whether or not this trace is normal). It does this through a collection of highly interconnected processing elements (called neurones) which are trained to produce reliable output. In Figure 9.12, the two outputs may be labeled "normal" and "requires investigation" (continuing the ECG example).

In the example in Figure 9.12, information flows forward only (i.e., from left to right). At each point processing occurs and this is fed forward into the next layer. It is not difficult to extrapolate this model into one with multiple hidden layers and/or that feeds backward as well as forward. The hidden layer is so called because this auto-tunes itself. To achieve this, data with known results are divided into two sets: a training set and a test set. The training set is used to tune the processing in the hidden layers. Once tuned, they are fixed and the test set is used to determine how successful the tuning has been. If the results are not as good as required (and the required level of success may depend on the problem being posed and the consequences of an incorrect output) then redesign is required: either in the processing neurones or the structure of the network. The process is then repeated. At no point must the test set be used to train the network or vice versa, and the importance of adequately dividing the data into training and test sets can clearly be seen.

Neural networks are very good at pattern recognition and, as such, have been applied to optical character recognition, signature recognition (where the data are notoriously noisy and a level of confidence is required more than an absolute result), and aircraft engine monitoring. Medical applications include diagnosis, biochemical analysis, and image analysis.

One of the successful areas of AI has been the expert system. An expert system seeks to replicate the decision-making processes of a human expert and, as such, combines the principles described above: it is heuristic (some inputs produce a definitive output), can account for noisy data (some inputs produce a probabilistic output), and is adaptive (it can be trained). It comprises two parts: an inference engine and a knowledge base. The knowledge base is a set of rules in an IF...THEN structure, for example:

- IF the chest wall is moving THEN respiration is taking place.
- IF the identity of the germ is not known with certainty AND the germ is gram-positive AND the morphology of the organism is "rod" AND the germ is aerobic THEN there is a strong probability (0.8) that the germ is of type enterobacteriaceae (Lewis, 2003).

Note the second example, from the MYCIN system, developed to assist in helping a doctor to decide whether a patient has a bacterial infection, which organism is responsible, which drug may be appropriate for this infection, and which may be used on the specific patient. This utilizes further logic (the AND clauses) to deliver a result.

The inference engine (as the name implies) infers new rules based on the ones already present in the knowledge base. It may do this by chaining together the outputs from several different rules (possibly in a network structure akin to the ANN) or by requesting clarification from the operator. A key output of the expert system as opposed to the ANN is that the logic used in reaching the result is displayed along with the result. In an ANN, the logic remains hidden.

Equipment Management Database Systems

An equipment management database system is, as the name implies, a database of

[53]Processing flows "forward" from inputs to outputs.

[54]www.doc.ic.ac.uk/~nd/surprise_96/journal/vol4/cs11/report.html

TABLE 9.3   Technologies for Implementing Device Tracking

Technology	Description	Pros and cons
Wi-Fi (wireless network)	This system uses an organization's Wi-Fi network (the same network that supports wireless computing devices). Tags transmit a Wi-Fi signal, which is picked up by Wi-Fi access points; if the signal is picked up by four or more access points, the location of the device can be triangulated in three dimensions. This may be supplemented by additional ultrasound "beacons" to more accurately locate devices within a particular room.	There is a claimed accuracy of location to 2 m (although there exists some skepticism about this as it is probably only possible with the addition of ultrasound beacons). Makes use of an existing Wi-Fi so no additional network is required. Fully scalable for patients, assets, and staff. It requires many more Wi-Fi access points than is needed just for wireless computing. Produces an additional load on the network. Tags are relatively expensive.
RFID (radio frequency identification)	A network of RFID receivers is positioned across the organization, connected to the local area network (LAN). Tags (either active or passive) communicate with these receivers, which can track their location. Passive tags respond to a signal from gateways (from where they also draw their power to do so), whereas active ones will transmit data to a nearby receiver and have onboard power. This is a proven technology, being used in many industries, including retail, where cheap passive tags are attached to stock, where their location/presence is tracked.	Tags are very cheap (passive tags are pennies each). The granularity of the mapping is not very precise. An additional network (of RFID transponders) is required.
RFID with IR (infrared)	This uses RFID technology, supplemented by infrared transmitters/detectors. IR can be used to track devices to a very granular level (bed space, or even a drawer), and information is then communicated centrally via an RFID network.	Accurate location of items at a very granular level. Fully scalable for patients, assets, and staff. Additional network needs to be installed of RFID transponders. The solution is quite complex in that there are several technologies involved. The risk of failure is therefore higher.

equipment around which a system has been constructed to manage that equipment. As such, it shares a lot of features with asset management database systems but contains additional features to aid in the management of equipment, which for the focus of this book means medical equipment.

An asset management database will contain such information as the equipment name, manufacturer, purchase cost and date, replacement cost (and planned replacement date if known), location, and a service history (if appropriate).

Reports may then be run against this data to produce information on capital spend and assistance in forecasting future spend. It may also be used to highlight unreliable equipment, commonalities of purchasing (to assist in contract negotiations and bulk buying), and equipment loss due to theft or vandalism.

An equipment management database system includes all of these features, but also includes information so that the equipment may be managed appropriately. Probably the most common example of this additional functionality is the PPM (planned preventative maintenance).[55] To implement PPM, all equipment managed must be assigned a service plan. In its simplest form, this is a set time frame: for example, a set of scales that requires calibration once a year. A more complex plan might include different tests or calibrations to be performed at different times. This type of service plan will generally still include a fixed time interval, but will also include a service rotation. On the first rotation, it might be visually inspected to ensure that all seals are still in place. On the second rotation a calibration might be added, and on the third a portable appliance test (PAT) renewal might be undertaken in addition to the inspection and calibration. The system will keep track of all work that is due (producing worklists for a particular time period), the service rotation that is due, parts and labor used in the maintenance, and so on.

Inspecting, calibrating, and testing all of a hospital's medical equipment is not a short task: most such services spread the load across the entire year, which the management system is also able to assist in planning by reporting on peaks and troughs in the planned workload, adjusting for planned staff absences, and potential peaks and troughs in emergency repairs (from historical data).

Further functions that such a system can assist with are the management of contracts (where equipment is maintained by a third party), the recording of repairs undertaken (thus determining when equipment has reached the end of its economic life), and the recording of medical device training.

In this last example, all staff of the hospital are recorded along with the equipment they have been trained to use. Thus training can be kept current and a mechanism for preventing the unsafe use of equipment implemented.

To achieve all this functionality, an equipment management database system needs to be interfaced to other systems, such as a capital asset management system, a human resources system, and a contracts management system. (See the section "Links to Hospital Administration Systems" for information on interfacing.)

Device Tracking Systems

Electronic tracking devices are a growing element of hospital practice and their role is considered here in terms of potential, rather than system design. The benefits of tracking devices are numerous, from the reduction in time spent looking for equipment (estimated as being 2.5% of nurse time by University College London Hospitals in 2011) to theft prevention. Other potential benefits are improved servicing due to reliable equipment assetting and the ability to run a more efficient equipment library.

Broadly, there are three types of technology to consider, as described in Table 9.3. Each requires the asset to have a tag attached, which communicates with the central system. The tags may vary in size, depending on the amount of data required and the distance from the transponder that it is required to communicate with. This means that such a system may be deployed for multiple purposes. For example, it may describe bed occupancy, the stocking of a pharmacy cabinet, or track the whereabouts of babies and/or vulnerable patients, as well as staff in lone working situations.

References

Bennett, S., McRobb, S., Farmer, R., 2010. Object-Oriented Systems Analysis and Design using UML. McGraw-Hill.

British Computer Society (BCS), 2001. Standard for Software Component Testing. (Draft from which BS7925-2 was produced). [online]. Available: <http://www.testing-standards.co.uk/Component%20Testing.pdf> (accessed 27.04.12).

---

[55]Sometimes simply referred to as "servicing."

Burns, A., Wellings, A., 2009. Real-Time Systems and Programming Languages. Addison-Wesley.

Hobbs, C., 2012. Build and Validate Safety in Medical Device Software. Medical Electronics Design [online] Available: <http://www.medicalelectronicsdesign.com/article/build-and-validate-safety-medical-device-software> (accessed 26.04.12).

Kernighan, B.W., Pike, R., 1999. The Practice of Programming. Addison-Wesley.

Knuth, D.E., 2011. The Art of Computer Programming. Volumes 1–4a. Addison-Wesley.

Lawlis, P.K., 1997. Guidelines for Choosing a Computer Language: Support for the Visionary Organization [online]. Available: <http://archive.adaic.com/docs/reports/lawlis/content.htm> (accessed 05.07.12).

Lewis, P., 2003. Knowledge Representation, University of Southampton [online]. Available: <http://users.ecs.soton.ac.uk/phl/ctit/ho1/node2.html> (accessed 23.12.11).

Miles, R., Hamilton, K., 2006. Learning UML 2.0 – A Pragmatic Introduction to UML. O'Reilly.

Pao, Y.C., 1986. A First Course in Finite Element Analysis. Allyn and Bacon, Inc..

Poole, D., Mackworth, A., Goebel, R., 1998. Computational Intelligence: A Logical Approach. Oxford University Press, New York.

Roylance, D., 2001. Finite Element Analysis. [online]. Available: <http://ocw.mit.edu/courses/materials-science-and-engineering/3-11-mechanics-of-materials-fall-1999/modules/fea.pdf> (accessed 17.08.12).

Turing, A., 1950. Computing Machinery and Intelligence, Mind LIX [online] (236): pp 433–460. Available: <http://mind.oxfordjournals.org/content/LIX/236/433> (accessed 23.12.11).

Web Development

Paul S. Ganney*, Sandhya Pisharody†, and Ed McDonagh**

*University College London Hospitals NHS Trust, London, U.K., †Varian Medical Systems, U.K.,
**Diagnostic Radiology/PACS/ Royal Marsden Hospital NHS Trust, London, U.K.

Many of the software applications used in hospitals now use web technology to enable the sharing of data and resources. The technology is described in this chapter, dealing first with hosting strategies and then with web programming.

WEB TECHNOLOGY

Hosting Strategies

Content or applications accessible via a web browser can be served from almost any computer device, from a mobile phone to a huge server farm with multiple locations across different continents. The computers providing these web services can be running almost any operating system and there is also a large choice of web server software.

However, the majority of websites are hosted on UNIX-like operating systems (mainly various flavors of Linux), with almost all the rest being hosted on Microsoft Windows. An even greater majority of websites are hosted using the Apache web server software (which runs on both UNIX-like operating systems and on Windows servers), with Microsoft's IIS software and nginx (pronounced engine-X) being the next two most popular.[1] Most web servers also have some sort of server-side scripting, from using the Common Gateway Interface (CGI) to execute Perl scripts or compiled binaries to having entire websites generated in languages such as PHP, Python, Perl, or Ruby. Further information on this topic is provided in the section "Web Programming."

The common use of Linux, Apache, MySQL (a database), and PHP to provide web hosting is commonly referred to as a LAMP stack. The P in LAMP can be used to refer to Python, Perl (the original "P"), or even Ruby! All of

---

[1]See *http://news.netcraft.com/archives/category/web-server-survey/* and *http://w3techs.com/technologies/overview/operating_system/all* for further details of the trends in market share along with the sampling methodology.

these elements are open source, which has helped them to gain such a large market share by being both free to use and collaborative in their development. A popular alternative to the LAMP stack is a WAMP stack, with the same software being used to run on Microsoft Windows operating systems, rather than using the Internet Information Services (IIS) application Microsoft provides with its server operating systems and, in a limited way, to its desktop systems.

Both Windows and Linux based web servers can be deployed at any level, from providing a few web pages from your desktop computer to utilizing the flexibility of cloud computing. Windows servers are usually slightly more expensive than Linux servers due to the licensing costs. The following is a description of different hosting strategies available, along with a summary of advantages and disadvantages of each.

Desktop Computer

The easiest way to experiment with creating a web server and providing a web service of some kind is to simply install the software on your own PC. This can be done on any of the more popular operating systems at no financial cost, and with relative ease depending on the applications chosen.

Advantages
- Cheap—the only cost is your time.
- Quick to install and get going, particularly if using the XAMPP web server software stack on Windows or if using Linux where all the software is readily available (if it isn't already installed).
- Secure, sort of—the web server won't normally be accessible from beyond your home or corporate network unless your router is configured with network address translation (NAT) to forward traffic from the router to your PC.

- Easy to prototype new ideas for services.
Disadvantages
- You need to know something of web server administration.[2]
- You need to have authorization to install software on your PC, and you may need to alter firewall settings.
- The server isn't normally accessible beyond your local area network.
- The solution does not scale—you will need to transfer the service off your PC if it becomes too useful, to gain uptime/availability, to reduce competition from resources, etc.

Local Server—Real Iron

This is a physical server (see "Local Server—Virtual Machine" next for a comparison). However, servers and desktop PCs that have reached the end of their useful life are often good (and free) candidates for setting up web servers due to the low processing power required. This can be done by installing a Linux server edition (free) or using the existing Windows Server license if applicable with IIS or Apache.

When reusing computers, however, always keep in mind the information governance requirements for any data that might have been on the computers' disks.

Advantages
- May make use of an old server that is being decommissioned.
- Can run other services for the department, such as a source control management system.
- No competition for resources.
- Security—not accessible beyond the corporate network.
- Easy to prototype new ideas for services.
Disadvantages

---

[2]There are many guides on the Internet.

- Expensive if purchasing hardware for the purpose.
- The hardware, the operating system, and the web server software will need to be maintained.
- The cost of power and the cost and availability of cooling.
- Space requirements.

Local Server—Virtual Machine

If a virtual machine host server is available, either within the department or from central IT services, this can be a very attractive web hosting option.

Advantages
- Low cost of deployment (depending on the funding model of the provider of the service).
- Can increase and decrease the resources available within the limits of the virtual server environment.
- Security—not accessible beyond the corporate network.
- Easy to prototype new ideas for services.
- Can run other services for the department such as a source control management system.
- No need to worry about the hardware.

Disadvantages
- Competing for resources with other virtual servers on the same host.
- Cost, depending on funding model of the provider of the service.
- The operating system and the web server software need to be maintained.

Managed Hosting

Managed hosting is where an Internet-based company provides a web server, usually with only email, ftp, and http access, along with web-based tools to configure and manage the server. There will normally be tens or even hundreds of other websites served from the same server, all sharing the same IP address (but different domain names).

Advantages
- Low cost.
- Not necessary to know anything about server operating systems.
- Not necessary to know anything about web server software.
- Easy-to-use web-based management tools to set up email, install software such as content management systems, create templated websites (functionality varies depending on provider).
- No need to worry about resources, hardware, resilience, or redundancy.
- Quick to set up.
- Security of the server is mainly the provider's concern—you only need to worry about your content and the web applications you are running on top of the service they provide.
- The service is available to anyone with an Internet connection.

Disadvantages
- Usually no shell/command line access.
- Cannot control how web server software is built/configured (e.g., PHP safe mode).
- Cannot control which version of software is installed (e.g., PHP, MySQL).
- Cannot install services not provided by the supplier, such as source control management.
- Vulnerable to resources being used up by other users of the same server.
- Vulnerable to being blacklisted due to other users abusing the same server with the same IP address and root hostname.
- No influence over availability/resilience/ redundancy.
- Limited resources—lack of flexibility to scale service.
- Might be "white label" reseller hosting— your provider might not have access to fix things.

- Security—you are responsible for the content that is on the server, its appropriateness, and its security.
- Minimum term contracts make testing new ideas more costly.

Virtual Dedicated Server

Different providers either call this hosting solution "virtual dedicated" or "virtual private" server but they both refer to the same thing. This is where you are supplied with a virtual machine to build and run whichever services you like on it, in a very similar way to the local virtual machine hosting discussed previously. The difference is that this machine has a static routable IP address.

Advantages
- Complete control of the operating system and web server software and how it is built and configured.
- Can install software and run other services such as source control management or voice over IP (VoIP).
- Choice of resources—processors, memory, hard disk.
- Resources are usually protected.
- Static and dedicated IP address(es).
- Some suppliers will install and configure the operating system and web server software for you.
- Some suppliers will provide web server management software (at extra cost).
- No need to worry about the hardware.
- The service is available to anyone with an Internet connection.

Disadvantages
- Considerably more expensive than managed hosting.
- The operating system and the web server software will need to be maintained—you will need server software administration knowledge.
- No help desk for anything other than hardware or connectivity issues.

- Shared resources that could be subject to other systems on the same hardware dominating them, though some level of protection can be expected.
- Resilience in terms of backups are your responsibility.[3]
- Content—you are responsible for the content that is on the server, its appropriateness, and its security.
- Security—you are responsible for the security of the operating system and the software running on the server.
- Minimum term contracts make testing new ideas more costly.

Dedicated Server

A dedicated server is very similar in almost all respects to the virtual dedicated server, except it is a whole real computer dedicated to you. It features the same advantages and disadvantages of the virtual dedicated server mentioned previously, with the following additions:

Advantages
- Dedicated resources—the only runaway processes that can bring down the server are yours.

Disadvantages
- More expensive again than the virtual dedicated server.

Cloud Provisioned Server

This is effectively the same as the virtual dedicated server but with much more flexibility and choice of resources, location, and functionality. The increase in scale and developments in technology should reduce the chance of other users of the service being able to compromise your server. Usually these services are billed on a pay-as-you-use basis.

[3]Unless the site captures data, the simplest solution is to develop the site on a local PC and then upload it. This provides a development environment and backup in one.

Advantages

- Complete control of the operating system and web server software and how they are built and configured.
- Can install software and run other services such as source control management or voice over IP (VoIP).
- Much more flexible provision of resources—user can change the number of processors, memory, and disk space as required.
- Resources can be chosen to suit different tasks—cheap and slow archive storage versus expensive and fast transactional storage, for example.
- Easier to replicate server or services for redundancy or resilience.
- Pay per resource per unit time used—from very cheap to very expensive.
- Easy to run up new servers to test ideas without committing to contracts—pay for your testing time then destroy the machine.
- Provider might provide resilience across data centers to quickly bring your server up in a different location should disaster strike.
- The service is available to anyone with an Internet connection.

Disadvantages

- More expensive than managed hosting, can be very expensive if lots of resources are required.
- The operating system and the web server software will need to be maintained—you will need server software administration knowledge.
- No help desk for anything other than hardware or connectivity issues.
- Resilience in terms of backups are your responsibility.
- Content—you are responsible for the content that is on the server, its appropriateness, and its security.
- Security—you are responsible for the security of the operating system and the software running on the server.

For all the servers that are not owned and located within your institution, extra special attention must be given to the information governance of the data held on that server. For example, you must know where the server is physically located (even for virtual servers), as there is a distinction between data held within the EU and data that is held outside the EU. You should also consider the speed of access to the server; distance does make a difference, though there are many other very important factors in this too.

Summary

Choice of hosting strategy will depend on several factors: the task you wish to achieve, the audience, the sensitivity of the data, the budget, your skills, how mission critical it is or might become. The points discussed can be used to guide and inform the decision-making process.

PROGRAMMING

Web Programming

The World Wide Web is an information-sharing concept that resides on the Internet. The roots of it are in work done by Sir Tim Berners-Lee in 1989 when he worked at CERN in Geneva. The basic concepts are:

- The separation of content server from content display.
- The use of markup to describe but not enforce the layout (HyperText Markup Language, or HTML).
- The use of links to jump from one piece of information to another.

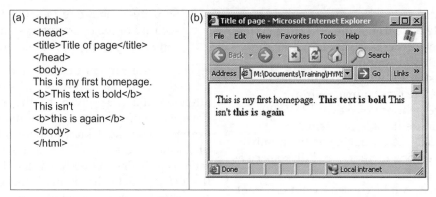

FIGURE 10.1   (a) The HTML code; (b) the display.

Web pages reside on web servers and are delivered to web browsers, where they are displayed. A basic page consists of the content and a series of tags describing how that content should be displayed. Originally these were not prescriptive and left the browser to determine how they should be displayed, but later additions to HTML have allowed such prescription.

A very simple web page might consist of the code and display shown in Figure 10.1.

Note the following:

- Each tag begins with < and ends with >.
- For each tag <tag > there is an opposite </tag > denoting the beginning and end of that tag's influence. (There are some exceptions to this.)
- Although the source has line breaks, the display does not—it needs to be told where the breaks should appear.
- There are two main parts to the page: the <head > and the <body>. Most of the display content resides in the <body>.

Some key formatting tags are:

<b >   Bold text

   Line break—this is an empty tag so does not require a </br >[4]

<i >   Italic text
<h1 >   Heading style 1—there are six predefined heading styles
<p >   Paragraph break—a larger break than
[5]

Further complexity is added by the ability to program the page. This can be achieved in multiple ways and using various languages. For brevity, only JavaScript is considered here. Such programming adds interactivity to the page, for example, a simple calculator, as shown in Figure 10.2.

Note that in this code the bulk of the JavaScript is contained in the <head > section of the page. This separates the code from the display and means that the page renders quicker. The second piece of JavaScript is in the line

```
< input type = "button" value = "Calculate"
onclick = "self.docalc();" >
```

This calls the function in the <head > section when the button is clicked. JavaScript in the <head > section is usually for calling later, whereas that in the <body > section is for immediate execution.

---

[4]In XHTML the <br > tag must be properly closed, like this: <br />.

[5]Although most browsers will interpret <p > without a corresponding </p > it is contrary to the standard and the use of attributes means that the browser does not know where to stop applying them.

(a)
```
<html>
<head>
<script type="text/javascript">
function docalc()
{
 var f=self.document.calc
 f.week.value=f.packs.value*f.cost.value*7;
 year=f.packs.value*f.cost.value*365;
 year=year/100;
 year=Math.round(year);
 f.year.value=year;
 f.cds.value=Math.round(year/15); // assumes cost of CD is £15
}
</script>
</head>

<body>
<h1>Chips calculator</h1>
<form name="calc">
How many packets of chips do you buy per day? <input type=text name="packs">
<p>
And how much does a packet cost (pence)? <input type=text name="cost"></p>
<p>
<input type="button" value="Calculate" onclick="self.docalc();"></p>
<p>
This costs you <input type=text name="week" disabled> pence per week
and <input type=text name="year" disabled> pounds per year</p>
<p>
Instead of this you could have bought <input type=text name="cds" disabled> CDs per
year</p>
</form>
</body>
</html>
```

(b)

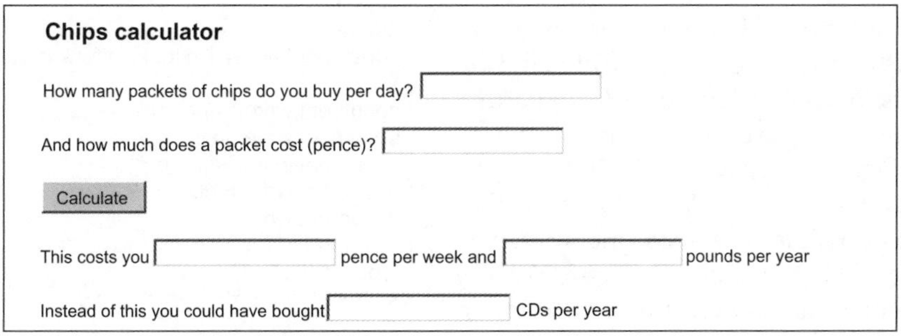

Chips calculator

How many packets of chips do you buy per day? [＿＿＿＿＿]

And how much does a packet cost (pence)? [＿＿＿＿＿]

[Calculate]

This costs you [＿＿＿＿＿] pence per week and [＿＿＿＿＿] pounds per year

Instead of this you could have bought [＿＿＿＿＿] CDs per year

FIGURE 10.2   (a) The HTML code; (b) the display.

It is also worth noting the use of the variable f (assigned the value self.document.calc) which is then used as a prefix to all the input fields within the form. Without this, the prefix "self.document.calc" would have to be used on all input fields to identify them.

Such functionality provides a web page with a degree of interaction, but programming web pages can go far beyond this. Because a web page as served to a browser is simply a stream of text, it is possible to program the entire system in languages such as C++, Python, and PHP so that the page that is served depends completely on the choices made on previous pages or visits. PHP is a server-side scripting language (unlike JavaScript that is client-side) and is therefore preprocessed, the text passing through the PHP preprocessor before being sent to the browser.

PHP files consist of standard HTML (and therefore can also contain JavaScript for executing on the browser) with specific PHP commands embedded. For example, this code:

```
<html> <body> <?php echo "Hello
World";?>
<p>"...again"</p> </body> </html>
```

Is sent to the browser as:

```
<html> <body> Hello World
<p>"...again"</p> </body> </html>
```

Figure 10.3 shows some examples in different languages which produce the same output. Note that, apart from Figure 10.3(a), these are code snippets and require some additional code to produce the output in Figure 10.3(d) and that the comments in Figure 10.3(a) also apply to the code in Figures 10.3(b) and 10.3(c). See the next section "Forms and Data" for a further example. It is possible to embed one system in another. The C++ version in Figure 10.3(b) could, of course, just mimic the JavaScript in Figure 10.3(a) (prefixed by "cout <<" on each line). This might be appropriate for JavaScript in the <head> section, for example.

(a)

```
<html>
<head>
</head>
<body>
<script type="text/javascript">
//If the time on your browser is less than 10,
//you will get a "Good morning" greeting.
var d=new Date()
var time=d.getHours()

if (time<10)
{
document.write("<b>Good morning</b>")
}
else
{
document.write("Good day!")
}
document.write("<p>")

//You will receive a different greeting based
//on what day it is. Note that Sunday=0,
//Monday=1, Tuesday=2, etc.

theDay=d.getDay()
switch (theDay)
{
case 3:
  document.write("Woeful Wednesday")
  break
case 5:
  document.write("Finally Friday")
  break
case 6:
  document.write("Super Saturday")
  break
case 0:
  document.write("Sleepy Sunday")
  break
default:
  document.write("I'm looking forward to this weekend!")
}
document.write("</p><p>")
document.write(time)
document.write("</p><p>")
document.write(theDay)
document.write("</p>")
</script>
</body>
</html>
```

FIGURE 10.3 (a) JavaScript; (b) C++; (c) PHP; (d) the browser rendering.

(b)

```
time_t now=time(NULL);
struct tm *localtm=localtime(&now);
int hours=localtm->tm_hour,day=localtm->tm_wday;
cout << "<html>\n<head>\n</head>\n<body>";
if(hours<10) cout << "<b>Good morning</b>";
else cout << "Good day!";
cout << "<p>";
switch(day)
{
case 3:
  cout << "Woeful Wednesday";
  break;
case 5:
  cout << "Finally Friday";
  break;
case 6:
  cout << "Super Saturday";
  break;
case 0:
  cout << "Sleepy Sunday";
  break;
default:
  cout << "I'm looking forward to this weekend!";
}
cout << "</p><p>" << hours << "</p><p>" << day << "</p>";
cout << "</script>\n</body>\n</html>";
```

(c)

```
<html>
<head>
</head>
<body>
<?php
 $now=getdate();
 if ($now[hours]<10) echo  "<b>Good morning</b>";
else echo "Good day!";
echo "<p>";
switch ($now[wday])
{
case 3:
  echo "Woeful Wednesday";
  break;
case 5:
  echo "Finally Friday";
  break;
case 6:
  echo "Super Saturday";
  break;
case 0:
  echo "Sleepy Sunday";
  break;
default:
  echo "I'm looking forward to this weekend!";
}
echo "</p><p>";
echo $now[hours];
echo "</p><p>";
echo $now[wday];
echo "</p>"
?>
</body>
</html>
```

(d)

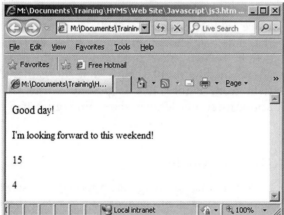

FIGURE 10.3   Continued

Forms and Data

Displaying information is the primary use of web technology, but it is increasingly being used to collect data also. The example in Figure 10.2 uses a specific form and it is worth investigating its structure.

As with other HTML, the form is enclosed between tags, in this case <form> and </form>. Within the form lie the input fields

(a)

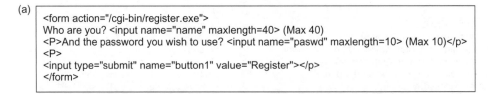

```
<form action="/cgi-bin/register.exe">
Who are you? <input name="name" maxlength=40> (Max 40)
<P>And the password you wish to use? <input name="paswd" maxlength=10> (Max 10)</p>
<P>
<input type="submit" name="button1" value="Register"></p>
</form>
```

(b)

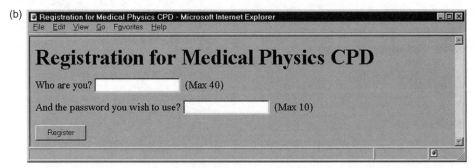

FIGURE 10.4   (a) The HTML code for the input form. (b) The code in (a) as rendered by the browser.

(named "packs" and "cost") and the output fields (in this case they are the disabled input fields named "week," "year," and "cds"). Note that the form itself is given a name (within the <form> tag) meaning that multiple forms can exist within a web page, can be nested, and the processing of one form may address data from another.

The most important part of a form, however, is its action. In the form in Figure 10.2, this action is initiated by the input button (labeled "Calculate"), which calls the embedded JavaScript. In contrast, the form in Figure 10.4 uses probably the most commonly used action, that of calling an external routine described in the <form> tag itself and initiated by clicking a "submit" type of input element. This external routine is a Common Gateway Interface (CGI), which by default usually resides in the cgi-bin folder of the web server.

We illustrate this by considering an example, that of registering a user into a CPD (continuing professional development) database system. The back-end database for this example (PostgreSQL) comes complete with C and C++ libraries (although it should be noted

that libpq++ is a merely a wrapper for libpq and therefore both need to be included at the linking stage), which can be used to simply produce CGI scripts.

The registration for the CPD database takes place in two parts: an HTML form to gather the data, as shown in Figure 10.4, and the action (which also produces a web page) shown in Figure 10.5. Note that the gathered data from Figure 10.4 is posted to the CGI application register.exe, shown in Figure 10.5.

Dynamic Content

A dynamic web page (as opposed to a static one) is a page that changes as a result of user interaction with it. The simplest form of such interaction is access—the page is generated for display in real time, but to all other intents is a static page. See Figure 10.3 for such an example. At the other end of the scale are continuously updating pages: they may use RSS feeds or continually refresh. In-between these are the buttons and forms that we saw in the previous section, "Forms and Data."

(a)
```
#include <stdlib.h>
#include <libpq++.h>
#include <strstream.h>
#include "parse.h"
#include "parse.cpp"

void main()
{
    char *query_str=getenv("QUERY_STRING"),buf[4];
    Parse list(query_str);
    int id;
    PgDatabase data("cpd");
    ostrstream os;

    cout << "Content-type: text/html"<<endl<<endl<<endl;
    cout << "<html>"<<endl<<"<head>"<<endl<<"</head>"<<endl<<"<body>"<<endl;

    if(data.ConnectionBad()) cout << "<h1>Error:Bad connection</h1>" << endl;
    else if(data.ExecTuplesOk("select * from config") && data.Tuples()==1)
    {
        id=atoi(data.GetValue(0,"next_id"));
        os << "insert into person values ('" << list.get_item_n("name") << ", " << id
", '" << list.get_item_n("paswd") << "')" << ends;
        if(data.ExecCommandOk(os.str())    &&    data.ExecCommandOk("update    configs
next_id=next_id+1"))
        {
                cout << "<h2>CPD Registration</h2>"<<endl;
                cout << "name    : "<<list.get_item_n("name")<<"<p>";
                cout << "password: "<<list.get_item_n("paswd")<<"</p><p>";
                cout << "Has been registered on the Medical Physics CPD system</p><hr>";
        }
        else cout << "<h1>Error: Registration failure</h1>" << endl;
        os.rdbuf()->freeze(0);
    }
    else cout << "<h1>Error:Next ID not collected</h1>";
    cout << "</body>"<<endl<<"</html>"<<endl;
}
```

(b)
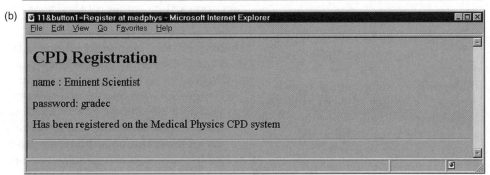

FIGURE 10.5   (a) The C++ code for processing the data and returning a confirmatory web page. (b) The confirmatory page from the code in (a) as rendered by the browser.

The techniques for implementing dynamic content may be either client-side or server-side, as we saw in the section "Web Programming." We shall therefore consider just two techniques here: the page refresh and the RSS feed.

A page refresh does as the name implies: it refreshes the page. In the case of the sample code in Figure 10.3, this would have the effect of keeping the page current as time progresses, moving through the phases of the day and days of the week. JavaScript to accomplish such an effect (and therefore is client-side) would be:

```
<html>
<head>
<script type = "text/JavaScript">
<!--
function Refresh(timeout) {
        setTimeout("location.reload
(true);",timeout);
}
// -->
</script>
</head>
<body onload = "JavaScript:Refresh
(10000);">
<script type = "text/javascript">
var d = new Date()
var time = d.getSeconds()
document.write("<b>time passing...<b>")
document.write(time)
</script>
</body>
</html>
```

This page refreshes every 10 seconds. This uses the "onload" event to call the function Refresh(). The parameter value "10000" equals 10 seconds.

In itself, this is interesting but not especially useful. However, consider a web page that itself displays other web pages in frames. Each frame has a separate page address and therefore a different <body> tag, meaning that a web page can have several parts which refresh at different rates (a weather forecast hourly, a webcam image every minute, etc.). A refresh attribute does take control away from the user, however. Therefore it must be part of the functionality of the page to display changing content, otherwise it will contravene the World Wide Web Consortium's (W3C) accessibility guidelines. It may be better in such cases to have a "refresh" button on the page that would call a function such as the Refresh() one above.

While the JavaScript:Refresh() parameter can be as low as 1, causing the page to refresh every millisecond, this is not a good way to implement real-time content. A far better way is with an RSS feed.

RSS (Rich Site Summary[6]) is a format for delivering regularly changing web content. Many news-related sites, weblogs, and other online publishers syndicate their content by providing an RSS feed for incorporation into another website. This is not the same as embedding YouTube videos, which are really just a dynamic frame. An RSS feed is a URL and can be embedded into a page in the same way as any other, as demonstrated in Figure 10.6.

Interfacing with Databases

The first question that must be answered when interfacing with databases is "which database engine?" In most settings, this will have been determined prior to the project's commencement due to the hosting structure and preexisting applications. Three very common databases are:

- MySQL (the community edition is free to install): Originally only available for Linux, this is now available for many platforms.
- PostgreSQL: An open-source system available on Linux, UNIX, Windows, and Mac OS platforms, this has some very useful programming libraries in languages such as C++, making interaction simpler.

[6]Also known as "Really Simple Syndication."

(a)
```
<html>
<head>
<title>Weather RSS demo</title>
<frameset cols="25%,75%">
<frame src="./title.htm">
<frame src="http://open.live.bbc.co.uk/weather/feeds/en/7284876/3dayforecast.rss">
</frameset>
</head>
<body>
</body>
</html>
```

(b)
```
<html>
<head>
<title>Title of page</title>
</head>
<body>
<h1>a simple weather feed from the BBC</h1>
</body>
</html>
```

(c)

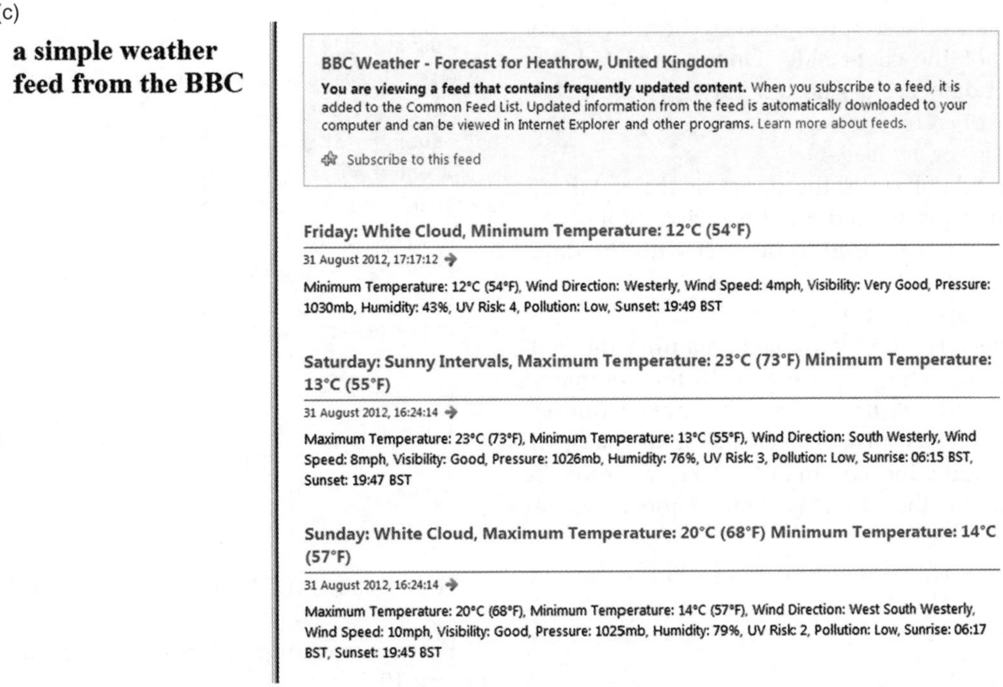

a simple weather feed from the BBC

BBC Weather - Forecast for Heathrow, United Kingdom
You are viewing a feed that contains frequently updated content. When you subscribe to a feed, it is added to the Common Feed List. Updated information from the feed is automatically downloaded to your computer and can be viewed in Internet Explorer and other programs. Learn more about feeds.

Subscribe to this feed

Friday: White Cloud, Minimum Temperature: 12°C (54°F)

31 August 2012, 17:17:12

Minimum Temperature: 12°C (54°F), Wind Direction: Westerly, Wind Speed: 4mph, Visibility: Very Good, Pressure: 1030mb, Humidity: 43%, UV Risk: 4, Pollution: Low, Sunset: 19:49 BST

Saturday: Sunny Intervals, Maximum Temperature: 23°C (73°F) Minimum Temperature: 13°C (55°F)

31 August 2012, 16:24:14

Maximum Temperature: 23°C (73°F), Minimum Temperature: 13°C (55°F), Wind Direction: South Westerly, Wind Speed: 8mph, Visibility: Good, Pressure: 1026mb, Humidity: 76%, UV Risk: 3, Pollution: Low, Sunrise: 06:15 BST, Sunset: 19:47 BST

Sunday: White Cloud, Maximum Temperature: 20°C (68°F) Minimum Temperature: 14°C (57°F)

31 August 2012, 16:24:14

Maximum Temperature: 20°C (68°F), Minimum Temperature: 14°C (57°F), Wind Direction: West South Westerly, Wind Speed: 10mph, Visibility: Good, Pressure: 1025mb, Humidity: 79%, UV Risk: 2, Pollution: Low, Sunrise: 06:17 BST, Sunset: 19:45 BST

FIGURE 10.6 (a) The HTML for a page containing frames. Note that the first frame contains a local file, the second an RSS feed. Note also the absence of a body. (b) The HTML for the file title.htm. (c) The code rendered by a web browser.

- Microsoft SQL Server: A Windows-only installation, this scales well and is suitable for large, mission-critical databases.

The most common method for interfacing is via CGI (see the section "Forms and Data") so that queries (generally written in SQL—see "Structured Databases") can be posted to the database engine and the results returned to the web page for display. The examples here use PHP and MySQL.

Assuming the database has already been created (we will use the structure from the "Structured Databases" section), the PHP script connects via

```
mysql_connect(localhost,$username,
$password);
```

where $username and $password have been defined earlier in the script. This is not a security risk as the script is processed on the server side, not the client side. Once connected, the correct database is selected via

```
@mysql_select_db($database) or die( "Unable
to select database")[7];
```

where $database is the name of the database, similarly predefined in the script. SQL commands are then used to interact with the database, via commands of the form

```
mysql_query($query);
```

$query is a text variable containing the SQL command. This is preferable to the command being written directly as the mysql_query() parameter as processing can take place in formulating the command. Data can thus be added to the database via $query values such as

```
$query="INSERT INTO Item { Item_ID,
Artist_ID, Title_ID,
Track_ID, Track_No} VALUES {4, 1, 1, 4, 3}";
```

and

```
$query=" INSERT INTO Track {Track_ID,
Track} VALUES {4, 'The Way'}";
```

(Obviously there needs to be a "mysql_query($query);" command between these two in the final script.) Note the use of single quotes to delimit text strings.

Let us assume that we wish to display the list of artists in our database. This would be accomplished via

```
$query="SELECT * FROM Artist";
    $result=mysql_query($query);
$num=mysql_numrows($result);
```

The second command places the results into an array and the third command returns the number of rows in that array. The full script to produce this information is thus:

```
<?
$username="username";
$password="password";
$database="music";
mysql_connect(localhost,$username,
$password);
@mysql_select_db($database) or die( "Unable
to select database");
$query="SELECT * FROM Artist";
$result=mysql_query($query);
$num=mysql_numrows($result);
mysql_close();
echo "<h1>List of Artists</h1>";
$i=0;
while ($i<$num) {
$artist=mysql_result($result,
$i,"Artist");
$id=mysql_result($result,$i,"ID");
echo "<p><b>$artist</b> ID: $id</
p>";
$i++;
}
Echo "$num listed";
?>
```

From our example, this gives the output in Figure 10.7.

---

[7]A failure to connect therefore outputs the text "Unable to select database" to the user and execution halts.

FIGURE 10.7 The screen produced by the PHP database code.

SECURITY

Limiting Access

Although most web pages and content are designed to be viewable by all, some pages require, by design, a restriction in access. Examples of this may be research work as yet unpublished and telephone lists, as well as the commercial applications such as banking and paid-for content (such as journals). In clinical computing terms a common access restriction is for data repositories: collection via an Internet interface is desirable due to its ease of use, yet it must be secure enough to prevent unauthorized access to that data.

There are various ways of restricting access, all of which require some form of authentication. For UNIX systems running an Apache web server, the .htaccess file is one method.

This is a plain-text file that resides in the same directory as the page that it protects.[8] The following five lines in this file:

```
<Limit GET POST>
order deny,allow
deny from all
allow from .clineng.ac.uk
</Limit>
```

will only allow access to the pages to users from the University of Clinical Engineering. Any other users attempting to access these pages will receive a "403 Forbidden" error. These lines:

```
AuthUserFile /security/web/passfile
AuthGroupFile /dev/null
AuthName ByPassword
AuthType Basic
<Limit GET POST>
require valid-user
</Limit>
```

mean that access is via a username and password combination, validated against the file/security/web/passfile. The disadvantage of this system is that the password file probably has to be maintained on the UNIX system, although this may also be viewed as an advantage, depending on the website's functionality.

Other methods of restricting access depend a lot on the language the system has been written in. PHP provides functionality to require a password, as in this example:

```
<?php
if ( $PHP_AUTH_USER != "user" || $PHP_AUTH_PW
 != "please" ) {
header("HTTP/1.1 401 Unauthorized");
echo "Failed to log in.";
exit();
}
else {
echo "You are logged in successfully as: ".
$PHP_AUTH_USER;
echo "Welcome";
```

[8]It also protects subdirectories from this point down.

```
}
?>
```

This has a hard-coded password, which is not ideal, but does illustrate the technology. It is clearly possible to replace this with a database look-up (see the section "Interfacing with Databases").

On an ASP.NET-based server, the relevant controls would be placed in the web.config file. For example, the following section would allow only the users Clinical and Engineering to access the site:

```
<authorization>
<allow users = "Clinical, Engineering"/>
<deny users = "?"/>
</authorization>
```

Finally, it is possible to code the password into JavaScript at the head of each page that requires restrictions (Figure 10.8). However, this is a very insecure method and can be bypassed simply by loading the page and then viewing the HTML source (unlike the PHP which is server-side so does not reach the browser). It is useful, however, for proof-of-concept and demonstration work where setting up a dedicated web server is not appropriate.

Public and Private Key Encryption

In an increasingly online world, it is imperative that software systems remain operational in the event of error or malicious access attempts. In a clinical setting, there are several instances where digital security is applicable. For example, when drug lists or drug information is stored in a central database and accessed electronically, it needs to be ensured that the data are not tampered with either accidentally or intentionally. Similarly, electronic access to patient records requires strong user authentication measures

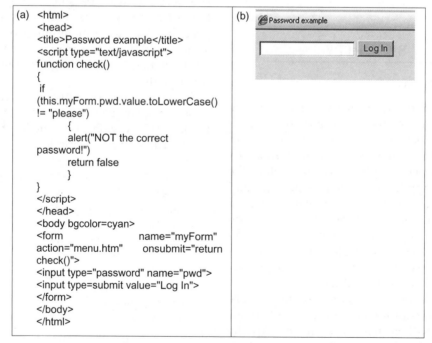

FIGURE 10.8   (a) JavaScript for generating a password prompt. (b) The code rendered by a web browser.

and any patient data that passes through an open network needs to be encrypted. In this section we aim to introduce the basic concepts of encryption and some of the key encryption techniques.

Introduction

Cryptography is an algorithmic process of converting a plain text (or clear text) message to a cipher text (or cipher) message based on an algorithm that both the sender and receiver know, so that the cipher text message cannot be read by anyone but the intended receiver. The act of converting a plain text message to its cipher text form is called *enciphering*. The opposite action, converting cipher text form to a plain text message, is called *deciphering*. The terminology enciphering and deciphering are more commonly referred to as *encryption* and *decryption*, respectively.

To start off with, it is useful to define some common terminology associated with encryption.

- *Cryptosystem* or *cipher system* is a method of disguising messages so that only certain people can see through the disguise.
- *Cryptography* is the art of creating and using cryptosystems.
- *Cryptanalysis* is the art of breaking cryptosystems—seeing through the disguise even when you're not supposed to be able to.
- *Cryptology* is the study of both cryptography and cryptanalysis.
- *Plaintext* is the original message input to the encryption system.
- *Ciphertext* is the disguised message output from the encryption system.
- The letters in the plain text and cipher text are often referred to as *symbols*.
- *Encryption* is any procedure to convert plain text into cipher text.
- *Decryption* is any procedure to convert cipher text into plain text.

- *Cryptology primitives* are basic building blocks used in the encryption process. Examples of these include block ciphers, stream ciphers, and hash algorithms.

Block and Stream Ciphers

One of the first known examples of encryption is Caesar cipher which involved replacing every letter by the letter 3 places ahead of it in the alphabet; for example, Caesar became Fdhvdu. This is known as a *substitution cipher*—substituting a letter for another throughout the text. It is also one of the easiest to decrypt given that certain letters occur much more frequently in English words than others.

Another encryption option is the *stream cipher* where a string of characters known as the *keystream* is repeatedly combined with the plain text to form the cipher. For example, one of the earliest known stream ciphers, the Vigenère stream cipher, considers each letter of the alphabet to correspond to a number such as $A = 0$, $B = 1$, ... , $Z = 25$, and simply adds the position numbers of the plain text symbols and substitutes the corresponding letter in the cipher text. So if we used "Julius" as the key, in this case, the letter "C" becomes "M" ($C = 3 + J = 9 \rightarrow 12 = M$). Continuing this through, "Caesar" would be encrypted as "Muqauj." To crack this type of cipher requires a long enough extract of cipher text on which frequency analysis of phrases may be used to guess the key.

Unlike stream ciphers that encrypt a symbol at a time, *block ciphers* encrypt blocks of plain text at a time using unvarying transformations on blocks of data. One of the simplest examples of block ciphers is the Playfair cipher where the plain text is divided up into blocks of 2 letters and a random arrangement of alphabets in a 5×5 block is used to encrypt each of these blocks individually. Given a large enough extract of cipher text, it might be possible to reconstruct the coding matrix. One way around this is to have large block sizes and iterative encryption. Some of the common encryption technologies used today such as the Advanced Encryption Standard (AES) used worldwide, use a form of block cipher incorporating large (64–128-bit) blocks and pseudorandom encryption keys.

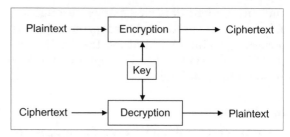

FIGURE 10.9   Symmetric key cryptography.

Public-Key Cryptography

Both stream ciphers and block ciphers are types of *symmetric key* cryptography, that is, the same key is used both for encryption and decryption (Figure 10.9).

Asymmetric key cryptography, on the other hand, uses two separate keys; one of these keys will encrypt the plain text and the other will decrypt it (Figure 10.10). This prevents the need for and associated risks with transmitting keys either with or without the message.[9]

In public key cryptography, one of the keys is made public, typically the encryption key. So when a message is transmitted, it is encrypted using the public key of the recipient. The matching decryption key is kept private

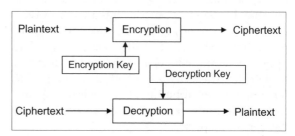

FIGURE 10.10   Asymmetric key cryptography.

and known only to the recipient who can then use it to decrypt the cipher text.

The primary advantage of an asymmetric cryptosystem is the increased security since private keys never need to be transmitted, unlike the symmetric key system where the secret key must be communicated between the sender and receiver as the same key is required for both encryption and decryption. Interception of this secret key can compromise the security and authenticity of future data transmissions.

Digital signatures are an application of public-key cryptography often used in e-commerce applications. These are provided via a *certification authority*. The certification authority manages a repository of public keys that it signs with its own private key to confirm authenticity. A sender can then digitally "sign" a message using his own private key and the recipient can confirm authenticity of

[9]Consider an example of a situation where a box of valuables needs to be securely sent from one location to another. Sending the key to unlock the box with it or separately would both be risky. So the sending official puts a padlock on the box to which only he has the key. At the receiving end, the official there double locks the box by adding his own padlock and sends it back. The sending official can now remove his padlock using his key leaving the box locked with just the receiving official's lock on it. At the other end, the receiver can then unlock the box and retrieve the contents. In this case, the need to send any keys is completely eliminated. This technique of sending, receiving and sending again is known as the three-pass protocol for secure transmission of messages without key exchange. Here both the encryption and decryption keys are private.

the sender by validating the signature against the public key authenticated by the certification authority.

A drawback of using public-key cryptography for encryption is speed. The multistep validation and verification process as well as the hashing algorithms typically used with public-key cryptography tend to be slower than symmetric key algorithms for large amounts of data. *Digital envelopes* are a way of combining the best of both by using the more secure public-key encryption to exchange secret keys and then using the faster secret-key encryption on the message itself.

The choice of encryption methodology depends on the application. In circumstances where it is possible to securely exchange secret keys, for example by meeting in person, there is then no need to use public-key encryption. This is also the case for users protecting their own documents with a password. Public key cryptography is advantageous in a multi-user scenario across open networks.

This section has touched only a small area in the vast field of cryptography. But it has hopefully introduced some of the basic concepts and key terminology.

Further Reading

Clay Mathematics Institute, The RSA algorithm [online]. Available: <http://www.claymath.org/posters/primes/rsa.php> (accessed 07.06.12).

DoH, The e-GIF and the NHS - a policy statement [online]. Available: <http://www.dh.gov.uk/en/Publicationsandstatistics/Publications/PublicationsPolicyAndGuidance/DH_4130864> (accessed 17.08.12).

Exelis, I.D.L., [online]. Available: <http://www.exelisvis.com/language/en-US/ProductsServices/IDL.aspx> (accessed 17.08.12).

FDA, U.S. Food and Drug Administration [online]. Available: <http://www.fda.gov/> (accessed 07.09.12).

HL7, [online] Available: <www.hl7.org> (accessed 07.06.12).

Haskell, The Haskell Programming Language [online]. Available: <http://www.haskell.org/haskellwiki/Haskell> (accessed 05.07.12).

ImageJ. [online] Available: <http://rsbweb.nih.gov/ij/docs/intro.html> (accessed 17.08.12).

Information Commissioner's Office (ICO), Data Protection Principles [online]. Available: <http://www.ico.gov.uk/for_organisations/data_protection/the_guide/the_principles.aspx> (accessed 16.12.11).

Information Commissioner's Office (ICO), Freedom of Information Act [online]. Available: <http://www.ico.gov.uk/for_organisations/freedom_of_information.aspx> (accessed 22.12.11).

Kirsch, C.M., Sengupta, R., 2006. The Evolution of Real-Time Programming [online]. Available: <http://www.cis.upenn.edu/~lee/10cis541/papers/kirsch.pdf> (accessed 04.09.12).

MathWorks, MatLab [online]. Available: <http://www.mathworks.co.uk/products/matlab/> (accessed 17.08.12).

Materialise, Mimics [online]. Available: <http://biomedical.materialise.com/mimics> (accessed 17.08.12).

McConnell, S., 2004. Code Complete: A Practical Handbook of Software Construction. Microsoft Press.

Medicines and Healthcare Products Regulatory Agency (MHRA) [online]. Available: <http://www.mhra.gov.uk/> (accessed 07.09.12) Specifically: <http://www.mhra.gov.uk/Howweregulate/NewTechnologiesForums/DevicesNewTechnologyForum/Forums/CON084987> and <http://www.mhra.gov.uk/Howweregulate/Devices/MedicalDevicesDirective/index.htm>.

NEMA. The DICOM standard. [online]. Available: <http://medical.nema.org/standard.html> (accessed 18.05.12).

Netcraft, August 2012. Web Server Survey [online] Available: <http://news.netcraft.com/archives/category/web-server-survey/> (accessed 07.09.12).

Ringholme, HL7 Message examples: version 2 and version 3. Available: <http://www.ringholm.de/docs/04300_en.htm> (accessed 07.06.12).

Sibtain, A., Morgan, A., MacDougall, N. (Eds.), 2012. Physics for Clinical Oncology. Oxford University Press.

SQA., An independent website that presents information about SQA, SQC and Software development. [online]. Available: <http://www.sqa.net> (accessed 31.08.12).

Stergiou, C., Siganos, D., Neural Networks. Imperial College [online]. Available: <http://www.doc.ic.ac.uk/~nd/surprise_96/journal/vol4/cs11/report.html> (accessed 23.12.11).

W3schools, A large collection of references and examples covering HTML, JavaScript, PHP, XML etc. [online]. Available: <http://www.w3schools.com/>.

W3Techs., Usage of operating systems for websites [online] Available: <http://w3techs.com/technologies/overview/operating_system/all> (accessed 07.09.12).

Wikipedia, the free encyclopedia [online]. Available: <http://en.wikipedia.org>.

Wellings, A., Book Web Page for Concurrent and Real-Time Programming in Java. A collection of sample programs and an introduction to Java for those familiar with C++/C and OOP. [online] Available: <http://www.cs.york.ac.uk/rts/books/CRTJbook.html> (accessed 05.09.12).

CLINICAL INSTRUMENTATION AND MEASUREMENT

Thomas Stone, Richard G. Axell, Paul A. White, Christine Denby, and Elizabeth M. Tunnicliffe

Overview

Part III provides a summary of medical electronics, basic circuit design, and the importance of differential amplifiers and instrumentation required for clinical measurement. It delves into the theory and reviews the concepts of accuracy and precision of measurements and measurement errors, as well as identifying the importance of calibration, sensitivity, and specificity of measurements. Examples of clinical measurement are observed in various organ systems such as in cardiology, respiratory, and neurosciences. In each of these organ systems basic clinical measurements are described in addition to data processing, clinical interpretation, and reviewing key safety issues of clinical measurement systems. In addition to describing physiological measurements from particular organ systems, they are also described in terms of measurement type, namely electrophysiological and the measurements of pressure and flow. Clinical examples are provided for each measurement type.

Medical Electronics

Thomas Stone

Cambridge University Hospitals NHS Foundation Trust, Cambridge, U.K.

ELECTRONIC COMPONENTS

There is a huge, and some would say bewildering, range of electronic components available to the designer. It is vital to have a well-constructed specification for your design to inform your decisions. To know what to specify it is important to have a grasp of the variety of factors that define a component, from the operational factors (electronic function) to the physical factors (size, mounting, etc.). We start with the passive devices, those that store energy and resist current flow, before moving on to the active devices, those that can "actively" change the electrical signal by amplification.

Resistor

A resistor is a passive device with two terminals (Figure 11.1). It has an ideal voltage–current relationship of

$$v(t) = Ri(t) \tag{11.1}$$

where R is the resistance in ohms and $v(t)$ and $i(t)$ are voltage and current, respectively.

Resistors in a parallel network are equivalent by the definition

$$R_{eq} = R_0 + R_1 + R_2 + \cdots + R_n \tag{11.2}$$

Resistors in series are equivalent to

$$\frac{1}{R_{eq}} = \frac{1}{R_0} + \frac{1}{R_1} + \frac{1}{R_2} + \cdots + \frac{1}{R_n} \tag{11.3}$$

Resistors are specified by several parameters, however, many will have specific parameters if they are designed for specific tasks:

- Resistance in ohms
- Tolerance in percent: This means that the actual value of resistance is within this percentage of the value quoted as the resistance
- Power in watts: The power rating of the resistor is the power ($P = i^2R$) above which the resistor will fail; it is a bad idea to operate anywhere near this value
- Working voltage: The appropriate voltage range that should be applied to the resistor
- Temperature coefficient of resistance (TCR) in ppm/°C: The change in the actual resistance with a change in temperature;

FIGURE 11.1 A resistor symbol and a variety of devices.

a value of 10 ppm/°C would imply a change of 10 Ω in every 1 MΩ for every 1°C

- Voltage coefficient in ppm/V: The change in the resistance with the change in voltage (within the working voltage range); described as a change in parts per million

Resistors are constructed in many ways. Each method has certain benefits and weaknesses and will impact on design consideration.

Metal and Carbon Film Resistor

Carbon film resistors are created by laying a thin carbon film over a nonconductive substrate. The carbon is then etched away (Figure 11.2) and the pitch of the etch will dictate the resistance of the device.

Metal film resistors are created by depositing a thin film of a metal or metal oxide over a ceramic core. The resistance is again controlled by cutting a helical path into the deposited metal.

Thin Film and Thick Film

Thin film resistors are created using a technique called sputtering which permits a very thin and controlled layer of a material, tantalum nitride or Nichrome, to be deposited onto a substrate. The environment in which the film is laid and without the necessity for high power laser trimming ensures a very stable and accurate resistor.

Thick film resistors are constructed from a resistive paste which is printed onto an

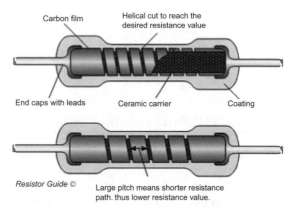

FIGURE 11.2 Metal and carbon film resistor.

insulating substrate and then fired. The paste includes glass particles that when fired help fix the resistive material in place; however, it also causes random imperfections within the printed material that can change the conductive properties of the resistor. Thick film resistors are also laser trimmed, as the conductive paste does not have an accurate ohmic property, to achieve the correct resistance value (Figure 11.3). Laser trimming can have the undesired effect of causing microfractures in the fired paste which cause imperfections in the conductance properties of the resistor.

Away from the Ideal

There are several reasons why a resistor may not act in the way described earlier in this section. There are times when a resistor may stop acting as an ideal resistor. For example, if

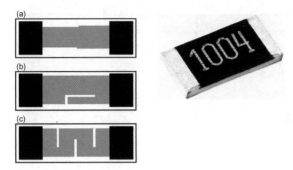

FIGURE 11.3 Trim methods: (a) edge, (b), L-cut, (c) S-cut.

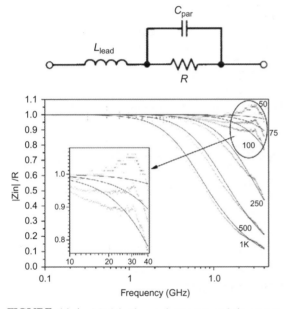

FIGURE 11.4 Model of a real resistor and frequency response.

we were to move away from a direct current (DC) application and into higher-frequency applications (AC) the helical cut in a carbon film resistor would increasingly act as an inductor, which is essentially what it is. In addition, a resistor has inherent capacitance, most significantly between the terminals of the device. The model of a resistor that includes all these less than ideal characteristics is shown in Figure 11.4.

Network and Circuit Analysis

Kirchhoff's voltage laws state that the sum of voltages, both positive and negative, around a closed circuit must equal zero. Equally, Kirchhoff's current laws state that the sum of all currents, both positive and negative, flowing into a node must equal zero. From these principles we can analyze circuits in terms of loop currents (Figure 11.5):

$$\begin{cases} i_1 R_1 + R_2(i_1 - i_2) + R_3(i_1 - i_3) = 10 \\ R_2(i_1 - i_2) + i_2 R_6 + R_4(i_2 - i_3) = 0 \quad\quad (11.4) \\ R_3(i_1 - i_3) + R_4(i_3 - i_2) + i_3 R_5 = 0 \end{cases}$$

Using Gaussian elimination, the result for each current can be isolated.

Thévenin equivalent circuits use the principle that any combination of voltage sources and resistors can be reduced to a single voltage source and single series resistance (Figure 11.6).

Equivalently, the Norton equivalent circuit is a combination of a current source and a parallel resistance (Figure 11.7).

We calculate the Thévenin equivalent circuit by calculating the open circuit voltage of the circuit and dividing it by the short circuit current.

$$V_{TH} = V \,(open\ circuit) \quad\quad (11.5)$$

$$R_{TH} = \frac{V_{TH}}{I\,(short\ circuit)} \qu\quad (11.6)$$

Capacitor

A capacitor is a two-terminal device described schematically in Figure 11.8.

Capacitance can be defined by

$$C = \frac{\varepsilon_0 \varepsilon_r A}{d} \quad\quad (11.7)$$

where ε_0 is the permittivity of a vacuum, ε_r is the relative dielectric constant of the material between the plates, A is the area of the plates, and d is the distance between the plates.

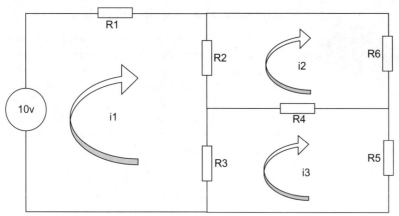

FIGURE 11.5 Loop currents, network analysis.

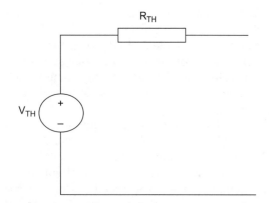

FIGURE 11.6 Thévenin equivalent.

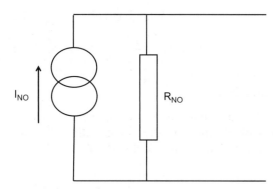

FIGURE 11.7 Norton equivalent.

The charge on a capacitor is described as

$$q(t) = Cv(t) \qquad (11.8)$$

FIGURE 11.8 Capacitor symbol.

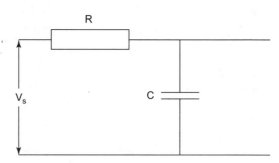

FIGURE 11.9 RC circuit.

where C is the capacitance in coulombs. We can define current as

$$i = \frac{dq(t)}{dt} \Rightarrow i = C\frac{dv(t)}{dt} \qquad (11.9)$$

which is the ideal voltage–current relationship of a capacitor.

As mentioned, Thévenin's law states that the sum of voltages around a closed circuit must equal zero. If we want to know the charge on a capacitor at some time t for the circuit in Figure 11.9 we can write

$$\left(V_s - \frac{Q}{C}\right) - IR = 0 \qquad (11.10)$$

$$\therefore \left(V_s - \frac{Q}{C}\right) - \frac{dQ}{dt}R = 0 \qquad (11.11)$$

This is a linear first-order differential equation and can be solved by the separation of parts; it is of the form $\frac{dy}{dx} + P(x)y = Q(x)$ with $Q(x) = 0$.

$$\Rightarrow \frac{dQ}{dt}R = \left(V_s - \frac{Q}{C}\right) \qquad (11.12)$$

$$\frac{dQ}{dt} = \frac{1}{R}\left(V_s - \frac{Q}{C}\right) \qquad (11.13)$$

$$\frac{dQ}{dt}C = \frac{1}{R}(V_s C - Q) \qquad (11.14)$$

Separate by parts and take the definite integral for both sides:

$$\frac{dQ}{(V_s C - Q)} \equiv -\frac{1}{RC}dt \qquad (11.15)$$

$$\int_{Q=0}^{Q=Q(t)} \frac{dQ}{(V_s C - Q)} \equiv -\frac{1}{RC}\int_{t=o}^{t} dt \qquad (11.16)$$

Recalling $\ln(m/n) = \ln(m) - \ln(n)$

$$\ln\frac{Q(t) - CV_s}{-CV_s} = -\frac{t}{RC} \qquad (11.17)$$

$$\therefore Q(t) = CV_s(1 - e^{-t/RC}) \qquad (11.18)$$

which gives the charge on the capacitor after time t. Remembering that $q(t) = Cv(t)$ we can also derive the voltage across it.

The energy stored on a capacitor is defined by

$$E = \frac{1}{2}CV^2 \qquad (11.19)$$

Capacitors in a parallel network (Figure 11.10) are equivalent by the definition

$$C_{eq} = C_0 + C_1 + C_2 + \cdots + C_n \qquad (11.20)$$

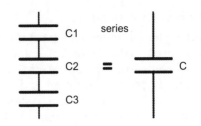

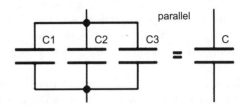

FIGURE 11.10 Capacitor in series and in parallel.

Capacitors in series (Figure 11.10) are equivalent to

$$\frac{1}{C_{eq}} = \frac{1}{C_0} + \frac{1}{C_1} + \frac{1}{C_2} + \cdots + \frac{1}{C_n} \qquad (11.21)$$

Capacitors are specified by a variety of parameters:

- Capacitance in farads (F), defined mathematically previously
- Operating voltage in volts (v), the highest voltage that can safely be applied across the terminals of the device
- Equivalent series resistance (ESR) in ohms (Ω)

Capacitors are also defined by their construction, which tailors them to their application. What was not mentioned in the list of parameters is the material of the dielectric, which defines the energy storage characteristics of the device.

Ceramic

Ceramic capacitors are constructed of layers of ceramic separated by layers of metal. Ceramic capacitors have the lowest equivalent series resistance.

Electrolytic Aluminum

Electrolytic aluminum capacitors can be manufactured with large capacitances. The plates of the capacitor can be coated with a very fine layer of aluminum oxide, which has a very high dielectric constant (~ 10). Therefore, if we look at Equation 11.21 we can see that we are able to construct a large dielectric without taking up too much space and increasing the distance between the plates too much; this equates to larger potential capacitances.

Electrolytic capacitors suffer from higher dissipation factors than other capacitors. The dissipation factor is the combination of all the energy losses within a capacitor under nonideal operation. The dissipation factor is defined as

$$D = \frac{R_s}{X_c} \qquad (11.22)$$

where R_s is the series equivalent resistance and X_c is the series equivalent reactance, which is equivalent to

$$X_c \equiv \frac{1}{j2\pi fC} \qquad (11.23)$$

It is for this reason that ripple current is a limiting factor for both aluminum and tantalum capacitors (see next). An alternating current resulting from the ohmic and dielectric losses causes temperature rises in the capacitor that could significantly reduce the capacitor's operational life. Therefore, electrolytic capacitors are often given a rated ripple current that should not be exceeded.

Tantalum

Capacitors have a similar construction to electrolytic aluminum in terms of a deposited fine layer of an oxide of tantalum, which has a high dielectric constant (approximately 27). Therefore, tantalum capacitors tend to have the same benefits of large capacitances as electrolytic aluminum capacitors do. In addition, however, they tend to have lower ESR which increases the stability of the capacitor. As a tradeoff for a high capacitance in a small space but with good stability, tantalum capacitors tend to have a higher cost.

Film

Film capacitors are the jack-of-all-trades in the capacitor world. They provide a low-cost stable capacitor with good leakage characteristics. They are formed by using layers of plastic film between layers of conductor. Modern capacitors have more complex layering techniques which produce equivalent series capacitance. By this process the equivalent series resistance can be significantly reduced in this type of capacitor. Film capacitors are flexible nonpolarized capacitors that are widely used.

Inductor

Inductors, like capacitors, store energy. An inductor is a two-terminal device with the symbol shown in Figure 11.11.

The inductance of a coil can be calculated as

$$L = \frac{\mu_0 N^2 A}{l} \qquad (11.24)$$

where μ_0 is the permeability of the core, N is the number of turns of wire, A is the cross-sectional area of the coil, and l is the length of the coil.

The voltage–current relationship (Figure 11.12) of an inductor is given by

$$\psi(t) = Li(t) \qquad (11.25)$$

$$v(t) = \frac{d\psi(t)}{dt} \qquad (11.26)$$

$$\Rightarrow v(t) = L\frac{di(t)}{dt} \qquad (11.27)$$

FIGURE 11.11   Inductor symbol.

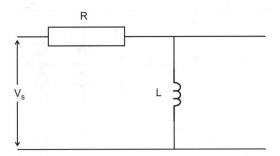

FIGURE 11.12 Inductor symbols and common electronic packages.

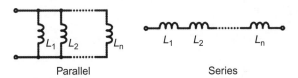

FIGURE 11.13 Inductors in parallel and in series.

In a similar fashion to the capacitor we can derive the current flowing in the inductor for any time t. Recalling Thévenin's rule

$$L\frac{di(t)}{dt} + Ri = V_s \qquad (11.28)$$

Separation of parts gives:

$$\frac{di(t)}{V_s - i} \equiv \frac{R}{L}dt \qquad (11.29)$$

Integrate both sides and take the natural logarithm:

$$\frac{i - V_s}{V_s} = e^{-tR/L} \qquad (11.30)$$

$$\therefore i = V_s(1 - e^{-tR/L}) \qquad (11.31)$$

The energy stored in an inductor is given by

$$E = \frac{1}{2}LI^2 \qquad (11.32)$$

Inductors placed in a parallel network are equivalent by the definition (Figure 11.13)

$$\frac{1}{L_{eq}} = \frac{1}{L_0} + \frac{1}{L_1} + \frac{1}{L_2} + \cdots + \frac{1}{L_n} \qquad (11.33)$$

Inductors placed in series are equivalent by the definition

$$L_{eq} = L_0 + L_1 + L_2 + \cdots + L_n \qquad (11.34)$$

Diode

A diode is a nonlinear passive device indicated by the symbol in Figure 11.14.

Diodes only permit current flow in one direction, and that direction is indicated by the arrow in the symbol. Diodes have a voltage–current relationship as described in Figure 11.15.

The diode will conduct when the anode is approximately 0.6 V above the cathode.

Transistor

The transistor is an active device, that is, it can perform tasks such as amplification. The bipolar junction transistor, or BJT, is shown as the symbol in Figure 11.16.

Bipolar Transistor

The bipolar transistor has two formations: NPN type and PNP. Transistors permit charge to flow from the collector to the emitter depending on the current flow within the base:

$$I_c = \beta I_B \qquad (11.35)$$

where β is the current gain of the transistor.

The flow of current is governed by an increase in free electrons as the forward-biased base and emitter junction reduces the depletion layer between these two regions. Having gathered sufficient energy within the electric field, the excess electrons cross the narrow base layer and are swept across the reversed biased base and collector junction generating a net current flow.

A bipolar transistor has maximum values for the current that the base I_B and the collector I_c can sink. In addition, a maximum voltage between the collector and the emitter V_{ce} will be defined.

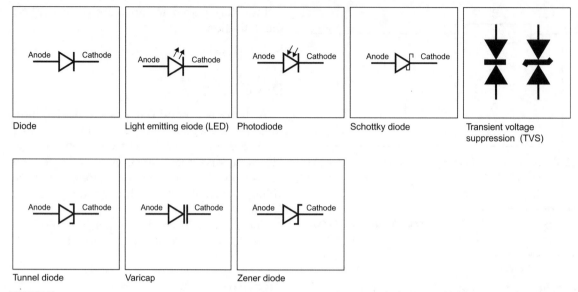

FIGURE 11.14 Diodes and their symbols.

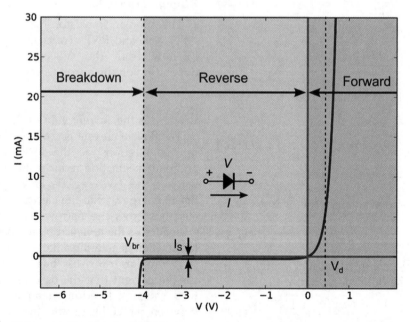

FIGURE 11.15 Voltage relationship of a diode.

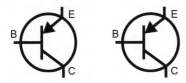

FIGURE 11.16 Symbols for bipolar transistors.

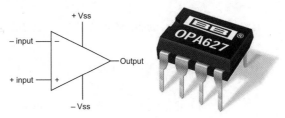

FIGURE 11.18 An op-amp.

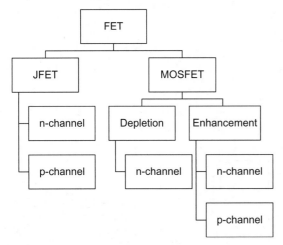

FIGURE 11.17 Transistor family.

Field Effect Transistor

Whereas the bipolar transistor can be considered a current-controlled device (the current in the base), the field effect transistor is a voltage-controlled device. The field effect transistor, or FET, has the symbol shown in Figure 11.17 shows the family of transistors.

The terms *base*, *collector*, and *emitter* are now replaced by *gate*, *source*, and *drain*, respectively. The voltage applied to the gate now controls the current through the drain. The electric field generated from the drain can be thought of as "pinching" the flow of charge carriers through the source.

Operational Amplifier

The operational amplifier is a three-terminal device (excluding power and balance terminals). It has an inverting input and a noninverting input defined by − and +, respectively. The op-amp is a high gain, high input impedance, low output impendence device which can be used, among a vast number of other possibilities, for amplifying a signal or buffering it from another part of a circuit (Figure 11.18).

In simple terms the output of an op-amp will try and do whatever it can, if fed back to the input, to make the voltage difference between the two inputs zero. Ideally the inputs draw no current.

AC SIGNALS' COMPLEX OHM'S LAW

When an alternating signal is applied to a capacitor or an inductor we have to consider the impact of the varying voltage and current. We take the capacitor as our first example.

We know that the current relationship for a capacitor is

$$i(t) = C\frac{dv(t)}{dt} \tag{11.36}$$

Let the voltage be a sine wave

$$v(t) = V\sin(\omega t) \tag{11.37}$$

$$\therefore i(t) = \omega V\cos(\omega t + \phi) \tag{11.38}$$

These can be rewritten in the complex polar form

$$i(t) = Re\{I_0 e^{-j\omega t}\} \tag{11.39}$$

$$v(t) = Re\{Ve^{-j\omega t}\} \quad where \; V = V_0e^{-j\phi} \quad (11.40)$$

$$i(t) = C\frac{dv(t)}{dt} = C\frac{d}{dt}e^{-j\omega t} = Cj\omega e^{-j\omega t} \quad (11.41)$$

$$i = Cj\omega e^{-j\omega t} \equiv Cj\omega v \quad (11.42)$$

$$\frac{1}{Cj\omega}v \equiv \frac{V}{R} \; \therefore \; \frac{1}{Cj\omega} \equiv R \; is \; the \; impedance \; \mathbf{Z}_{cap}$$

$$(11.43)$$

Equivalently for an inductor

$$v = Lj\omega e^{-j\omega t} \equiv Lj\omega i \quad (11.44)$$

$$Lj\omega i \equiv IR \therefore Lj\omega \equiv R \; is \; the \; impedance \; \mathbf{Z}_{ind}$$

$$(11.45)$$

The impedance for a resistor of resistance R remains R and there is no phase lag in the resistor.

BASIC CIRCUIT DESIGN IN INSTRUMENTATION

Voltage Divider

The ubiquitous voltage divider (Figure 11.19) has a voltage output that is a fraction of the voltage in

$$I = \frac{V_{in}}{R_1 + R_2} \quad (11.46)$$

The output voltage is the voltage across R_2, therefore

$$V_{out} = IR_2 = \left(R_2 \cdot \frac{V_{in}}{R_1 + R_2}\right) = \frac{R_2}{R_1 + R_2}V_{in}$$

$$(11.47)$$

We can produce a Thévenin equivalent of this circuit, as shown in Figure 11.20.

We know that the Thévenin equivalent voltage is given by

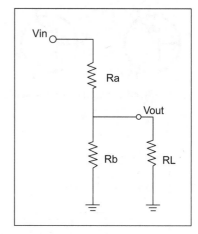

FIGURE 11.19   Voltage divider.

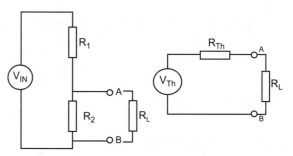

FIGURE 11.20   Thévenin equivalent voltage divider.

$$V_{TH} = open \; circuit \; voltage = V_{in}\frac{R_2}{R_1 + R_2} \quad (11.48)$$

In addition we know that the Thévenin equivalent resistance R_{TH} is given by

$$R_{TH} = \frac{V_{TH}}{short \; circuit \; current}$$
$$= V_{IN}\frac{R_2}{R_1 + R_2} \cdot \frac{R_1}{V_{IN}} \quad (11.49)$$
$$= \frac{R_1R_2}{R_1 + R_2}$$

If we add a load across the output of this equivalent circuit we find

$$V_{out} = V_{TH}\frac{R_L}{R_{TH} + R_L} \quad (11.50)$$

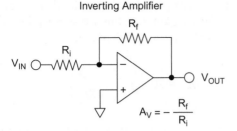

FIGURE 11.21   An inverting amplifier.

The load resistance causes the output of the voltage divider to drop. This is an important consideration in the design of measurement equipment. We must ensure that the load resistance is very large compared to the source resistance $R_L \gg R_{source}$. Equivalently, the source resistance should be as low as possible; there are only a few exceptions, for example, if the signal is a current, in which case we want a small input impedance.

Simple Signal Amplifier Built Using an Op-Amp

Because the noninverting input is grounded and because the output acts to equalize the difference between the inputs, the inverting input is also at ground (known as a virtual ground), as shown in Figure 11.21. As already mentioned, the inputs draw no current, so the current flowing through R_1 and R_2 is the same. Therefore

$$I = \frac{V_{out}}{R_2} = -\frac{V_{in}}{R_1} \Rightarrow \frac{V_{out}}{V_{in}} = -\frac{R_2}{R_1} \equiv voltage\ gain$$

(11.51)

The noninverting amplifier is shown in Figure 11.22.

The inverting input has a voltage

$$V_{inv} = V_{out} \frac{R_1}{R_1 + R_2}$$

(11.52)

$$V_{in} = V_{inv} \Rightarrow V_{in} = V_{out} \frac{R_1}{R_1 + R_2}$$

(11.53)

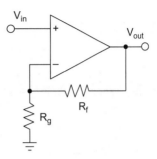

FIGURE 11.22   A noninverting amplifier.

$$\frac{V_{in}}{V_{out}} = \frac{R_1}{R_1 + R_2} \Rightarrow \frac{V_{out}}{V_{in}} = \frac{R_1 + R_2}{R_1} = 1 + \frac{R_2}{R_1}$$

(11.54)

which is the gain of the noninverting amplifier.

Importantly the input impedance of this circuit is, in theory, infinite and in practice very high ($10^{12}\Omega$).

Difference Amplifier

There are many applications where the important information within a signal is contained in the presence of significant and relatively high amplitude noise. We might think of the electrical signal from a muscle amplifying speech in a noisy environment. The information can be extracted if we improve the signal-to-noise ratio. This can be done by recording a signal from two spatially different points and removing the signal that is *common* to both. The amplifiers seen thus far have only a single input so we need something different (Figures 11.22 and 11.23).

The operational amplifier seen earlier is essentially a difference amplifier; the difference between the two inputs is reflected by a change in the output. However, because of the huge common mode gains of an operational amplifier very small differences in voltage between the input terminals produce huge

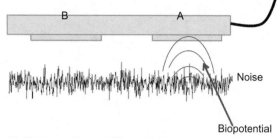

(A–B)* Some Gain = Differential Output

FIGURE 11.23 Differential amplifier operation. The noise is common to both electrodes; the biopotential is present predominately at one. *Source: Thomas Stone, Measuring the Motor System, Biomedical Lecture Series Birmingham University 2009.*

swings in the output, so it does not make a practical device for instrumentation. A solution is shown in Figure 11.25.

In this circuit

$$V_{out} = (V_1 - V_2)\frac{R_2}{R_1} \qquad (11.55)$$

Unfortunately this is not an ideal solution as V_2 "sees" an input impedance of $R_1 + R_2$ but V_1 only sees an input impedance of R_1. This imbalance of impedance means that if we were to apply a voltage to one terminal and ground the other then reverse the configuration, different currents would flow through V_2 and V_1 whereas ideally they would be the same. This affects the common rejection ratio as similar signals would not appear so and would not be rejected.

The common mode rejection ratio (CMRR) is defined as

$$CMRR = A_{diff}\left(\frac{V_{cm}}{V_{out}}\right) \qquad (11.56)$$

That is, it is a ratio of the common signal present at both terminals (V_{cm}) and the output (V_{out}) multiplied by the differential gain (A_{diff}). The logarithmic expression of the CMRR, the common mode rejection (CMR), can be defined; you may also see this expressed as a logarithmic term $CMR = 20_{log10}CMRR$.

The CMR in dB can be defined in terms of the amplifier gain and the n resistor tolerances, Kr, or total fractional mismatch.

$$CMR = 20 \ log_{10}\left(\frac{1 + \frac{R_2}{R_1}}{nKr}\right) \qquad (11.57)$$

It can be seen that if typical values of resistors were used where the tolerances were in the order of 0.1%, the CMR would be very poor.

It should also be noted that the input impedances of the inputs mentioned earlier are really rather low. If we remember what happened to the voltage divider network when $R_L \ngtr R_{source}$ we can see that we need to maintain a high input impedance so as not to force a reduction in the divider output, which is the signal we are trying to measure. It should be noted that if the signal is acting as a current source, the inverse is true and we want a small input impedance.

The disparity between input currents can be overcome by introducing unity gain (gain = 1) buffer amplifiers to the inputs of the circuit (Figure 11.26). This essentially produces a very high impedance input to the differential amplifier network; because op-amps typically have very high input impedances, they ideally draw no current through their inputs. The buffer simply provides feedback of the output to the noninverting input which *follows* the inverting input; as such, it merely buffers the input.

The gain of this differential amplifier is set similarly to the inverting amplifier, except the voltage at the noninverting input is not ground. The gain is therefore the ratio of R_1 to R_2; however, to change the gain we must remember that the circuit must remain balanced, and small changes in the resistor pairs will produce different gains on the inputs. How do we solve these problems? The instrumentation amplifier is one solution.

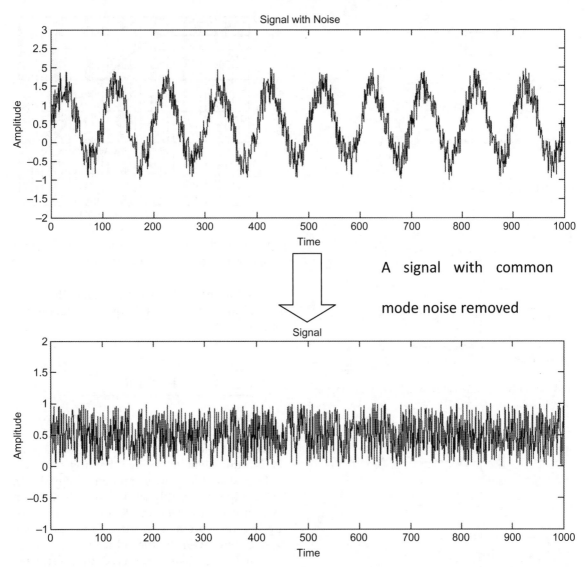

FIGURE 11.24 Signal with common mode noise. *Source: Thomas Stone, Patterns of muscle activity during walking after stroke, Ph.D Thesis, 2006, Bournemouth University.*

Instrumentation Amplifier

The instrumentation amplifier is a precision differential amplifier that is prepackaged in a monolithic device. The circuit shown in Figure 11.27 can be adapted as described in the following.

The gain of this circuit is defined by R_{gain}. Because the inputs of the two buffer op-amps draw no current, the voltage drop across R_{gain}, which is proportional to the differential voltage V_1 and V_2, produces a current that runs entirely through the resistors R. This produces

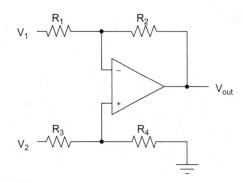

FIGURE 11.25 Differential amplifier.

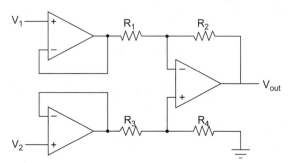

FIGURE 11.26 Differential amplifier with buffered inputs.

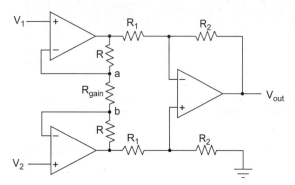

FIGURE 11.27 Instrumentation amplifier.

a voltage that forms the input to the differential amp we saw previously.

$$V_{out} = (V_1 - V_2)\left(1 + \frac{2R}{R_{gain}}\right)\frac{R_1}{R_2} \qquad (11.58)$$

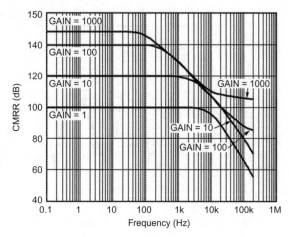

FIGURE 11.28 Frequency response of the AD8221 instrumentation amp with respect to CMRR.

It is important to remember that the function of an amplifier is frequency dependent (Figure 11.28).

FILTER

A basic filter can be constructed from a capacitor or an inductor (Figure 11.29). As we saw earlier, the frequency response of a capacitor is equivalent to saying that it is increasingly resistant to lower-frequency signals. In fact, a capacitor will block DC signals and can be useful for preventing DC offsets from disrupting measurement devices. The same can be said, but inversely, for inductors.

Signals we have considered thus far have been considered in the *time domain*, that is, they vary with time; we can just as easily consider a signal within the *frequency domain*. If this is the case, the signal now varies with respect to frequency (Figure 11.30).

We use the frequency domain to understand the response of a signal to a filter; specifically we can use the concept of a *transfer function*. A transfer function can define the response of a linear system given knowledge of its input; nonlinear

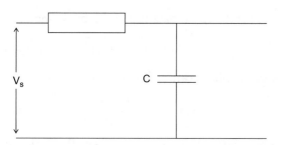

FIGURE 11.29 Low pass filter resistor capacitor network.

systems are more complex and require different treatment. Importantly, the input can be arbitrary. The transfer function is defined as the Laplace transform of the system's output over the Laplace transform of the system's input:

$$H(s) = \frac{Y(s)}{X(s)} \qquad (11.59)$$

So, for example, if the input were a sine wave, the output would be the Laplace

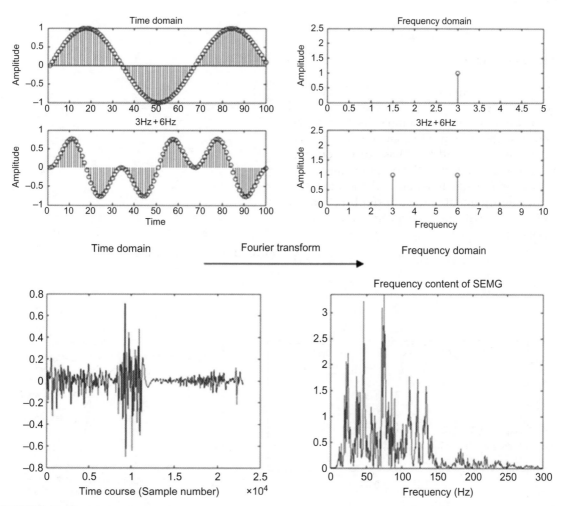

FIGURE 11.30 Time and frequency domain. *Source: Thomas Stone, Patterns of muscle activity during walking after stroke, Ph.D Thesis, 2006, Bournemouth University.*

transform of the sine wave input multiplied by the transfer function of the system.

$$Y_s = H_s \mathcal{L}\{A\sin\omega t\} \tag{11.60}$$

$$Y_s = H_s \left(\frac{A\omega}{s^2 + \omega^2} \right) \tag{11.61}$$

For the simple capacitor resistor first order network shown in Figure 11.29, the relationship of the input to the output as a frequency response can be defined by using Kirchhoff's voltage analysis and taking the Laplace transform of the result.

$$V_{out} = \frac{1}{RCs + 1} V_{in} \tag{11.62}$$

which means that

$$H(s) = \frac{1}{RCs + 1} \tag{11.63}$$

The transfer function allows us to understand how the system will change with respect to the magnitude and the phase of the output signal. The magnitude of the output is the modulus of the transfer function and the phase is the inverse argument of the transfer function.

We can express the transfer function H(s) in its polar form:

$$H(s) = |H_s|e^{j\phi(s)} \tag{11.64}$$

Therefore, the magnitude can be defined as

$$|H_{(s)}| = \sqrt{\Re\{H_s\}^2 + \Im\{H_s\}^2} \tag{11.65}$$

and the phase can be defined as

$$\phi(s) = \tan^{-1} \left(\frac{\Im\{H_s\}}{\Re\{H_s\}} \right) \tag{11.66}$$

where $\Re\{H_s\}$ and $\Im\{H_s\}$ represent the real and imaginary parts, respectively, of the transfer function H(s).

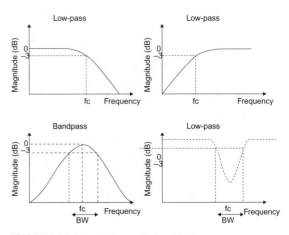

FIGURE 11.31   Different forms of filters.

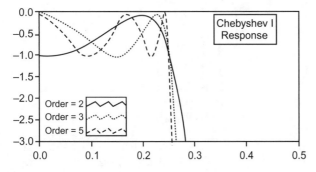

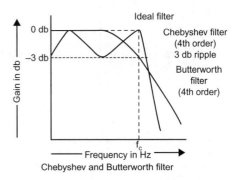

FIGURE 11.32   Butterworth filter compared to a Chebyshev filter.

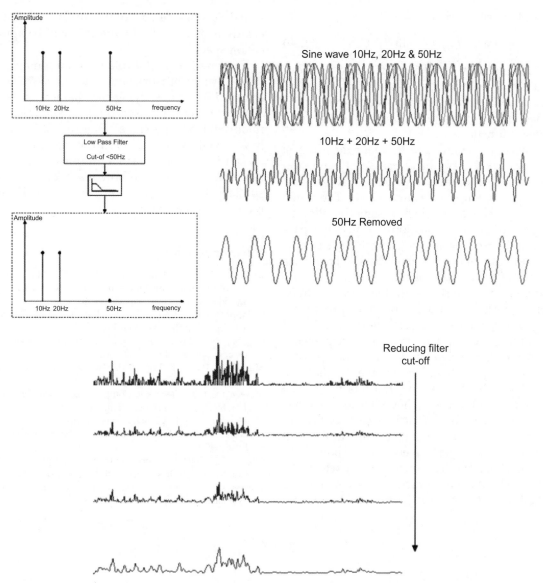

FIGURE 11.33 Effect of filtering on a surface electromyography (sEMG) signal with increasingly lower cut-off frequency. *Source: Thomas Stone, Patterns of muscle activity during walking after stroke, Ph.D Thesis, 2006, Bournemouth University.*

It is possible to replace s with $j\omega$. Thus, complex representation of the transfer function (Eq. 11.63) can be

$$H_{(j\omega)} = \frac{1}{1 + R^2C^2\omega^2} - j\frac{RC\omega}{1 + R^2C^2\omega^2} \qquad (11.67)$$

Therefore, the magnitude is

$$\left|H_{(j\omega)}\right| = \sqrt{\frac{1}{(1 + R^2C^2\omega^2)^2} - j\frac{R^2C^2\omega^2}{(1 + R^2C^2\omega^2)^2}}$$

$$(11.68)$$

and the phase is

$$\phi_{(j\omega)} = \arg(H_{(j\omega)}) = -\tan^{-1}(RC\omega) \qquad (11.69)$$

These results can be plotted to provide a phase and magnitude response with respect to frequency. Often a special format of plot called a Bode plot is used to visualize the nature of a transfer function.

Filters can be grouped into four main types: low pass, high pass, band pass, and band stop (Figure 11.31). Filters attenuate a signal and this attenuation is measured in dB. The range of frequencies over which a signal is allowed to pass without attenuation is called the pass band. The frequencies over which the signal is stopped is called the stop band. There are many different types of filters, which provide different characteristics in their pass bands. For example, the Butterworth filter provides a very flat pass band where the signal is not modulated very much over these frequencies. In contrast, the Chebyshev filter has a very oscillatory pass band which means that the signal will be distorted within this band. The advantage of the Chebyshev filter is that it has a much sharper *knee*. That is, signals of the frequency within the stop band are attenuated much more aggressively, as shown in Figure 11.32. The frequency beyond which the signal should be sufficiently attenuated is called the *cut-off frequency*. The level at which a signal is considered to be attenuated sufficiently is often 3 dB, though this will be application specific.

Filtering has a vital role in conditioning signals during measurement and great care is required when specifying the characteristics of a filter to ensure useful information is not removed from the signal. Figure 11.33 shows how a rectified surface electromyogram can be turned into a more visually intuitive envelope representation but also how many of the more subtle characteristics of this signal are removed in the process.

Further Reading

Brown, W.F., Bolton, C.F., Aminoff, M.J., 2002. Neuromuscular Function and Disease: Basic, Clinical, and Electrodiagnostic Aspects. W. B. Saunders.

Chiappa, K.H., 1997. Evoked Potentials in Clinical Medicine. Lippincott-Raven Press.

Glisson, T.H., 2010. Introduction to Circuit Analysis and Design. Springer Limited, London.

Horowitz, P., Hill, W., 1998. The Art of Electronics. Cambridge University Press, Cambridge.

Metzer, D., 1981. Electronic Components, Instruments and Troubleshooting. Prentice Hall, New Jersey.

Mississippi, A.A.L.A.P.N.U., Mississippi, V.C.T.N.R.U., 2000. Atlas of Electromyography. Oxford University Press, USA.

Oh, S.J., 2003. Clinical Electromyography: Nerve Conduction Studies. Lippincott, Williams & Wilkins.

Ziemer, R., Tranter, W., Fannin, R., 1998. Signals and Systems. fourth ed. Prentice Hall, New Jersey.

Clinical Measurement

Richard G. Axell

Clinical Scientist, Medical Physics and Clinical Engineering, Cambridge University Hospitals NHS Foundation Trust, Cambridge, U.K. and Honorary Visiting Research Fellow, Postgraduate Medical Institute, Anglia Ruskin University, Chelmsford, U.K.

Measurement underpins science and good measurement equals good science. A measurement tells us a property of the quantity to be measured and by assigning that property a number and expressing it in the correct SI unit within a healthcare environment it allows us to interpret the patient's condition to provide a diagnosis or intervention.

The measurement may be obtained by using invasive and noninvasive methods, with blood pressure being an ideal example of both. The gold standard technique for the accurate measurement of blood pressure is to use an invasive catheter. Invasive aortic blood pressure is measured by cutting into the peripheral vessel and inserting a catheter that has a pressure sensor mounted at the tip into the vessel and advancing the catheter to the exact point in the aorta to be measured. The most commonly used technique to measure blood pressure within a clinical environment is to perform the measurement noninvasively. Noninvasive blood pressure can be measured by placing a cuff around the patient's arm and listening for the Korotkoff sounds due to the turbulence of the blood underneath the cuff to determine the blood pressure. Although the most accurate method of measuring blood pressure is to perform an invasive measurement using a central line, this is at an increased risk to the patient than having the measurement performed noninvasively. It is the responsibility of the clinician to decide on the condition of the patient and whether the patient will benefit most from the most accurate and potentially harmful invasive measurement, of if the slightly less accurate noninvasive measurement of blood pressure is sufficient.

ACCURACY AND PRECISION

Firstly, to understand the concept and scientific requirement of equipment calibration to perform clinical measurement, we must consider the accuracy and precision of the measurement. Resolution is the smallest incremental value that can be measured. The digital measurement display on an ultrasound scanner with a larger number of digits ($1/1000$ mm) has a

211

higher measurement resolution than one with fewer digits (1/100 mm). Similarly, an analog measurement display on a sphygmomanometer with a tick mark every 1 mmHg has a higher measurement resolution than one with a tick mark every 5 mmHg. However, a higher resolution does not imply higher measurement accuracy; it will give you a high precision. Measurement precision is the repeatability of obtaining the output values from repeated measurements under identical input conditions over a period of time.

Figure 12.1(a) shows low precision where the measurement values are scattered all over the target. Figure 12.1(b) shows high precision where the measurement values are tightly clustered on the target. The measurement accuracy is given by the difference between the true value and the measured value divided by the true value. Figure 12.2(a) shows that although the measurement values are tightly clustered with high precision, they have low accuracy to the true value. Figure 12.2(b) shows high accuracy and high precision. When using electronic devices to measure physiological systems, the foremost consideration is selecting the

measurement device with the highest possible precision, repeatability, and accuracy.

Although good repeatability would suggest you have a valid result, a second person should repeat the experiment using different equipment to see if they observe the same outcomes. There are many different factors that can cause an inaccuracy or error in your results. These inaccuracies or errors can typically be characterized in two groups (1) errors intrinsic with the measured system, or measurement device errors, and (2) errors intrinsic to the measuring system, or measurement errors. A measurement device error could include mechanical, electronic, software, scale, environmental, and human errors. A measurement error could be classified as systematic or random. To improve measurement accuracy you should follow a strict measurement protocol to ensure all measurements are collected in the same manner and that the measurement equipment you use is calibrated against a national standard. However, even with following this measurement protocol there will be some uncertainty with the measurement.

FIGURE 12.1   (a) Low precision and (b) high precision.

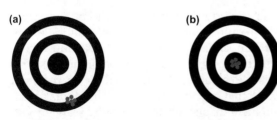

FIGURE 12.2   (a) Low accuracy and (b) high accuracy.

MEASUREMENT DEVICE ERRORS

A mechanical, electronic or software fault with the measurement device may cause the relationship of the input signal to output reading to be altered by some fundamental characteristic built into the measurement device. For example, a digital filter may have been set up with an inadequate sampling frequency so a component of the input data is being lost on the output from the device.

To perform a clinical measurement, the measurement device in question must have a visual display of the output to allow the observer to perform the measurement. As previously alluded to, this could be in the form of an analog scale or a digital display. Going back to the analog measurement display on a sphygmomanometer with a tick mark every 5 mmHg, the total error with the reading can be no better than the intrinsic error from the manufacturer marking the scale on the display combined with the human interaction from reading the scale. A reading error is one that arises from the pointer-scale-reader interaction, which makes the reading differ from the actual indication on the display. This is two-fold: A pointer positioned between two points marked on a scale means the observer has to make some form of assumption on the true value of the indicated point; secondly, if the observer was to take a reading from any point other than normal to the plane of the scale there would be a parallax error with the measurement. Parallax errors can also occur from the incorrect use of measurement equipment; for example, if a thermal camera was set up so that the object to be measured was not parallel to the field of view, the measurand would not be entirely covered by the focal region of the thermal camera and errors with the accuracy of the temperature measurement would be induced.

A digital display can only measure to the precision of its most significant digit. So a scale with a precision of 1 g cannot measure to the precision of a set of scales to 0.001 g. There may also be some fluctuation with the value displayed on the scale, and so, again, the observer has to make some form of assumption on the true value indicated on the display.

An environmental error is one that arises from factors external to the instrument caused by the environment that the measurement device is being operated in. Some of the most common types of environmental error are temperature, humidity, pressure, and sound. A temperature change in an environment in which an electrocardiogram (ECG) machine is being operated may cause the internal amplifier electronic components to have undesired slow changes in the voltage output, called drift.

A thermally induced voltage across the amplifier may also cause undesired random voltage to be measured across the component, called noise.

MEASUREMENT ERRORS

A systematic measurement error is one that relates to the magnitude of the output signal for the entire measurement output range of the device. If a measurement device output was 3% higher than the true value, or if the scale was offset by 2 units higher than the true value, the error is termed *systematic*. This error could arise if the zero point of the measurement device scale shifts or if the calibration of the measurement device scale is incorrect. Since a systematic error is constant across the entire output range of the measurement device, it would normally be easy to detect and correct for.

A random measurement error is one that stems from fluctuation in the conditions within a system being measured which has nothing to do with the true signal being measured. The complexity of any biological system will mean the measurement conditions are continually varying with time, and these variations will appear to the measurement device as differences in the input signal and be displayed as such on the output as noise. This will only significantly impact the accuracy of the reading if the magnitude of the noise level is a considerable proportion of the output signal itself. The signal-to-noise ratio (SNR)—the ratio of signal magnitude to noise magnitude—is used to define the quality of the measurement device output. For example, to be able to detect and measure an electroencephalography (EEG), you would need high SNR to be able to distinguish the relatively small signal from background noise.

CALIBRATION

The basic concept of measurement calibration is that the test equipment should be periodically compared against a standard of higher accuracy. The organisation performing the calibration against the standard of higher accuracy will then provide a calibration certificate. If errors are identified the measurement device may be adjusted so that the measurement is more accurate or correction factors will be specified in the calibration certificate. These correction factors will allow the user to continue using the measurement device and the user can correct for the systematic error ensuring the measurements recorded are within an acceptable tolerance of measurement accuracy. The calibration certificate will be provided by an accompanying statement of uncertainty. This is where uncertainty analysis is performed on the calibration to specify the uncertainty of the measurement. Within the healthcare environment a label will then be attached to the measurement device that indicates the calibration certificate number and the date that it is due for recalibration. A measurement device that is uncalibrated may also be used, however, it should be clearly indicated as such, with a label such as "for indication only."

If the measurement device is linear in its operation, its output can be set to zero for zero input. A one-point calibration can then be used to define the calibration curve that plots output versus input. If the linearity of the measurement device is unknown, a two-point calibration must be performed with these two points plus the zero point used to plot the calibration curve to guarantee linearity. However, if the resulting calibration curve from the two-point calibration is nonlinear, a multipoint calibration must be plotted to obtain the calibration curve. Due to the possibility of environment effects influencing the performance of the measurement device, calibration curves should be calculated for a range of different temperatures that would be expected during normal use conditions to determine temperature drift effects on the zero point and the gain.

TRACEABILITY

For a calibration to be considered traceable there must be an unbroken chain of measurements back to a national or international standard. This will ensure that the measurement taken is an accurate representation of the true value by way of its traceability back to a known standard. The National Metrology Institute (NMI) exists to maintain primary standards of measurement to provide traceability for the calibration of measurement devices. Within the United Kingdom, the National Physics Laboratory (NPL) is the accredited national measurement institute and, as such, measurements performed using a calibrated device should be traceable back to their reference standard (accurate to 0.002%). However, the chain of calibrations is only traceable if the correct calibration calculations and uncertainties are correctly applied at every step of the chain back to the reference standard; and with every step along the chain away from the reference standard the uncertainty increases. For instance, NPL's reference standard may be calibrated to an accuracy of 0.002%. A calibration laboratory's "laboratory standard" device may then be calibrated against the reference standard to an accuracy of 0.01%; a calibration laboratory "field use" device may then be calibrated against the "laboratory standard" device to an accuracy of 1.0%; a calibration laboratory would then use the "field use" device to calibrate the clinical measurement device used within a clinical environment to an accuracy of 5.0%.

UNCERTAINTY

With any clinical measurement there will always exist some doubt about the true value of the measurement. A correct understanding of uncertainty is fundamental to good quality measurements for: (1) calibrating a measurement device; (2) correctly interpreting a calibration certificate for the measurement device; and (3) using a measurement device to perform a measurement. Any potential source of error whose value is unknown is a potential source of uncertainty. As such, the overall uncertainty in the measurement will be calculated from the combination of all the individual sources of uncertainty within the measurement device. By performing this analysis of the clinical measurement it will allow you to work at reducing the source of greatest uncertainty in your measurement. However, it is important to understand that a mistake when performing the measurement is not a source of measurement uncertainty.

To carry out an uncertainty analysis for a clinical measurement, you must first define all the potential sources of error, then perform the measurements following a strict measurement protocol. If an operator is inadequately trained, or the operator performs each measurement differently, or an inappropriate measurement device is used to carry out the test, there will be no scientific validity to the result. After using a strict measurement protocol to record the measurements you then have a choice of using two different methods to express the uncertainty. Statistical or Type A uncertainty is used to express the uncertainty in a measurement by calculating the standard deviation of many repeated measurements using the equation

$$u = \frac{s.d}{\sqrt{n}}$$

where u is the standard uncertainty, $s.d$ is the standard deviation, and n is the number of measurements performed.

Type B uncertainty is used to express uncertainty from supplementary information provided for the measurement device. This could include calibration certificates, specifications,

TABLE 12.1 k Factor

Required Confidence Level (%)	k Factor
99.7	3
99.0	2.58
95.0	2
68.0	1

and previously published information on the use of the system. By expressing the individual uncertainties in consistent units, the individual uncertainties calculated for both Type A and Type B can be combined using the "root sum of all squares" equation:

$$Combined\ Uncertainty = \sqrt{a^2 + b^2 + c^2 + \cdots + n^2}$$

Finally, the combined uncertainty should be expressed as a confidence interval to indicate how confident you are that the true measurement is within the stated range. This is done by multiplying the combined uncertainty by a k factor using the equation

$$Expanded\ Uncertainty\ U$$
$$= Coverage\ Factor\ k$$
$$\times Combined\ Standard\ Uncertainty$$

The k factor is specified depending on the confidence level, as shown in Table 12.1.

SENSITIVITY AND SPECIFICITY OF MEASUREMENT TECHNIQUE

We have already determined that clinical measurement is subject to error, be it random, systematic, or human. The random variation with the measurement will mean that if you use an automated blood pressure monitor to record a subject's blood pressure more than once, the values obtained for systolic and diastolic blood pressure will differ slightly. These differences could be due to variations in the measurement method, the observer, an equipment fault, or physiological differences in the subject. The variations may also increase further by introducing a second observer to perform the measurement and interpret the results.

Sensitivity and specificity can be used by the operator to determine the probability of the clinical measurement test to correctly identify the pathology being tested for in the patient. When a clinical measurement is performed there are two possible test results: (1) the test result is positive, indicating the presence of pathology, or (2) the test result is negative indicating the absence of pathology. There are also two possible clinical outcomes: (1) the patient has the pathology, or (2) the patient is pathology free. Table 12.2 shows the test result in columns and the true status of the patient under test in rows.

Sensitivity can be defined as the probability that the clinical measurement indicates the patient has the pathology when in fact they do have the pathology. Sensitivity is a measure of how likely the clinical measurement is to detect the presence of pathology in a patient who has it. The sensitivity is given by

$$P(T^+|S^+) = \frac{a}{a+b}$$

Specificity can be defined as the probability that the clinical measurement indicates the patient does not have the pathology when in fact they are pathology free. The specificity is given by

$$P(T^-|S^-) = \frac{d}{c+d}$$

The ideal clinical measurement test should have a high sensitivity and high specificity, but in the clinical environment there may have to be some tradeoff between sensitivity and specificity. Usually you can use a clinical measurement test with a relatively high sensitivity and specificity, however, you will still get false positives and false negatives. A false positive is when the clinical measurement test reports a positive result for a patient that is, in fact, pathology free. The false positive rate is given by

$$P(S^-|T^+) = \frac{c}{a+c}$$

In an ideal world the number of patients who test positive to pathology that are in fact pathology free would be zero ($c = 0$). In reality, due to the errors associated with clinical measurement techniques, this would be impossible to achieve as soon as the test population increases in number. For small study numbers you may find random chance dictates you observe no false positives within your result.

Conversely to false positives, a false negative is when the clinical measurement test reports a negative result for a patient that has pathology. The false negative rate is given by

$$P(S^+|T^-) = \frac{b}{b+d}$$

The clinical implications of a false positive or a false negative depend on the type test being performed. A false positive would indicate that you have informed the patient that they have pathology when in fact they do not, causing them unnecessary worry and concern and they may also be given inappropriate intervention or

TABLE 12.2   Sensitivity and Specificity Testing

	Clinical Measurement/Test Result (T)	
True Status of Pathology (S)	Positive (+)	Negative (−)
Pathology (+)	a	b
No Pathology (−)	c	d

TABLE 12.3 Hypothesis Testing

True Status of Pathology (S)		Clinical Measurement/Test Result (T)	
		Positive (+)	Negative (−)
	H_a true (+)	a	b (Type II)
	H_0 "true" (−)	c (Type I)	d

treatment. Conversely, a false negative would indicate that you have informed the patient that they do not have pathology when in fact they do. This could lead to withholding intervention or treatment, and the absolute worse-case scenario is the patient dies from this.

TYPE I AND TYPE II ERRORS

Significance tests are used to test a hypothesis, and sensitivity and specificity relate to the correct decisions, whereas a false positive and false negative relate to Type I and Type II errors, respectively. When a hypothesis is tested there are two possible outcomes: (a) the test result is positive, a statistically significant rejecting of the null hypothesis, or (b) the test result is negative and is not statistically significant, so we decide not to reject the null hypothesis. Table 12.3 shows that if the true status of the pathology is positive[1] then the alternative hypothesis is true, whereas if what you are testing for is not true[2] then the null hypothesis is true.

Test result a represents a correct decision to reject the null hypothesis, as the alternative is really true. A high value of a results in a high probability of rejecting a null hypothesis, corresponding to a highly sensitive test. Test result b represents an incorrect decision not to reject the null hypothesis when in fact it is not true. This represents Type II error and corresponds to a false negative. However, a low value of b would represent a low probability of Type II error. Test result c represents an incorrect decision to reject the null hypothesis when in fact it is true. This represents Type I error and corresponds to a false positive. However, a low value of c would represent a low probability of Type I error. Test result d represents a correct decision not to reject a "true" null hypothesis when it should not be rejected. A high value of d results in a high probability of not rejecting a "true" null hypothesis, corresponding to a test with high specificity.

Further Reading

Webster, J.G., 2008. Bioinstrumentation. John Wiley & Sons.

Webster, J.G., 2009. Medical Instrumentation Application and Design. John Wiley & Sons.

International Organization for Standardization (ISO), 2010. ISO 21748:2010 Guidance for the use of repeatability, reproducibility and trueness estimates in measurement uncertainty estimation.

[1]The pathology is present; hence what we are testing for is true.

[2]There is no pathology.

13

Cardiology

Richard G. Axell

Clinical Scientist, Medical Physics and Clinical Engineering, Cambridge University Hospitals NHS Foundation Trust, Cambridge, U.K. and Honorary Visiting Research Fellow, Postgraduate Medical Institute, Anglia Ruskin University, Chelmsford, U.K.

ANATOMY AND PHYSIOLOGY

The human heart is about the size of a clenched fist and is located in the thoracic cavity between the sternum and the vertebrae. It is the pumping mechanism that uses the blood vessels as a transportation system to deliver essential minerals to the body and to carry away waste products. The basic structure of the heart is illustrated in Figure 13.1. The heart is made up of four chambers: the upper chambers, or atria, receive the blood returning to the heart and transfer it to the lower chambers, ventricles, which then pump the blood from the heart. The two chambers on the right side send deoxygenated blood to the lungs (pulmonary circulation). The two chambers on the left side supply oxygenated blood through arteries to the rest of the body (systemic circulation). When the blood reaches the capillary beds the nutrients are exchanged for waste products and then the blood is transported along the venous system

back to the right side of the heart and the cardiac cycle starts again.[1]

The circulatory system is made up of pulmonary and systemic circulation. The blood flows in one fixed direction and four one-way heart valves aide the unidirectional blood flow. The valves are positioned so that they open and close passively due to the pressure differences between the heart's chambers and arteries. Deoxygenated blood flows back from the body and enters the right atria via large veins called the vena cava (superior and inferior). This deoxygenated blood flows from the right atria through the tricuspid valve into the right ventricle which pumps it out through the pulmonary (or semilunar) valve and into the pulmonary artery and into the lungs. As the blood flows through the lungs it loses the carbon dioxide and collects a fresh supply of oxygen before flowing down the pulmonary veins and into the left atria. The oxygenated

---

[1]A cardiac cycle is a series of events that occur during a single heartbeat.

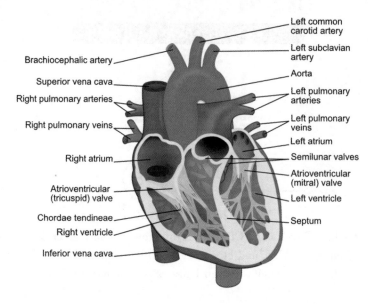

FIGURE 13.1   Basic structure of the heart.

blood flows from the left atria through the bicuspid (or mitral) valve into the left ventricle which is subsequently propelled through the aortic (or semilunar) valve and into the aorta to be circulated around the body.

The tricuspid[2] and bicuspid[3] (atrioventricular or AV) valves are positioned between the atrium and ventricle on the right and left sides of the heart, respectively. When the atrial pressure exceeds ventricular pressure the valves are forced open to allow the blood to pass into the ventricle during ventricular filling. As the pressure within the ventricle rises and exceeds the atrial pressure, the AV valves close to prevent the back flow of blood into the atria during ventricular emptying. When the ventricular pressure exceeds the arterial pressure, the semilunar valves open and allow blood flow into the arteries. As the ventricles relax the ventricular pressure falls below

arterial pressure and the valves close to prevent the blood flowing from the arteries back into the heart.

The ventricles contract during systole and relax during diastole. During diastole the pressure within the left ventricle is low; hence the aortic valve is closed. The blood flows from the left atrium into the left ventricle and then the left atrium starts to contract after being excited by the sinoatrial (SA) node to further fill the left ventricle. The ventricle contracts increasing ventricular pressure causing the mitral valve to close. The pressure within the left ventricle continues to increase until it exceeds the aortic pressure, forcing the aortic valve to open and blood to flow into the aorta. The blood continues to flow into the aorta until the pressure in the left ventricle falls below arterial pressure and the aortic valve closes.

The timing of the cardiac cycles is controlled by the heart's pacemaker, the sinoatrial node. The SA node is located in the wall of the right atrium and consists of specialized muscle tissue which combines the characteristics

[2]Named because it consists of three cusps or leaflets.

[3]Likewise named because it consists of two cusps or leaflets.

of muscle and nerves; when it contracts it generates electrical impulses similar to those produced by nervous system. Nodal tissue is self-excitable and can contract without any signal from the nervous system. When the SA node contracts it generates a wave of excitation which travels through the walls of the heart; the impulse spreads rapidly causing both atria to contract simultaneously. The ventricles are separated from the atria by nonconductive tissue except for a second patch of nodal tissue positioned at the bottom of the wall separating the two atria, the atrioventricular (AV) node. As the wave of excitation reaches the AV node it is delayed by approximately 0.1 seconds to allow the atria to fully contract before the ventricles. After the delay, the excitation wave is conducted along the bundle of His to the apex of the ventricles, to spread upward through the ventricular walls along the Purkinje fibers.

ECHOCARDIOGRAPHY

Diagnostic ultrasound is the use of high frequency sound waves (1–10 MHz) to generate diagnostically useful information. Echocardiography is specifically the use of diagnostic ultrasound to investigate the cardiac system, providing information on the cardiac hemodynamics as well as structural images of the heart and vasculature. The sound waves are produced and detected by a transducer with an array of piezoelectric crystals which deforms under a mechanical (pressure) and electrical load, termed the *piezoelectric effect*. A voltage signal is passed across the piezoelectric crystal to generate a mechanical pressure wave. Electrolytic coupling gel is used to transmit the ultrasound signal from the transducer and into the body. As the mechanical pressure wave passes through the body, the difference in acoustic impedance of the tissue interfaces cause a proportion of the ultrasound wave to be reflected back to the transducer. This echo is then received by the transducer and the piezoelectric transducer converts the mechanical wave back into an electronic pulse to be processed by the ultrasound scanner. As the assumed speed of sound in soft tissue is $1540\,\mathrm{ms^{-1}}$, the "time of flight" is then used to determine the depth of the tissue interface and is displayed as such on the monitor.

The interaction between ultrasound and the tissue it propagates through is fundamental to the diagnostic ability of the clinical measurement technique. This can happen by reflection, scattering, refraction, and absorption. For tissue interfaces larger than the ultrasound wavelength, the ultrasound will either undergo specular or diffuse reflection,[4] or Raleigh scattering when the tissue interface is smaller than the wavelength. Reflection is caused by the difference in acoustic impedance between two tissues at a boundary. The greater the acoustic impedance mismatch the greater the proportion of the transmitted pulse that is reflected back to the transducer. In specular reflection the strongest reflection occurs when the tissue interface is perpendicular to the transducer. A diffuse reflection results in weaker reflections being transmitted in multiple directions. If the ultrasound pulse was incident to a tissue boundary at an angle other than 90° and the two tissues were of a different speed of sound, the transmitted pulse would be refracted as it passes into the second tissue as governed by Snell's law and the resulting echo would be incorrectly displayed on the image:

$$\frac{\sin\theta_1}{\sin\theta_2} = \frac{v_1}{v_2}$$

where θ_1 is the angle of the incident pulse, θ_2 is the angle of the refracted pulse, v_1 is the

---

[4]For smooth tissue boundaries the ultrasound pulse will undergo specular reflection. For rough tissue interfaces the ultrasound pulse will undergo diffuse reflection.

speed of sound in the first tissue, and v_2 is the speed of sound in the second tissue.

Absorption is caused by the viscoelastic properties of tissue leading to the deposition of the mechanical energy in the pulse into thermal energy in the tissue. This is the primary mechanism for attenuation, described by the attenuation coefficient and assumed to be approximately $0.5 \, \text{dB} \, \text{cm}^{-1} \, \text{MHz}^{-1}$ in soft tissue. Therefore, higher frequency sound waves are attenuated more, so there is a tradeoff between penetration depth and image resolution.

B-mode imaging uses the received timing and echo intensity to generate a 2D cross-sectional image of the heart. A group of piezoelectric crystal elements are excited with an electrical signal, this finite pulse is transmitted into the body, and each reflection provides information on the reflection strength for a specific time of flight. This information is used to create a scan line from the pulse-echo sequence, where the pixel brightness at each depth represents the strength of the reflector at that distance from the transducer, hence the name B-mode (brightness mode). The next group of elements are then sequentially activated to generate a 2D cross-section of the anatomy, or scan plane. The 2D image can then be used to visualize the cardiac anatomy, in addition to performing calliper measurements of the different parts of the heart.

Doppler ultrasound can be used to measure the frequency change in the Doppler pulse as it is reflected by a moving target, and the magnitude and direction of this frequency shift can be used to determine the velocity of the reflector as governed by the Doppler equation:

$$\Delta f = \frac{2vf\cos\theta}{c}$$

where Δf is the Doppler shift frequency, v is the velocity of the moving target, f is the frequency of the transmitted ultrasound pulse, θ is the angle between the ultrasound pulse and the target to be insonated, and c is the speed of sound of the transmitted ultrasound pulse.

There are two main types of Doppler, continuous wave (CW) and pulsed wave (PW), with PW Doppler being used in both spectral and color Doppler. CW Doppler requires one element to continuously transmit the ultrasound pulse and one element to continuously receive the reflected ultrasound echo, and all reflector velocities along the line of sight are detected. PW spectral Doppler transmits an ultrasound pulse, similar to B-mode, and reflected ultrasound echoes are only measured if they fall within a set time period from transmission. The operator uses a range gate or sample volume indicator on the B-mode image to determine where the PW Doppler signal will be measured. The spectral Doppler trace is displayed as a graph of Doppler shift frequency (reflector velocity) versus time. As with any pulsed signal, Nyquist's theorem states that the sampling frequency must be at least twice the maximum frequency in the signal to be sampled, otherwise aliasing will occur. Since there is an upper limit to the pulse repetition frequency (PRF) there is a maximal blood velocity ($v_{\max}$) that can be detected, given by

$$v_{\max} = \frac{c^2}{8df\cos\theta}$$

where d is the depth of the target.

Similarly to spectral Doppler, color Doppler transmits a pulse along a series of scan lines and reflected ultrasound echoes are only measured if they fall within a set time period from transmission defined by the position of the color box. Each scan line is divided up into a number of sample volumes and the mean Doppler frequency is calculated for each of them and then color-coded on the image. The color variation is then used to visualize the speed and direction of the blood flow. In general, red indicates flow toward the transducer and blue indicates flow away. However, the colors can be inversed and modern ultrasound scanners may have many different color options that can be selected by the operator. Color Doppler also suffers from aliasing, where high frequency content will be incorrectly displayed as the wrong color.

Although ultrasound is considered to be a safe imaging modality due to its nonionizing nature, there are potential risks to the patient in both the thermal and mechanical form. Since the ultrasound pulse is attenuated as it passes through the tissue, the dissipation on the pressure wave leads to localized heating which may have the potential to cause permanent tissue damage. This potential for thermal harm is indicated to the operator by way of the thermal index (TI):

$$TI = \frac{W_i}{W_0}$$

where W_i is the incident ultrasound power and W_0 is the power required to raise the tissue temperature by $1°C$.

The mechanical pressure wave will also interact with small gas bubbles and fluids as it passes through the tissue in the form of streaming and cavitation (noninertial and inertial). Streaming is caused by the ultrasound pulse radiation force causing the forced flow of liquid within the body, and can be seen within the testes. Noninertial cavitation is caused by the oscillations in the pressure wave causing gas bubbles to expand and contract in synchronism, creating local shear stresses between the gas bubble and the surrounding tissue. Noninertial cavitation is caused by a prolonged exposure to an oscillating pressure waveform, hence would only be observed while using CW Doppler. Inertial cavitation is caused by a sufficiently large rarefactional pressure that causes the bubble to expand rapidly and violently collapse, leading to localized areas of high temperature and pressure, which may cause cell necrosis. This potential for mechanical harm is indicated to the operator by way of the mechanical index (MI):

$$MI = \frac{P_r}{\sqrt{f_{awf}}}$$

where P_r is the attenuated peak rarefactional pressure and f_{awf} is the acoustic working frequency.

As with standard medical device management, an electrical safety test should be performed at acceptance and during any six-monthly or annual quality assurance (QA) or planned preventative maintenance testing. Firstly a visual inspection should be performed to check the physical integrity of the device. This allows the operator to check for any signs of physical damage and to make sure all the controls and ergonomic adjustments function correctly. Ultrasound QA is then performed using phantoms or test objects of known acoustical property. Imaging and Doppler phantoms are used by a trained operator to ensure the accuracy and validity of an image or measurement taken using the device. For example, an imaging phantom can be used to check the accuracy of caliper measurements, imaging resolution, contrast sensitivity, and penetration depth. A Doppler phantom may be used to check the accuracy of velocity measurements, sample volume dimension and position, detectable velocity limits, and the accuracy of waveform estimations.

The clinical examination can be performed using 2D phased array and 3D matrix array transthoracic probes and 2D and 3D transesophageal probes while the patient is under sedation to investigate cardiac hemodynamics and image the heart and vasculature. For a routine cardiac scan the patient will be positioned to lay left lateral decubitus with their right arm by their side and their left arm by their head. This obtains an optimal image of the heart as gravity acts to force it toward the front of the chest cavity, and by positioning the left arm above the head the intercostal space between the ribs is increased to obtain a better view. Since the acoustic impedance of bone is much greater than that of soft tissue, ultrasound images cannot be acquired through bone, so by increasing the intercostal space the operator can scan the heart between the rib bones. Three silver chloride electrocardiograph (ECG) electrodes are positioned on the patient's back so the patient's 3-lead ECG can be synchronized to the ultrasound scanner. An image or Doppler measurement is normally stored over a single beat, however, if the heart rate is over 60 bpm or the patient is in atrial fibrillation the information will be collected over two beats or more. A patient in atrial fibrillation will have an irregular heartbeat and by averaging the data collected over a series of heartbeats the parameters collected should better represent the patient's current physiology.

The operator will start the examination from the parasternal window and then move on to the apical window (Figure 13.2). Then, if clinically indicated, the patient will be moved to the supine position to examine the heart from the suprasternal and subcostal windows. By positioning the patient in this way you can scan the heart along standard planes and axis so that the information collected can be compared to normal data for diagnosis. A typical scan will involve taking measurements of the thickness of the aortic valve and mitral valve; the area and volume of the left and right atria; the size and function of the tricuspid valve; and Doppler is used to check for regurgitation. The severity of mitral regurgitation is determined by measuring the proximal isovelocity surface area (PISA) at the mitral valve during peak regurgitant flow. A PISA greater than 1.0 cm is severe and less than 0.4 cm is mild. The regurgitant flow is given by

$$Regurgitant\ Flow = 2\pi r^2 v_n$$

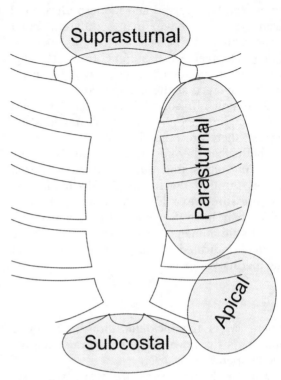

FIGURE 13.2 Echocardiography scanning windows for transducer positioning.

where r is the PISA radius and v_n is the normal velocity. The effective regurgitant orifice area (EROA) is the regurgitated flow divided by the maximum velocity calculated from CW Doppler through the mitral valve. The EROA is given by

$$EROA = \frac{2\pi r^2 v_n}{v_{max}}$$

The total regurgitant volume can be calculated from the EROA multiplied by the integral of the velocity-time curve:

$$Regurgitant\ Volume = EROA \int velocity - timecurve$$

ELECTROCARDIOGRAPHY

We have already determined that the heart is excited by electrical signals which cause the heart muscle to contract during systole. This electrical signal is transmitted from the heart and into the thorax cavity and the voltage across the resistive tissue can be detected by electrodes positioned on the skin and recorded as an electrocardiograph (ECG). The ECG trace provides the observer with information on the electrical characteristics of the heart, the extent of damage to the heart, and the effects of drug or surgical intervention on cardiac function. As with the measurement of neurophysiologic signals, ECG signals are measured using silver/silver chloride electrodes utilizing a driven right leg circuit. This provides a high common mode rejection ratio to remove the unwanted capacitive coupling induced electrical signals common across the body. The ECG is used to measure the voltages that arise from the differing contraction times of parts of the myocardium during the cardiac cycle. While at rest, myocardial cells have a negative potential across their membrane. An electrical signal is initiated at the sinus node; it travels to the atrioventricular (AV) node and into the ventricular myocardium. The electrical signal floods positive ions into the cell creating a positive potential that depolarizes the myocardium and causes it to contract. The excess positive ions are then pumped back out of the cell and it repolarizes and relaxes. The depolarization and repolarization of the myocardium occur at different times in different sections of the heart and this variation in potential is detected by positioning leads across the body in an ECG. On a conventional ECG a positive deflection indicates that the depolarized wave is traveling toward the electrode.

A typical ECG trace is shown in Figure 13.3. The P wave corresponds to the depolarization (contraction) of the atria. The PR interval is the

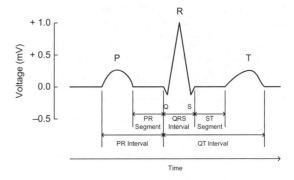

FIGURE 13.3 The normal PQRST ECG waveform.

time taken for the impulse travels through the AV node, bundle of His, and bundle branches to the Purkinje fibers. The QRS complex represents the ventricular depolarization (contraction); the Q wave corresponds to the septum depolarizing from left to right and the R wave is the main ventricular depolarization. The T wave is caused by the repolarization (relaxation) of the ventricles. The repolarization of the atria is covered up by the QRS complex. The apex of the T wave toward the ascending slope is the vulnerable period of the cardiac cycle and if an electrical impulse arrives during this period fibrillation can occur. Fibrillation is the involuntary recurrent contract of cardiac muscle, which disrupts the normal sinus rhythm of the heart and results in deficiency in the propulsion of blood from the heart chamber. Atrial fibrillation is indicated by the randomized contraction of the atria, causing an irregular and rapid heart rate. Ventricular fibrillation is indicated by randomized contraction of the ventricles due to the random repetitive excitation of the ventricular muscle fibers without coordinated ventricular contraction. Ventricular fibrillation will lead to the rapid onset of ischemia if a normal cardiac rhythm cannot be reestablished by defibrillating the heart by way of passing a large electric current across the heart to put the entire myocardium in a state where no impulse can occur.

TABLE 13.1   The 12-Lead ECG Definitions

Lead	Measurement
Lead I	Right arm to left arm (bipolar)
Lead II	Right arm to left leg (bipolar)
Lead III	Left arm to left leg (bipolar)
aVR	Right arm (unipolar ref. left arm and left leg)
aVL	Left arm (unipolar ref. right arm and left leg)
aVF	Left leg (unipolar ref. right arm and left arm)
Chest 1	Chest position 1 (unipolar ref. three limb electrodes)
Chest 2	Chest position 2 (unipolar ref. three limb electrodes)
Chest 3	Chest position 3 (unipolar ref. three limb electrodes)
Chest 4	Chest position 4 (unipolar ref. three limb electrodes)
Chest 5	Chest position 5 (unipolar ref. three limb electrodes)
Chest 6	Chest position 6 (unipolar ref. three limb electrodes)

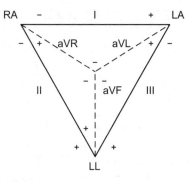

FIGURE 13.4   Einthoven's triangle.

A 12-lead ECG can be used to monitor the heart's electrical activity from 12 different angles by selecting different pairs of electrodes, or combinations of electrodes, through a resistive network to give an equivalent pair, which is referred to as a lead. The 12 leads are summarized in Table 13.1. The first three leads are taken from electrodes placed on the left arm (LA), right arm (RA), and left leg (LL). These three electrodes form Einthoven's triangle (Figure 13.4). Lead I is from RA to LA, Lead II is from RA to LL, and Lead III is from LA to LL. The electrical signal on each lead of Einthoven's triangle can be represented as a voltage source, thus we obtain Einthoven's equation:

$$I - II + III = 0$$

Leads I, II, and III are bipolar as they measure the voltage between two specific points. All other leads in a 12-lead ECG are unipolar as they measure the voltage at one electrode relative to a reference electrode, which is taken from the average of two or more other electrodes. The three unipolar limb leads, known as the *augmented limb leads*, are the voltage difference between one limb electrode and the average of the other two electrodes. Lead aVR is the voltage at RA referenced to the average of LA and LL; lead aVL is the voltage at LA referenced to the average of RA and LL; lead aVF is the voltage at LL referenced to the average of RA and LA. The first six leads I, II, III, aVR, aVL, and aVF are collectively known as the *limb leads*. The remaining six leads in the 12-lead ECG are known as the *chest leads*. The chest leads provide local information and are measured relative to the average of the limb electrodes. The accurate placement of the chest ECG leads is crucial to obtain reliable and repeatable ECG recordings. The incorrect placement of the chest electrodes can lead to artefacts within the measurement, ultimately causing serious misdiagnosis errors within the trace.

The frequency response required to accurately construct an ECG signal lies between 0.05 and 150 Hz. Thus it is not advisable to filter mains frequency as it would remove useful information. DC offsets are removed by using

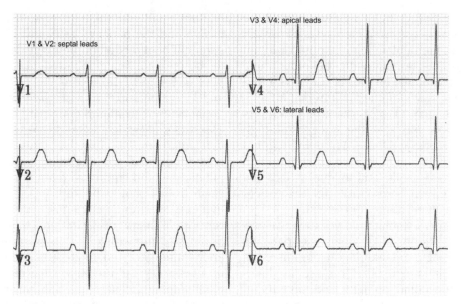

FIGURE 13.5   A simulation of normal chest lead appearance in a 12-lead ECG.

a 0.05 Hz high-pass filter and, likewise, high frequency noise from monitors and radio interference are removed by using a 150 Hz low-pass filter. The most common types of artefacts encountered during ECG measurement are mains noise, baseline wander, and jitter. Although mains noise is reduced by good instrumentation design (driven right leg circuit) it can be loaded on the system from poor connections within the circuit. This could be from electrical connections or high impedance contacts between the electrode and the patient. Baseline wander can be caused by respiration,[5] cable motion, and faulty electrodes. Jitter is caused by unwanted muscle movement causing muscle noise on the ECG trace.

A resting 12-lead ECG is used extensively within a healthcare environment. It is a standard, noninvasive test that can be performed by a healthcare practitioner and then interpreted by a medically trained professional to provide information on the electrical characteristics of the heart. To continuously monitor the patient's ECG in a clinical ward environment, a 1-lead (lead II) can be taken and continuously displayed on a bedside monitor. This is used to monitor trends in the cardiac rhythm and to alarm the clinician to a too slow or fast heart rate. To avoid artefacts caused by mains, motion, and muscle noise, the frequency response would have to be reduced to 0.05−40 Hz. A simulated trace from the six chest leads is shown in Figure 13.5.

Ambulatory monitoring, more commonly referred to as Holter monitoring, is the use of a portable device to record the patient's ECG for a prolonged period of time. This is normally performed over a 24-hour period; however, it can be performed up for up to a 7-day period if clinically indicated. It is used to provide the clinician with long-term information on the patient's cardiac function to help identify the cause of the patient's symptoms, such as chest pain, angina, dizziness, and palpitations.

[5]This is most likely to be observed on the chest electrodes as the chest cavity moves up and down with respiratory swing.

Typically a 3-lead ECG monitor would be used. This will use four electrodes to obtain three leads; the neutral lead is positioned on the top of the sternum with the other three electrodes positioned on the chest to give modified bipolar views of chest electrodes V1, V3, and V5. The monitor is checked for good electrode contacts and that the trace is of high quality and artefact free. When the patient returns to the hospital the recorded trace can be transferred to a PC and analyzed digitally. A report can then be produced for the patient summarizing the findings of the test.

Exercise stress testing involves the measurement of a 12-lead ECG when the patient's heart is being stressed. It can be used in patients with known heart disease to determine the functional capacity and severity of the disease. Diagnostically it can be used to determine between ischemic heart disease and other causes of chest pain. A positive test result would indicate that cardiac ischemia is occurring on exercise and further investigations will be performed. The electrodes are positioned the same as with a resting 12-lead ECG, except the limb electrodes are positioned on the patient's torso to minimize artefacts generated from muscle motion. A standard clinical protocol will be followed where resting ECGs and blood pressure are recorded and then the patient is put on a treadmill. The treadmill will gradually increase in speed and incline and the patient's ECG and blood pressure will be recorded at set intervals according to the protocol being followed. Since the purpose of the test is to record the patient's ECG and blood pressure on a stressed heart, additional safety considerations must be observed. The stress on the heart could lead to a myocardial infarction, so exercise stress testing is contra-indicated for high-risk patients. Best practice would indicate that a defibrillator crash trolley should be kept in the room and that staff performing the tests are trained in cardiopulmonary resuscitation. A test would be stopped immediately if an exercise-induced arrhythmia occurred or if there were systems of altered cardiac function, such as a drop in blood pressure. There are also hazards to the patient from trips of falls while using the treadmill, and a test would be stopped if the patient's gait became unsteady or if they felt they had reached their maximum exertion.

Further Reading

Hoskins, P., Martin, K., Thrush, A., 2010. Diagnostic Ultrasound Physics and Equipment. Cambridge University Press.

Webster, J.G., 2008. Bioinstrumentation. John Wiley & Sons.

Pressure and Flow

Paul A. White* and Richard G. Axell**

*Cambridge University Hospitals NHS Foundation Trust, Cambridge, U.K. and Anglia Ruskin University, Chelmsford, U.K., **Clinical Scientist, Medical Physics and Clinical Engineering, Cambridge University Hospitals NHS Foundation Trust, Cambridge, U.K. and Honorary Visiting Research Fellow, Postgraduate Medical Institute, Anglia Ruskin University, Chelmsford, U.K.

INTRODUCTION

Flow of fluids in the body are essential to life, providing the body with fuel for respiration and transporting and excreting waste products. In the simplest model, flow in a tube such as a small blood vessel or bronchiole can be expressed in analogy to electrical flow:

$$Q = P/R$$

where Q is the volume flow rate, P is the pressure drop, and R is the resistance, given by Poiseuille's law:

$$R = \frac{8\eta l}{\pi r^4}$$

where η is the viscosity of the fluid, l is the tube length, and r is its diameter. Note the very strong dependence on r. This demonstrates the close relationship between pressure and flow: in physiological measurements often a pressure difference will be used to quantify a flow, or vice versa. The above model is very simple, but works reasonably well in laminar flow, which occurs when the Reynolds number is less than about 2000 or so:

$$Re = \frac{2\rho v r}{\eta}$$

where ρ is the fluid density and v is its velocity. Above 4000 the flow is fully turbulent; between these numbers the flow is partially turbulent and partially laminar, known as transitional flow.

In the body, reduced flow of blood or air into the lungs can be due to either an increase in resistance, for example because of obstructive lung disease, or a decrease in pressure gradient, for example in heart failure. This can ultimately lead to ischemia and death. However, absolute pressure also has physiological significance, particularly in the cardiovascular system, in which a high blood pressure can lead to coronary heart disease, stroke, and hypertensive nephropathy.

Physiological measurements of pressures and flows allow the diagnosis and enable the severity of the disease to be assessed.

Clinical Engineering.

BLOOD PRESSURE

Methods of Blood Pressure Measurement

Blood pressure is one of the commonest physiological measurements carried out, as it is straightforward and provides essential information about the health of the cardiovascular system. For most of the twentieth century, routine blood pressure measurements were made by the auscultatory method. This involves occluding the brachial artery by encircling the upper arm with an inflatable bladder attached to a mercury sphygmomanometer. A stethoscope is held over the antecubital fossa to note the cuff pressures at which the first and fifth Korotkoff sounds are heard. These are the systolic and diastolic blood pressures.

For patients in critical and intensive care situations, continuous monitoring is required so that rapid clinical decisions can be made as a result of change in status. Intra-arterial blood pressure is a commonly used as a monitoring tool in these environments. It is measured using a continuously flushed, fluid-filled catheter which is connected to a pressure transducer, often a strain gauge. The physics of fluid-filled catheters for pressure measurement is discussed in more depth in the next subsection. These measurement systems provide a continuous readout of arterial blood pressure, from which systolic, diastolic, and mean arterial pressures can be determined.

In the 1970s, the first automatic blood pressure monitor was put on the market, the Dinamap ("device for indirect noninvasive mean arterial pressure"). This worked on the oscillometric principle, measuring the amplitude of oscillations in the cuff pressure to determine mean arterial pressure. The cuff is pumped up to a certain pressure and the device waits for a predefined time to check for oscillations in cuff pressure. If none are noted, it steps down to the next pressure. Once it does detect oscillations in pressure due to the contractions of the heart, it always waits until it has two close readings for the oscillation amplitude before moving on to the next pressure. This is shown diagrammatically in Figure 14.1. The requirement for two similar oscillations to be detected is used in artefact rejection. An intrinsic and extrinsic artefact are shown in Figure 14.1.

In oscillometric measurements, only mean arterial pressure is well defined, as the pressure at which maximum cuff oscillations occur. Systolic and diastolic pressures are then calculated using proprietary algorithms from the oscillation amplitude at different cuff pressures. Unfortunately, there are two differing standards for oscillometric blood pressure monitors (also commonly known as noninvasive blood pressure (NIBP) monitors). The Association for the Advancement of Medical Instrumentation (AAMI) in the United States compares NIBP to an intra-arterial reference (FDA, 1997), with the requirement that the mean difference be $\ll 5$ mmHg and the standard deviation of the differences $\ll 8$ mmHg. In contrast, the British Hypertension Society advocates the referencing of NIBP to auscultatory measurements. To obtain an A-rating in this system, 60% of readings must be within 5 mmHg, 85% within 10 mmHg, and 90% within 15 mmHg of the reading taken by the auscultatory method (O'Brien et al., 2001).

In principle, these two standards should be complementary, but this assumes that the values of blood pressure measured intra-arterially and by the auscultatory method agree. There is a wide range of scatter in results of studies comparing auscultatory to intra-arterial blood pressure, but the general trend appears to be that the auscultatory method measures a systolic pressure that is too low and a diastolic that is too high compared with the intra-arterial method (Nielsen et al., 1983; Darovic, 1995). Thus, companies have tended to develop more than one

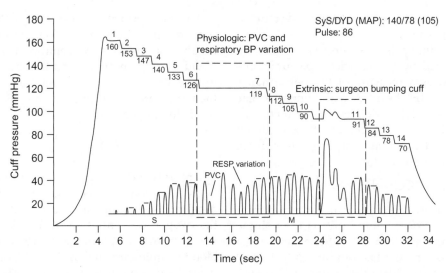

FIGURE 14.1 An oscillometric noninvasive blood pressure measurement with artefacts rejected by the automated algorithm. A premature ventricular contraction (PVC) results in a lower than normal blood volume and so is ignored by the algorithm, along with the following larger than normal compensatory beat. Similarly, the large pressure variation due to the surgeon bumping the cuff is ignored.

algorithm, one referenced to intra-arterial and the other to auscultatory methods. Safety issues specific to oscillometric automated blood pressure monitors are addressed in BS EN 60601-2-30 (BSI, 2000). As well as complying with general electrical safety standards, there are particular requirements due to the hazard of the pressurized cuff around the patient's arm. Specifically, excessive pressure could damage the tissues in the patient's arm and cause significant pain, and cuff pressure being maintained for too long could lead to venous blood pooling. The standard sets out limits for the maximum pressure and the maximum inflated time in both normal and single fault conditions for such devices.

Hemodynamic Monitoring

The physiological state of the heart, heart chambers, and heart valves can be assessed by the direct measurement of cardiac pressures. In the cardiac catheterization laboratory

catheters are inserted into the heart chambers to measure cardiac function to determine if intervention is required, and post intervention to evaluate the outcome of the procedure. A basic monitoring catheter will incorporate a strain gauge or set of strain gauges to measure the change in pressure by applying the Wheatstone bridge principles. A more advanced conductance catheter typically used in a clinical research setting also allows you to measure the volume of blood within the heart chamber, derived from measuring blood conductance. A catheter can also be used to introduce a stent or to inject a radiopaque dye to determine the location of a vascular constriction or plaque. A catheter is typically a flexible tube which can be inserted via a sheath into the narrow opening of an artery or vein and then fed along the vessel into the heart chamber of interest. The catheter is normally introduced into the femoral or internal jugular artery or vein, although a brachial approach may be desired. Cardiac pressure can be measured using a catheter-type[1] system or catheter tip[2] sensor.

Cardiac output is the effective volume of blood expelled by the ventricles per unit time and is used as a measure of the performance and health of the heart. A maintained cardiac output is essential to deliver all the metabolic agents to the body and remove the unwanted waste products. Cardiac output (CO) is given by the following equation:

$$(CO) = HR \times SV$$

where the heart rate (HR) is in beats per minute and the stroke volume (SV) is the volume of blood pumped from a ventricle in a single heartbeat (l). The cardiac output (CO) (l/min) is directly correlated to the size of the patient, by way of standardizing to the body surface area (BSA) (m^2). Cardiac index (CI) is given by the following equation:

$$CI = \frac{CO}{BSA}$$

The ejection fraction (EF) is the volumetric fraction of blood ejected in a single heartbeat, and is given by the following equation:

$$EF = \frac{SV}{EDV}$$

where the end-diastolic volume (EDV) is the volume of blood left in the ventricle at the end of diastole. End-systolic volume would be the volume of blood left in the ventricle at the end of systole.

The Fick principle states that if the oxygen concentration in arterial blood supplying an organ, the oxygen concentration in venous blood leaving an organ, and the rate of oxygen uptake by an organ per unit time is known, then the cardiac output can be calculated by the following equation:

$$CO = \frac{VO_2}{C_a - C_v}$$

where VO_2 is the oxygen consumption (l/min), C_a is the oxygen concentration of the blood sample taken from the artery (l/l), and C_v is the oxygen concentration of the blood sample taken from the pulmonary artery (l/l). The blood samples are analyzed in a blood-gas analyzer to determine the oxygen concentration, and the oxygen consumption is measured by using spirometry. By assuming a value for oxygen consumption, cardiac output can be closely approximated to the actual value without the need for the spirometry equipment. The assumed oxygen consumption is typically calculated from the LaFarge equation:

$$VO_{2\sim males} = 138.1 - (11.49 \log_e age) + (0.378HR)$$

and

$$VO_{2\sim females} = 138.1 - (17.04 \log_e age) + (0.378HR)$$

where $VO_{2\sim males}$ is the assumed oxygen consumption in males and $VO_{2\sim females}$ is the assumed oxygen consumption in females ($ml/min/m^2$).

The pulmonary artery catheter, more commonly known as the Swan–Ganz catheter, allows the direct measurement of right atrial, right ventricular, and pulmonary artery pressures and blood saturations. It also allows the indirect measurement left atrial or wedge pressure by inflating the balloon when positioned in the pulmonary capillary. By inserting a venous sheath into the internal jugular (approach) or femoral (approach) vein, the catheter can be passed along the vein and

---

[1]The blood pressure is measured using a fluid-filled catheter with an open tip. The catheter is of known length and diameter so an external pressure sensor can be attached to the end of the catheter outside the patient's body to measure the blood pressure within the heart chamber. This type of catheter is sensitive to damping and resonance which can be avoided by using the more accurate catheter tip sensor.

[2]The sensor unit is placed at the tip of the catheter and in direct contact with the blood to be measured.

positioned within the right atrium, passed into the right ventricle, then out into the pulmonary artery. After inflating the balloon when the catheter is in the pulmonary capillary, a wedge pressure (PCWP) can be recorded as a measure of the left atrial pressure derived through the atrial wall. The combination of these measurements allows the physician to measure a number of cardiovascular parameters including system vascular resistance index (SVRI), pulmonary vascular resistance index (PVRI), and oxygen consumption (VO_2).

The Swan–Ganz catheter is also used to perform the thermodilution method of determining cardiac output. The catheter is advanced to the pulmonary artery and the balloon is inflated to float the catheter tip into the pulmonary artery. When in situ the balloon is deflated. The tip of the catheter contains a temperature-sensitive thermistor which is used to measure the transient temperature change after injection of a small bolus (10 ml) of cold saline (0.9% NaCl). By measuring the blood temperature at a known distance from the tip of the catheter, cardiac output can be calculated by measuring the resistance change of the thermistor as a function of time as it responds to the pulmonary artery blood temperature change due to the injected bolus of saline. A greater temperature change will correspond to a lower cardiac output. Thermodilution-derived cardiac output measurement would normally be repeated 3 to 5 times and a mean measurement would then be taken as the true cardiac output.

The conductance catheter is a novel technique used predominately within a research setting. The catheter consists of eight electrodes and one pressure transducer (Figure 14.2). Electrodes 1 and 8 are used to generate an intracavity electric field by which the remaining electrode pairs (2-3, 3-4, 4-5, 5-6, 6-7) measure the potential difference across the electrodes and generate a time-varying conductance. This is then used with a measurement of

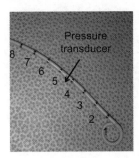

FIGURE 14.2 Millar conductance catheter: the two outermost electrodes (1 and 8) generate an intracavitary electric field; the remaining electrodes (2-3, 3-4, 4-5, 5-6, 6-7) measure the potential difference.

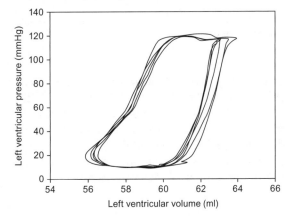

FIGURE 14.3   The pressure–volume loop.

the resistivity of the blood to calculate a volume of blood within the ventricle at that moment in time. The ventricular pressure can then be plotted against the ventricular volume in the form of a pressure–volume loop (Figure 14.3).

The catheter is inserted into both the left and right ventricle to measure beat-to-beat measures of load-independent contractility. By inflating a separate balloon in the inferior vena cava (IVC) to create a preload reduction, a series of pressure–volume loops can be recorded to represent the end systolic pressure–volume relationship (ESPVR), shown in Figure 14.4. The ESPVR describes the maximum pressure that can be generated by the ventricle for any given pressure. The slope of this line represents the end-systolic elastance (Ees) and is an index of myocardial

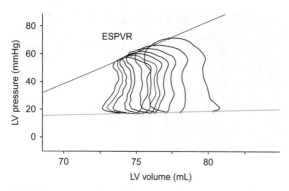

FIGURE 14.4 The end systolic pressure—volume relationship (ESPVR).

contractility. Arterial elastance (Ea) can be calculated from the slope of the line joining the end-systolic and end-diastolic points on the pressure—volume loop. Conductance quickly derives preload recruitable stroke work (PRSW), systolic power, maximum and minimum pressure time derivatives (dP/dT max,

dP/dT min), and isovolumic constant of relaxation (tau).

References

British Standards Institution (BSI). 2000 Medical Electrical Equipment—Part 2-30: Particular Requirements for Safety, Including Essential Performance, of Automatic Cycling Non-Invasive Blood Pressure Monitoring Equipment. BS 60601-2-30:2000.

Nielsen, P.E., Larsen, B., Holstein, P., Poulsen, H.L., 1983. Accuracy of Auscultatory Blood Pressure Measurements in Hypertensive and Obese Subjects. Hypertension. 5, 122—127.

O'Brien, E., Waeber, B., Parati, G., Staessen, J., Myers, M.G., 2001. Blood Pressure Measuring Devices: Recommendations of the European Society of Hypertension. BMJ. 322, 531—536.

Darovic, G.O. (Ed.) Hemodynamic Monitoring: Invasive and Noninvasive Clinical Application. second ed. Saunders, London, 1995.

U.S. Food and Drug Administration (FDA) Centre for Devices and Radiological Health, 1997. Non-Invasive Blood Pressure (NIBP) Monitor Guidance. <www.fda.gov/cdrh/ode/noninvas.html>.

Neurological Measurement

Thomas Stone* and Christine Denby**

*Cambridge University Hospitals NHS Foundation Trust, Cambridge, U.K.,
**Royal Liverpool University Hospital, Liverpool, UK.

Electrophysiological measurements are often used to support the diagnosis and nature of neurological conditions; they are used to infer the structure of the nervous system.

ELECTROMYOGRAPHY

Electromyography (EMG) measures the activity of the nervous system as it manifests during the contraction of skeletal muscles. Movements, voluntary or reflex, originating in the brain or spinal cord will, in the absence of impairment, result in the depolarization of a peripheral nerve which in turn will activate skeletal muscles. The biochemical process by which a muscle contracts results in an electrical field which can be measured either intramuscularly or from the surface of the skin.

The electrical signal measured during electromyography originates from the depolarization of the sarcolemma, a gated plasma membrane that responds to the release of the neurotransmitter acetylcholine across the synaptic cleft, the gap between the muscle fiber membrane and the nerve that innervates it.

The muscle or fascia is formed of many muscle fibers or fascicles, which are formed of many myofibrils, a series of contractile elements connected end-to-end. Nerve fibers contain efferent motor nerves. These are distinct from the sensory or afferent nerves that are also present, which originate within the gray matter ventral horn of the spinal cord and are contiguous with the alpha motor neuron—an all-or-nothing trigger that summates the huge number of inhibitive and excitory inputs from the spinal and supraspinal centers. The alpha motor neuron, the nerve, and the muscle fiber it innervates is termed the *motor unit*. This is an important insight into the complexity of the electromyographic signal because any fascia, or collection of fascicles, may have many motor units and many fiber innervations. Therefore, many moving electrical fields generated by, almost certainly, phasically dissimilar and geographically distributed sources form any single electromyography signal: the principle of superposition.

$$F(x_1 + x_2 + x_3 + \cdots + x_n) = F(x_1) + F(x_2) + F(x_3) \\ + \cdots + F(x_n)$$

Electromyogram can be measured using essentially two different types of electrode: in-dwelling and surface. Whereas a fine wire or needle electrode will be influenced by only a small number of motor units, possibly only one, the surface electrode will capture the resultant electrical field from the multitude of motor units that make up any one muscle. Thus one technique is very selective, reflecting localized activity, and the other provides a generic view of the various muscle types within the heterogeneous muscle.

Electrode Construction

Electrodes are used to transduce the electrical energy within the muscle to electrical energy within the recording system, the wires of the recording electrode. Electrodes for biopotential are typically constructed as nonpolarizable electrodes and commonly out of a silver/silver chloride (Ag/AgCl). Nonpolarizable electrodes allow the free passage of electrons across the electrode/electrolyte interface and do not rely on changes in charge distribution (capacitor-like action) which can be varied significantly by external factors such as electrode movement. Ag/AgCl forms a very stable half potential; the electrodes are constructed of a silver plate with an oxide layer that is in contact with an electrolyte impregnated sponge. With in-dwelling electrodes the intracellular fluid acts as the electrolyte.

A bipolar or differential amplifier amplifies the difference between its two inputs referenced to a common electrode. The differential amplifier has a low common mode gain, amplification of a signal appearing at both terminals at the same time, and a high differential gain. The ratio of common mode to differential gain forms the common mode rejection ratio; for a high quality biopotential amplifier for surface electromyography (sEMG) this should as high as possible.

$$CMRR = 20 \log_{10}\left(\frac{A_{diff}}{A_{com}}\right)$$

EMG and Nerve Conduction Studies

Nerve conduction studies are used to query the integrity of the peripheral nervous system, specifically by evoking an action potential within a nerve and recording its propagation to the muscle it innervates or to another part of the nerve.

Conduction studies can take the form of nerve, motor, or mixed studies. In the case of a motor study an innervating nerve is stimulated at two points proximal to the muscle. The difference ∂t between the time $t1$, for the stimulation to propagate from the most proximal point to the muscle, and the time $t2$, for the stimulation to propagate from the distal point to the muscle, is the conduction time. The conduction velocity, therefore, is the distance between the two stimulation points divided by the time difference.

Alternatively, it is possible to measure the conduction of the sensory nerve by stimulating a distal branch of a nerve and measuring the compound nerve action potential orthodromically from a proximal position. Equivalently, it is possible to make an antidromic measurement by stimulating proximally and measuring distally. As the nerve action potential is being measured directly and there is no delay introduced by the muscle depolarization, the conduction time can be calculated by dividing the time for the stimulus to propagate by the distance between stimulating and recording electrodes.

Nerve conduction velocities are highly repeatable measurements and can be related to standard data sets. However, there are several factors that can modify the conduction velocity, such as temperature.

Responses that are involved with the reflex arc can also be assessed using a combination of evoked response and biopotential measurement. Therefore, these techniques can extend the diagnostic potency from the peripheral nervous system to the spinal cord. The pattern of the electromyogram within quiescent or voluntarily active skeletal muscle can reveal the presence of disease.

Kinematic EMG

The use of electromyography in the analysis of isokinetics and kinetics requires a different set of analysis tools. The typical use is in biomechanics to understand muscle recruitment mechanisms, muscle fatigue, and activation patterns, however, this is still an interrogation of the nervous system.

Equipment for kinematic analysis, although very similar in principle, is often required to measure from a larger number of neurological signals though may still focus on a single motor unit. The systems must be worn by the patient during the measurement time and during activity such as walking. Therefore there are different challenges in capturing viable signals during this clinical scenario. The equipment can be formed of wireless sensors allowing more free movement of the limbs and reducing the chances of movement artefacts being induced by the movement of wires or wires tugging the sensors themselves. When wires are present they are often designed to be very thin and flexible and very well screened to reduce noise being induced onto the signal. Commonly these systems have a small patient-worn unit which digitizes the signal information and structures it for communication to a main signal processing and acquisition unit. In addition, these systems must allow the synchronization of other measurement systems

such as video or limb joint angle measurements.

Commonly the electromyography is presented as a rectified and filtered linear envelope. This has the advantage of being easier to interpret by eye but it also permits calculating an ensemble average and amplitude normalization of multiple dynamic signals.

Rectification for a set of n discrete samples creating a signal S_r is defined:

$$S_r = \sum_{i=0}^{n} |x_i|$$

Additionally, the root mean squared (RMS) value for a set of n discrete samples is defined by

$$S_{rms} = \sqrt{\frac{\sum_{i=0}^{n} x_i^2}{n}}$$
$$= \sqrt{\langle x \rangle^2}$$

where in this nomenclature $<.>$ denotes the mean.

Creation of the linear envelope can use many different filtering techniques but commonly a Butterworth or Bessel filter is used. These filters have maximally flat pass bands with rapid attenuation beyond the unwanted frequencies. Caution must be used if time lags are introduced and the signals are to be compared temporally.

Placement of electromyography sensors is of critical importance, as surface sensors' orientation and position with respect to the muscle will greatly impact the signal measured. The propagation of muscle action potential is along the length of the muscle fiber. As the differential amplifier has greater differential gain for a differential amplifier to operate appropriately, the line of electrodes should be parallel to the line of propagation of the motor unit action potential (MUAP).

At the innervation zone, the region where the spinal nerves innervate the muscle, the electrical field can be less predictable and thus electrodes should be kept away from these areas. A general rule of thumb is to place electrodes over the belly of the muscle, however, good guidelines have been developed by the International Society of Electrophysiology and Kinesiology (ISEK) and SENIAM (Surface Electromyography for the Non-Invasive Assessment of Muscles) to guide good electrode placement. Finally, it is inadvisable to place electrodes near the edge of a muscle, for two reasons: 1) this increases the chances of cross talk from other muscles significantly affecting the signal quality, and 2) in a related way, the electrical field generated by the MUAP is less consistent in this area.

Possibly the most important difference between isokinetic and kinetic analysis of electromyography is that the kinetic instance is likely to be a nonstationary signal. A stationary signal is one that has time invariant statistical properties. The EMG signal can be modeled on a stochastic process or, rather, a signal with values that can be defined probabilistically; they can be considered random (i.e., not deterministic). When the function that defines the probabilities remains constant for any time in the signal, the signal is stationary. However, if the probabilities vary over time, the signal is nonstationary. So, for example, a stochastic process is stationary when its mean (first moment) and variance (second moment about the mean) is a finite number that is not dependent of time:

$$\mu_x(t) = \mu_x(s + t)$$

The significance of this property becomes apparent in the analysis of frequency of a dynamic and isokinetic EMG signal.

Muscle fatigue manifests itself in a frequency shift in the EMG signal; the dominant frequency moves to a lower frequency bin. Frequency changes in EMG can reflect, among other things, a change in dominant action potential shape or change in muscle action potential conduction velocity; muscle types have different frequencies and speeds of depolarization and different muscle types are recruited in different ways in the response to movement and load. Pathology can adapt the frequency spectrum of an EMG signal due to the modification to the heterogeneity of muscle fibers and the predominance of often slower muscle fiber types.

Frequency analysis using familiar tools such as the Fourier series relies on some fundamental assumptions about the signal under analysis: continuous, infinite, and stationary. The reason for this is that the basic function of the Fourier method is a sinusoid and, thus, unless windowed in some way, it exists for all time: over the period of the whole signal under investigation. If a particular component of a signal exists at one frequency for all time then there is not a problem; however, what if a particular frequency component only exists for a finite time? Given this situation how are we to know when the "event" occurred if our basis function is infinite?

Observing the time series from an isokinetic exercise during testing on an isokinetic dynamometer we can measure the stochastic and wide-sense stationary[1] nature of the electromyogram. However, if we were to observe the EMG signal measured from the kinetic assessment of the quadriceps during walking, the requirements for a stationary signal are no longer met and we would be wise to use a different method for analyzing the frequency characteristics of the signal.

Signal analysis provides many opportunities to analyze the frequency characteristics of

---

[1]Many signals are stationary over their first two moments, i.e., mean and standard deviation about the mean, but not necessarily for any higher moments. These signals are considered wide-sense stationary. A truly stationary signal must be stationary over all moments.

these types of signals; two commonly applied techniques to electromyography are short-time Fourier transforms (STFT) and wavelet analysis.

EVOKED POTENTIALS

Electroencephalogram

Electroencephalography (EEG) is a technique that aims to detect the electrical activity of the active nerve cells in the brain, which is the top 1.5–4 mm of cerebral cortex. Measurements are made using surface electrodes (Ag/AgCl). The resultant voltage difference that can be measured on the scalp is of the order of 100 μV. Cortical measurements can be measured at approximately 1–2 mV on brain surface. In nerve cells, ionic potentials are present due to differences in the concentration of the ions. These are mainly sodium (Na$^+$), chloride (Cl$^-$), and potassium (K$^+$). The cell wall is a semipermeable membrane that is more permeable to K$^+$ and Cl$^-$ than Na$^+$. A sodium-potassium pump keeps Na$^+$ outside the cell and K$^+$ inside. The rate of sodium pumping is greater than potassium pumping resulting in a difference in ionic potential. The resting potential is −70 mV.

There are two different techniques in EEG measurements: unipolar and bipolar. Unipolar measurements are recorded with respect to a common reference. The main emphasis is on frequencies and amplitudes of the EEG signal. In bipolar measurements, recording is achieved from a series of electrodes in which the input of each amplifier is connected to the output of the next one. It is used mainly for visualizing phase reversals (loci of amplitude maxima) which is good for detecting epileptic foci. (See Figure 15.1.)

The EEG is typically described in terms of rhythmic activity and transients (Figure 15.2). Rhythmic activity is divided into 4 bands by

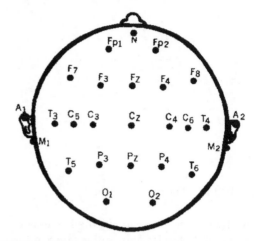

FIGURE 15.1 Electrode positions based on the International 10–20 Standard. Cz: Central vortex electrode, F: Frontal, C: Central, T: Temporal, P: Parietal, O: Occipital, M: Mastoid, N: Nasion.

frequency. Most of the cerebral signal observed in the scalp EEG falls in the range of 1–20 Hz; amplitudes in the adult are approximately 10–100 μV. Transient waves are classified as abnormal waveforms having specific wave shapes and have short time nonstationary properties. They are common in epilepsy (which has phenomena such as the single spike and the spike/wave complex). The 4 rhythmic bands are:

1. Delta waves: These are high-amplitude waves in the frequency range of up to 4 Hz. They are located frontally in adults and posteriorly in children. They occur during slow wave sleep in adults. In babies they are found during tasks that need continuous attention.
2. Theta waves: These are in the frequency range between 4 and 8 Hz. They are located in areas not related to the current task at hand. They occur in young children during drowsiness and lucid dreaming or arousal in older children and adults.

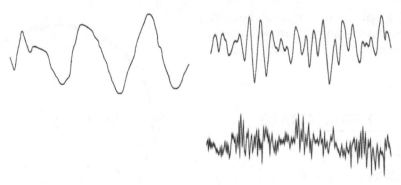

FIGURE 15.2   Rhythmic EEG bands.

3. Alpha waves: These are in the frequency range of 8–13 Hz. They are located in the posterior regions of the head, both sides, and central sites at rest. They are normally associated with relaxation, closing of eyes, and inhibitory activity in different locations across the brain. Pathologically, they are found in coma patients.
4. Beta waves: These are in the frequency of greater than 13 Hz. They occur on both sides of the brain, having a symmetrical distribution, most evident frontally, and are low amplitude waves. They occur when the subject is alert (e.g., working), when active, busy or anxious, and in active concentration. Pathologically, they are found in subjects with drug toxicity.

Pattern Shift Visual Evoked Potentials

Pattern shift visual evoked potentials (PSVEPs) are used to investigate the visual pathways and is commonly achieved through the reversal of black and white checkerboard images. While ensuring that other parameters such as luminescence, contrast, and color are not changed, pattern reversal is quite simply making all the white squares become black and all the black squares become white.

There are alternative approaches by using, for example, a strobe light or an image displayed for a short period of time; however,

due to better consistency of results and better discrimination of pathological results, pattern shift is most commonly used clinically.

The angle of the stimulus, the distance, and the location are very important factors affecting the resulting signal recorded from PSVEP. The PSVEP signal is recorded from the scalp at positions Fz, Oz, and Cz using A1 and A2 as reference points. These represent the standard locations for EEG electrode placement (Figure 15.1).

The signal produced during a pattern change reveals a strong positive response approximately 100 ms after the stimulus; this is called P100 using the naming nomenclature of the polarity of the wave and the latency. Diagnostically the latency and, to a lesser extent, the amplitude of the P100 are used; these factors can be dramatically affected if the angle subtended by the visual stimulus changes much beyond 10°. In fact, it has been shown that the amplitude of the P100 signal can be reduced by as much as 80% if the central 10° of the visual field is occluded.

PSVEP is most commonly conducted monocularly as differences in latencies between left and right can give diagnostic information. Even when both results from left and right eyes may be within normal ranges the difference can reflect damage to the nervous system. Systemic demyelization of the nerves, for example, within the optical system will slow down action potentials and produce increased latencies in the P100.

Brainstem Auditory Evoked Potentials

Brainstem auditory evoked potentials (BAEPs) use many hundreds, possibly 1000 or more, auditory "click" stimuli to elicit a response in the eighth cranial nerve and auditory centers of the midbrain.

The signal produced during BAEP characteristically has five peaks reflecting the different contributions of the eighth cranial nerve, cochlear nucleus, superior olivary complex, and colliculus and lemniscus. However, the true origins of the signals are still under conjecture.

Wave 1 is considered to be the response predominately from the eighth cranial nerve and is measured from the earlobe. Wave 2 and Wave 3 are most probably from activity originating in the cochlear nucleus and superior olivary complex; though Wave 2 is thought also to be produced by the proximal part of the eighth nerve. Waves 4 and 5 are considered to be a summation of the activity of multiple structures in the ascending lower auditory pathway. These waveforms are thought to include activity from the lower lemniscus and inferior colliculus. Waveforms 2−5 are best captured with electrodes placed on the vertex, Cz.

BAEP allows an analysis of the structures in the midbrain and areas close to the auditory tract. Each waveform is related to an anatomical region. Disruption of different areas by, for example, a tumor may be reflected as a disruption to one of the typical waveforms, though it is important to note that an abnormal BAEP does not relate itself to any specific cause; this is only elucidated by the wider clinical picture.

Short Latency Somatosensory Evoked Potentials

Somatosensory EPs are those sequentially generated by different neural structures in response to some stimulus. In the upper limb the medial, ulner, or radial nerve is often used; in the lower limb the tibial or peroneal nerve is used. The stimulus can take the form of muscle stretch, tap, or electrical stimulation; electrical stimulation being the preferred method. The stimulator provides a square wave stimulation pulse of frequency in the range $1\,Hz \leq f_{stim} \leq 100\,Hz$ with pulse widths of $100-200\,\mu S$ and a maximum current of $50\,mA$, though a current of $15-20\,mA$ is more common.

The latencies of the evoked response are measured at different points along the neural pathway.

Measuring locations are commonly the popliteal fossa, L1 and T12, and Erb's point (lateral route of the brachial plexus). Upper spinal and cortical regions are measured between C5 and Fz and Cz to C5. The areas of the premotor cortex (on the ipsilateral side of the stimulated limb) and the central scalp area over the primary motor cortex also provide a location for measuring the evoked response.

Up to 1000 repeated stimulation are ensemble averaged to form a single observation, and several observations are routinely made to provide knowledge of measurement reliability.

Respiratory

Elizabeth M. Tunnicliffe* and Paul A. White†

*University of Oxford Centre for Clinical Magnetic Resonance Research, John Radcliffe Hospital, Oxford, UK,

†Cambridge University Hospitals NHS Foundation Trust, Cambridge, U.K. and Anglia Ruskin University, Chelmsford, U.K.

INTRODUCTION

The lung function laboratory carries out a number of tests to assess pulmonary function and assist in diagnosing pulmonary disease. Among others, these include tests that assess the ability of the lungs to oxygenate blood under stress; for example, the six-minute walk and hypoxic challenge, and tests that assess the body's reaction to allergens, such as histamine provocation and skin allergy tests. The three tests carried out most frequently are spirometry, full body plethysmography, and gas transfer. These three tests are discussed in this chapter.

LUNG VOLUMES AND PHYSIOLOGICAL PARAMETERS

Measurement of lung volumes is important clinically as many pathological states change specific lung volumes or their relationships to each other (Cotes et al., 2006). The lung can be divided into four irreducible volumes, which can then be combined to form various different capacities, as shown in Table 16.1 and illustrated as part of a spirometric measurement in Figure 16.1.

Other parameters that are measured in lung function testing include FEV_1, the volume of air expired in 1 s at maximal effort (forced expiratory volume); FVC, the forced vital capacity (i.e., the VC measured when the breath is being forced out of the lungs); PEF, the peak expiratory flow in a maximal effort expiration; and $MMEF_{75-25\%}$, the mid-maximal expiratory flow, or average flow rate from 75% to 25% of the FVC (also sometimes known as $FEF_{75-25\%}$). In diffusion, measurements are made of $T_{1,CO,sb}$, the transfer factor for carbon monoxide in the lungs measured by the single-breath method; and $K_{CO,sb}$, the same transfer factor normalized by the alveolar volume, V_A.

VOLUME CONVERSIONS

Ambient air and that in the lungs are at different temperatures and aqueous vapor partial

243

TABLE 16.1   Definitions and Abbreviations of Lung Volumes and Capacities Measured During Lung Function Testing

Volume/Capacity	Abbreviation	Description
tidal volume	TV	volume inspired and expired in quiet breathing
expiratory/inspiratory reserve volume	ERV/IRV	volume that can be in-/expired above/below that contained in the lungs at the end of a normal tidal ins-/expiration
residual volume	RV	volume contained in the lungs at all times due to the elasticity of the alveoli
inspiratory capacity	IC	IRV + TV, not often used
functional residual capacity	FRC	ERV + RV, volume left in the lungs at the end of a tidal expiration, when the lungs are in equilibrium
vital capacity	VC	IC + ERV, maximum volume that can be inhaled (IVC) or exhaled (EVC) in one breath
total lung capacity	TLC	VC + RV, the total volume of the lungs

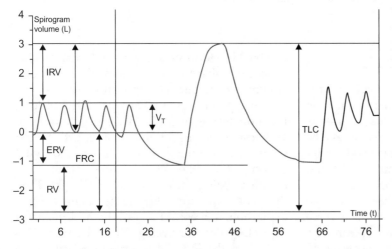

FIGURE 16.1   A screenshot of typical output for unforced spirometry. RV, and thus FRC and TLC, are estimates based on reference values.

pressures. Thus, when measuring volumes in lung function it is necessary to know the current conditions of the gas and which conditions the volume is usually reported at.

Air in the lung will be at body temperature and saturated with water vapor. At equilibrium, it will be at ambient pressure. This is referred to as BTPS (body temperature and pressure saturated). Ambient temperature and pressure is known as ATP, which includes the partial pressure due to the ambient humidity. This can be corrected for to give the dry ATP (ATPD). Volumes are sometimes quoted at "standard" temperature and pressure (STP), usually 0°C and 760 mmHg. The lung volumes referred to in Table 16.1 are almost universally quoted at BTPS. An equation, based on the ideal gas law, is used to convert between ATP and BTPS:

$$V_{\text{BTPS}} = \frac{310(P_{\text{amb}} - P_{H_2O,S}(t)H)}{T_{\text{amb}}(P_{\text{amb}} - P_{H_2O,S}(37))} V_{\text{ATP}},$$

where P_{amb} is ambient pressure, T_{amb} is ambient temperature (in K), H is relative humidity, and $P_{H2O,S}(t)$ is the saturated partial pressure of water at temperature t (measured in °C).

LUNG CONDITIONS

In the lung there are two main forces acting on the airways and alveoli. The first is the elastic recoil of the lung tissue, which alone would lead to the collapse of the lung. However, this is opposed by the outward recoil of the chest wall, which holds the lungs open. At the end of relaxed expiration (at FRC), these two forces are in equilibrium. During inspiration, the inspiratory muscles (primarily the diaphragm, along with the external intercostals) increase this outward force, leading to the expansion of the lung tissue and the inflow of air, at a rate controlled by the airway resistance, into the alveoli. In so-called "quiet" expiration, these inspiratory muscles relax and the lung returns to its equilibrium volume, driven by the elastic recoil of the lung parenchyma. In forced expiration, however, the expiratory muscles (the internal intercostals and abdominal muscles) are used to increase the force inward on the lungs to expel air more rapidly, and enabling expiration down to the residual volume (RV). In lung disease, these forces are altered, and lung function testing is used to measure these changes.

In most lung function laboratories four main conditions are most commonly encountered. These are asthma, emphysema, chronic bronchitis, and pulmonary fibrosis. Emphysema and chronic bronchitis are often seen together in the same patient and have similar symptoms, so are often grouped together as chronic obstructive pulmonary disease (COPD). Asthma, emphysema, and chronic bronchitis are examples of obstructive lung conditions, whereas fibrosis is an example of a restrictive lung condition. Table 16.2 shows the effect of these diseases on the frequently measured physiological variables.

The hallmark of an obstructive disease is an increase in the airway resistance, R_{aw}, due to a narrowing of the airways in the lung. This is reflected in the low values of FEV_1/FVC. In asthma, this is due to contraction of the smooth muscle lining the bronchioles. In chronic bronchitis, the narrowing occurs because of inflammation and excess mucous buildup. Emphysema causes airway narrowing because of the destruction of the alveolar walls. This reduces the elastic recoil of the lung and the bronchioles collapse slightly. Forced expiratory maneuvers exacerbate this effect and lead to dynamic compression, in which the inward pressure due to the expiratory muscles collapses the small airways because the reduced elastic recoil is not sufficient to keep them open. The reduced elastic recoil also leads to an increase in total lung capacity (TLC).

Restrictive diseases are characterized by a decrease in lung volumes, usually due to an

TABLE 16.2 Typical Impact of Different Conditions on Physiological Measurements of the Lung (N Indicates Normal, ↑ an Increase, and ↓ a Decrease)

	RV	FRC	TLC	FEV₁/FVC	$T_{l,CO,sb}$	R_{aw}
Asthma	↑	↑	↑	↓	N or ↑	↑
Emphysema	↑	↑	↑↑	↓	↓	↑
Chronic bronchitis	↑	↑	↑	↓	N or ↓	↑
Fibrosis	↓	↓	↓	N or ↑	N or ↓	N or ↓

increase in alveolar elastic recoil because of an excess of fibrous tissue in the lung. This extra elastic recoil can lead to an increase in FEV_1/FVC, and by holding the airways open, a decrease in R_{aw}.

REFERENCE VALUES

To compare whether the lung function values measured for a particular individual are normal or not, it is necessary to know what constitutes normal. For this purpose, reference values are used, which consist (for adults) of equations based on height and age, with separate equations for men and women. These are based on measurements taken on large healthy populations, usually excluding smokers. The primary tables in use have been published by the European Community for Coal and Steel (ECCS; Quanjer et al., 1983). However, ethnic origin also has an impact on the values, with non-Caucasian subjects having smaller lungs, on average, for a given height (Cotes et al., 2006). However, the variation with ethnicity is large, and also depends on factors such as diet, customary activity level, and habitual altitude. Thus, care is required when using reference values for non-Caucasian subjects. In practice, for routine clinical use, the software used to analyze lung function test data has a set of reference variables that provide a reasonable approximation for all non-Caucasians.

The upper and lower limits in these normal ranges are based on the mean $\pm 2 \times$ standard deviation. This means that one in 20 subjects is likely to fall outside this range and must be borne in mind when assessing test results.

SPIROMETRY

Spirometry comes from the Greek language and means the measurement of breath. Specifically in spirometry, lung volumes and flows are measured. In most modern equipment, spirometers actually measure flows which can then be integrated to obtain the lung volumes. There are two main types of airflow measurement device (pneumotachometer) used in spirometry: Fleisch- and Lilly-type pneumotachometers (Cotes et al., 2006). Both consist of resistive elements placed into the flow. Assuming laminar flow, the pressure drop across the resistive element is proportional to the flow rate.

In a Fleisch pneumotachometer, the resistive element consists of a bundle of fine capillary tubes. Their small radius reduces the Reynolds number so that the flow is laminar up to higher flow rates. The pressure is measured near the middle of the capillary bundle.

In a Lilly pneumotachometer, the resistive element consists of one or more mesh screens, with the pressure measured on either side of the screen(s). In this case, the range of flows that remain laminar is increased by tapering the tube outwards. For a given flow rate $Q = \pi r^2 v$, an increase in r will lead to a drop in gas speed v. Substituting back into the Reynolds number we find that, all other things being equal, $\text{Re} \propto 1/r$, so that an increase in the tube radius leads to a drop in the Reynolds number and the flow is more laminar. Generally the Fleisch-type are linear up to higher flow rates and more accurate (Cotes et al., 2006), but the Lilly-type are cheaper and easier to clean, making them more practical for hospital use. Pneumotachometers are heated typically to 32°C to avoid water vapor from the saturated breath condensing on the screen and disturbing the airflow.

A typical spirometric output is shown in Figure 16.1, with the lung volume plotted as a function of time.

As well as measuring the lung volumes and flow parameters, spirometers can provide a graphical output of flow rate plotted against lung volume, known as a flow-volume curve. Particularly useful are forced flow-volume

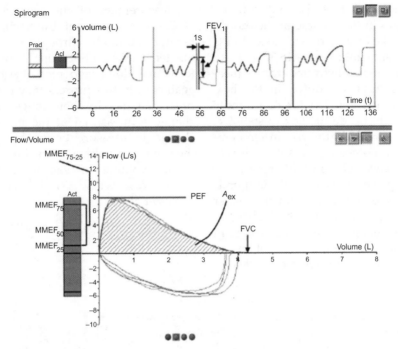

FIGURE 16.2 Flow-volume curves for normal lung function along with accompanying spirograms.

curves. A screenshot from the spirometry software is shown in Figure 16.2, illustrating the relationship between spirograms and the flow-volume curve. Forced flow-volume curves can provide useful feedback when assessing lung function as different conditions often have very distinct curve shapes.

Spirometry and other pulmonary function tests have to be carried out in a standardized fashion, as otherwise there could be large variations between centers, operators, and patients. European guidelines (Miller et al., 2005) exist to facilitate standardization. Lung function tests are unusual as they can be quite technically demanding for the patient, who needs enough coordination to carry out the tests successfully. The operator therefore acts almost as a coach, enabling the patient to perform the maneuver so that useful measurements are taken.

Spirometers are calibrated using a large syringe to inject a known volume of air into the instrument, and values of temperature, pressure, and humidity are entered into the system for volume correction purposes. Infection hazards are minimized by using separate mouthpieces for each patient, with a filter that is stated to remove 99.999% of bacteria and viruses. The pneumotachographs are taken apart and sterilized weekly.

WHOLE BODY PLETHYSMOGRAPHY

Whole body plethysmography is the measurement of changes in volume of the body due to differing amounts of air in the lungs. It enables the estimation of both the airway resistance and functional residual capacity (FRC) (and thus TLC). Measurements are carried out in an airtight box containing a pneumotachometer. Flows, or pressure changes at the mouth when the tube is shut off, are related to changes in the volume of the thorax. A screenshot of the airway resistance and FRC measurement is shown in Figure 16.3.

During a full body plethysmography measurement, the subject sits down in the box, which is then sealed and left for one minute for the temperature inside the box to stabilize. Any excess pressure from the rise in temperature due to the presence of the subject leaves via a high-resistance tube so that there is no pressure buildup. The resistance of the tube is such that the time constant is much greater than the breathing period. Then the higher frequency pressure changes due to the change in the subject's thoracic volume are not excessively diminished by the presence of the leak.

Measurement of the thoracic gas volume (TGV) is carried out by having the subject breathe normally through the pneumotachometer. At the end of a tidal expiration, a shutter comes down and the subject tries to pant against it. In this process they rarefy and compress the gas in their lungs. This pressure change is measured at the mouth, while the change in pressure of the box is related to the change in volume of the lungs.

The volume of gas is measured using Boyle's law where PV is a constant. Then (Cotes et al., 2006):

$$PV = (P + \Delta P)(V - \Delta V)$$
$$= PV + V\Delta P - P\Delta V - \Delta P\Delta V$$

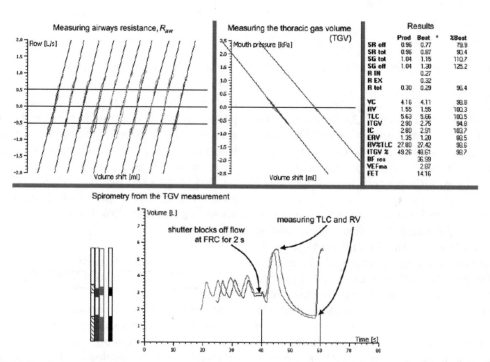

FIGURE 16.3   A screenshot of measurements of healthy volunteers using a full body plethysmograph. The linear fits are offset for clarity. Top left are the linear fits for the determination of airway resistance, top middle shows the linear fits for the TGV measurement, and top right are the numerical values. The TGV is sometimes called the ITGV, the intrathoracic gas volume. The horizontal axes are labeled "volume shift" and are equivalent to $P_{box}C_{box}$. The bottom panel shows the spirometric output from the pneumotachograph.

By cancelling, rearranging, and assuming that $\Delta P \Delta V$ is small, we obtain

$$V_L = P_{\text{atm}} \frac{\Delta V_L}{\Delta P_{\text{mouth}}}$$

$$= P_{\text{atm}} \frac{\Delta P_{\text{box}}}{\Delta P_{\text{mouth}}} C_{\text{box}} \left(\frac{V_{\text{box}} - M_{\text{body}}/1.07}{V_{\text{box}}} \right),$$

where $C_{\text{box}} = V_{\text{box}}/P_{\text{box}}$ is the compliance of the box, measured at calibration, and the term in parentheses corrects the volume of gas in the box compared to that at calibration, by subtracting the subject's estimated volume. Their mass, M_{body}, is known and 1.07 kg m^{-3} is the average density of the human body.

The changes in box and mouth pressure are monitored as the subject makes small panting attempts against the shutter. The gradient of the straight line plot of ΔP_{box} against ΔP_{mouth} is inserted into the above equation to obtain the volume of the lungs at the time of measurement, the TGV. This corresponds to the subject's FRC. Once the shutter is open again, the subject takes a breath into TLC, out to residual volume (RV) and back up to TLC. This enables measurement of all the lung volumes. Cancelling the term $\Delta P \Delta V$ depends on the volume and pressure differences being small. The subject attempting to pant against the shutter ensures this. The panting rate should not be too high ($>1 \text{ Hz}$), however, as otherwise the airway resistance means that the assumption that $P_{\text{mouth}} = P_A$, the alveolar pressure, does not hold. Even at lower breathing frequencies, this can be a problem for very obstructed patients and can lead to an overestimate of lung volume.

In a clinical study, the first measurement taken is that of the airway resistance. The subject breathes shallowly and quickly through the pneumotachometer, which is open to the air in the box, with a frequency of 30 to 60 breaths per minute. Data are collected until 10 consistent breaths are obtained.

If there was no airway resistance, there would be no change in pressure of the box during panting as air would flow instantly into the lungs as a result of the change in volume of the thorax. However, a pressure gradient is required to shift the air into the lungs, so that when the thorax expands to inspire, there is a pressure rise in the box. Using the relation between P_{mouth} and P_{box} measured during the TGV maneuvers and remembering that we assumed $P_{\text{mouth}} = PA$, the airway resistance is given by

$$R_{aw} = \frac{\Delta P_A}{Q}$$

$$= \frac{\Delta P_{\text{mouth}}}{\Delta P_{\text{box}}} \frac{\Delta P_{\text{box}}}{Q}.$$

By measuring the flow rate through the pneumotachometer, Q, and the change in box pressure, and taking the inverse of the gradient of the straight line obtained, we can calculate the airway resistance. The equation assumes that the air entering and leaving the lungs is all at BTPS. This is not strictly true, but is made closer to being so by panting with low flow rates, so most of the air remains in the pneumotachometer (Figure 16.4). In practice, corrections are made for this error electronically. R_{aw} also changes with lung volume (as the caliber of the airways changes) and with flow rate, so panting also helps to keep the curves linear.

A linear fit associates the Q–P_{box} curves to measure R_{aw}, referred to in Figure 16.3 as R_{tot}, in the top left pane and P_{mouth}–P_{box} curves. The resistances are therefore measured by simply plotting a straight line through the two volume extremes, and that gives R_{tot}. Multiplying R_{tot} by the volume at which the measurement is made gives SR_{tot}, the specific airway resistance. Calculating SR_{tot} removes almost all the dependence of resistance on lung volume. The inverse is SG_{tot}, the specific airway conductance. Another measure is to

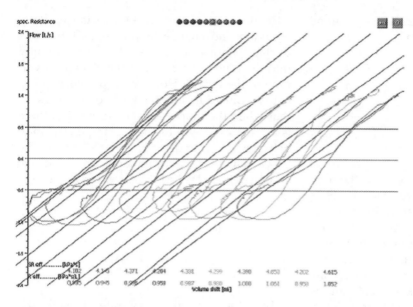

FIGURE 16.4 Showing the effect of early airway closure on the measurement of airway resistance.

take the average gradient over the whole loop, which gives $_sR_{eff} = 1/_sG_{eff}$, the effective specific airway resistance and conductance. This value relates to the actual work of breathing the subject experiences.

$_sR_{eff}$ and $_sR_{tot}$ usually only differ significantly if there is substantial noise present in the measurement, which artificially increases $_sR_{eff}$. Because of this susceptibility to error, $_sR_{tot}$ is usually the most reliable measure clinically.

The box used for full body plethysmography is transparent, making the test less intimidating for subjects. There is a handle on the inside of the box so the subject can open the door at any time should they feel the need. However, full body plethysmography cannot be carried out with very claustrophobic subjects. A vent in the bottom of the box is opened while the subject is in the box, except when tests are taking place. This helps to stabilize the temperature and ensures the oxygen concentration remains normal.

As mentioned earlier, a calibrated leak in the side of the box ensures no pressure buildup during tests. If the door handle should break, there are tabs on the door panel so that the frame can be removed for the subject to exit the box.

Calibration of the plethysmograph has two stages: Firstly, the leak decay time is checked by pumping a known volume of gas into the box. The decay in pressure is fitted to an exponential and ensures that any reduction in the pressure readings due to the leak can be accounted for. Secondly, a calibrating syringe pumps a small volume of gas sinusoidally in and out of the box while the pressure is monitored. This enables the compliance of the box to be measured. Each test is carried out three times and the median value for the time constant and compliance are used.

GAS TRANSFER

Gas transfer measures how well gases can diffuse into the bloodstream and bind with the hemoglobin in the erythrocytes. It is quantified using the transfer factor T_l, sometimes also known as the diffusing factor. The transfer factor can be expressed as (Cotes et al., 2006)

$$\frac{1}{T_1} = \frac{1}{D_m} + \frac{1}{\Theta V_c},$$

in which D_m is the diffusing capacity of the alveolar membrane, and ΘV_C quantifies how well the gas is taken up by the blood once it has diffused across the membrane. Diffusion across a membrane is governed by $D \propto A/t$, where A is its area and t its thickness. Therefore diseases that increase the thickness of interstitial lung tissue, such as fibrosis, or diseases that decrease the alveolar area, such as emphysema, will decrease D_m. The hemoglobin concentration in the blood will affect ΘV_C and so test results are usually corrected with a recently measured value of hemoglobin for that individual. For a given gas, G, the model used for T_1 is usually

$$T_1 = \frac{\text{uptake of gas per minute}}{P_{A,G} - P_{c,G}}$$

where $P_{A,G}$ is the partial pressure of gas in the alveoli and $P_{c,G}$ is the partial pressure of gas in the capillary blood. This treats the lungs as if they are a balloon with perfect gas mixing and contact with the diffusion membrane. In a real lung, there is regional variation in ventilation (the bottom of the lungs tends to fill before the top) and there is always some ventilatory unevenness within regions, even in healthy subjects.

A method to measure gas transfer is the single-breath method. This has the advantage that, because it is measured at total lung capacity, regional variations in ventilation are minimized. However, the uneven ventilation of the lung is not taken into account and this tends to lead to low values of the alveolar volume, V_A. This can give an estimate of gas trapping in the lung, however, which can be a useful diagnostic tool.

The test gas used is carbon monoxide (CO), for two main reasons. Firstly, CO is removed from blood plasma by binding with hemoglobin

very quickly. Thus, its partial pressure in blood plasma rises very slowly and does not saturate during the time the capillary blood is in diffusive contact with the alveoli. CO is therefore said to be diffusion limited. This makes it ideal for measurements of T_1, as gas will continue to be removed from the alveoli as long as it is in contact with the membrane. The other major advantage to using CO for T_1 measurements is that in most subjects, $P_{c,Co} = 0$ before the test begins. However, this is untrue for very heavy smokers or those who have recently smoked.

A spirogram of the test is shown in Figure 16.5. The test begins with normal tidal breathing. A larger breath is taken in and exhaled down to RV. Close to the bottom, the operator triggers the system so that the room air is shut off. When the subject begins to breathe in, a demand valve opens and the test gas mixture is inhaled up to TLC. The subject holds their breath for around 8 s and then exhales so that the test gas can be collected. The first ~ 750 ml of the expired gas are discarded to account for instrumental and anatomic dead space, and then between 0.5 and 1 l of gas are collected and analyzed. The breath holding time is defined from one third of the way into the inspiration to half of the way through the collection volume, labeled in Figure 16.5.

The test gas consists of 0.28% CO, 9.5% Helium (He), and 18% Oxygen (O_2), with the remainder made up of Nitrogen (N_2), which is supplied premixed. 18% O_2 is used because this leads to a roughly constant fraction of O_2 in the lungs however much gas is inspired. The uptake of CO in the lungs is proportional to its fractional concentration (assuming zero CO partial pressure in the capillaries), so that

$$\frac{dF_{A,CO}}{dt} = -\frac{F_{A,CO}(P_B - P_{H_2O,S}(37°C))T_{l,CO,sb}}{\kappa V_A}$$

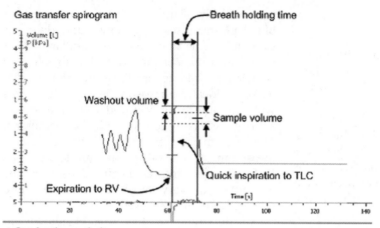

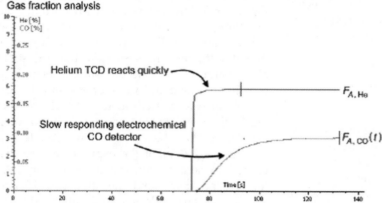

FIGURE 16.5 A screenshot for a gas transfer measurement, showing the spirogram for the test in the upper pane and the gas analysis in the lower pane.

where κ is a constant converting V_A from a volume into a quantity of gas, equal to $1/22.4$ mol/l at STP. Integrating and rearranging we find

$$T_{l,CO,sb} = \frac{\kappa V_A}{(P_B - P_{H_2O,S}(37°C))t} \ln\left(\frac{F_{A,CO}(0)}{F_{A,CO}(t)}\right).$$

where $F_{A,CO}(t)$ is the fraction measured in the expired sample. However, $F_{A,CO}(0)$ has to be calculated because the gas left in the lungs at the start of the maneuver will dilute it. The helium fraction is used to measure the level of dilution, and this is assumed to be the same for both gases. Helium is ideal for this as it is not lipid soluble so does not pass through the alveolar membrane. Then we have

$$F_{A,CO} = F_{I,CO}\frac{F_{A,He}}{F_{I,He}}$$

where F_I is the inspired fraction of gas. Helium dilution is also used to estimate V_A, the alveolar volume. This is the volume of the lungs that actually participates in gas transfer, so excludes the anatomic dead space. It is given at the ambient temperature and pressure saturated (ATPS) by

$$V_{A,eff} = (V_{I,ATPS} - V_{ID} - V_{DS})\frac{F_{I,He}}{F_{A,He}}$$

where, V_{ID} is the instrument dead space, including that of the mouthpiece used, and V_{DS} is the anatomic dead space, usually estimated using the empirical observation that dead space volume in ml equals the sum of the subject's age in years and body weight in pounds. This then requires conversion into BTPS (body temperature and pressure saturated). It is labeled as the effective V_A because, as mentioned earlier, it underestimates the true alveolar volume due to uneven ventilation. If the ratio $V_A/V_{A,eff}$, where V_A = inspired volume plus the RV measured by full body plethysmography, is less than about 0.8, this indicates significant gas trapping.

Note that in single-breath gas transfer measurements, the value measured directly is $K_{CO,sb}$, the gas transfer per unit volume. This is then multiplied by $V_{A,eff}$ to obtain $T_{1,CO,sb}$. $K_{CO,sb}$ gives a measure of how well diffusion occurs in ventilated parts of the lung and thus an indication of tissue thickening or loss, whereas $T_{1,CO,sb}$ gives a measure of the overall ability of the lungs to supply oxygen and remove carbon dioxide into and out of the bloodstream.

Gas analysis is shown in the lower half of Figure 16.5. The CO content of the expired sample is measured using an electrochemical cell, and the He content is measured using a thermal conductivity method (Cooper et al., 2003). In the electrochemical cell, carbon monoxide is oxidized to carbon dioxide and water at the cathode is reduced. A current will flow due to this reaction and this is what is measured. The response time is around 30 s to reach 90% of full signal due to the time needed for the CO gas to diffuse into the electrolyte in the cell. The thermal conductivity cell involves a Wheatstone bridge, illustrated in Figure 16.6.

The time response is improved by having a stabilized flow of gas across the circuit elements, so that it takes as little as 3 s to reach 95% of full signal. Thermal conductivity detectors for helium are very linear and stable.

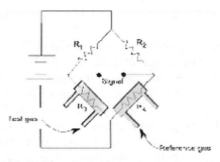

FIGURE 16.6 A thermal conductivity detector (TCD) for helium. The reference flow has a known thermal conductivity. A difference between the thermal conductivity of the test gas and reference gas will lead to different cooling in the two arms and unbalance the bridge.

The transfer factor is affected by the quantity of hemoglobin in the blood, so it is often standardized by this value to remove any variation that is simply due to anemia.

The equation for this corrected, $T_{1,CO,sb}$ is

$$T_{1,CO,c} = T_{1,CO} + 1.4(Hb_{stand} - Hb_{meas})$$

where Hb_{stand} is 14.6 g dl^{-1} for adult men and 13.4 g sl^{-1} for adult women.

As with spirometry, performance of the test is important to avoid errors. Errors can often be identified by inspecting the spirogram and including stepwise inspiration or expiration, inspiration not from RV, and an inspiration that is too slow. These either invalidate the assumptions made about the breath-holding time, or mean that the volume calculation by helium dilution is inaccurate.

Subjects should not have eaten recently, as this will decrease $T_{1,CO,sb}$, and should not have engaged in heavy activity, as this increases $T_{1,CO,sb}$. The test will also be inaccurate if subjects have recently smoked as there is likely to be an appreciable partial pressure of CO in the blood already. A four-minute interval is left in-between tests to enable the CO to diffuse out of the blood as a result of the previous test, and two tests agreeing to within 10% are a minimum for an acceptable determination of $T_{l,CO,sb}$.

The single-breath measurement of $T_{l,CO}$ cannot be carried out in patients who have a VC less than 1.5 l and those who cannot hold their breath for 4 s. If a gas transfer measurement was essential, these patients could be measured using a rebreathing method, although strictly speaking this is not directly comparable to single-breath measurements. The gas analyzers are calibrated twice daily. They undergo a two-point calibration, checking the zero and that the preprepared gas mixture reading agrees with the stated value. The primary risk involved in gas transfer measurements is related to the compressed gas bottles. This is minimized by using appropriate regulators, high pressure tubing, and connectors. The bottles are held in an enclosure so that they cannot fall and either cause injury or be damaged.

References

Cooper, B.G., Evans, A.E., Kendrick, A.H., Newal, C. (Eds.), 2003. Practical Handbook of Respiratory Function Testing, Part 1, second ed. Association for Respiratory Technology and Physiology.

Cotes, J.E., Chinn, D.J., Miller, M.R., 2006. Lung Function: Physiology, Measurement and Application in Medicine. Wiley-Blackwell, Oxford.

Miller, M.R., Hankinson, J., Brusasco, V., Burgos, F., Casaburi, R., Coates, A., et al., 2005. Standardisation of spirometry. Eur. Respir. J. 26, 319–338.

Quanjer, P.H., Temmeling, G.J., Cotes, J.E., Pedersen, O.F., Peslin, R., Yernault, J.-C., 1983. Standardised lung function testing. Bull. Eur. Physiopathol. Resp. 10 (S5), 1–95.

PART IV

REHABILITATION ENGINEERING AND ASSISTIVE TECHNOLOGY

David Long, Mike Hillman, Duncan Wood, Ian Swain, Dan Bader, Ladan Najafi, Donna Cowan, Jodie Rogers, Fiona Panthi, Will Wade, Robert Lievesley, Tim Adlam, David Ewins, Tom Collins, Martin Smith, Vicky Gardiner, Paul Horwood, Chris Morris, Tim Holsgrove, and Tori Mayhew

Overview

This section of 10 chapters describes how clinical engineers contribute to the provision of rehabilitation and assistive technology services. It covers a variety of subject areas, some of which overlap with each other and some of which are quite distinct. Many areas involve a patient-facing role for the clinical engineer, which provides additional challenges, particularly in the area of communication.

It is both an inspiring and a rewarding area of work where there is always the next challenge to overcome. Just when it would appear that all permutations of a condition or presentation have been observed, a previously unforeseen set of variables is presented.

The reader should be aware that while there is a quantity of material covered, it should be considered an introduction. Further knowledge will be obtained from supervisors, peers, and, importantly, through personal study, development, and experience. This is a changing and developing area of expertise and, as such, requires those involved to not only to keep up to date with the latest developments, but to contribute to those developments themselves.

17

Introduction: Medical Engineering Design, Regulations, and Risk Management

David Long and Mike Hillman[†]*

*Oxford University Hospitals NHS Trust, [†]University of Bath

INTRODUCTION

It is hoped that the reader of this section will be provided with some foresight into the world of rehabilitation engineering and assistive technology, particularly in relation to dealing with face-to-face patient contact, an area in which engineers are often less experienced. Much of the material presented may be applied broadly to the different subject areas, and some is more directly relevant to certain areas.

Titles for Engineers

The term *rehabilitation engineering* is not universally defined. In this book it is used to encapsulate a broad spread of clinical areas. Engineers working in these fields are referred to either as *rehabilitation engineers, clinical engineers, bioengineers,* or, less commonly these days, *technical officers*. Technicians may support the work of an engineer and it is generally the case that clinical engineers and bioengineers will carry out clinical assessment independently, but in some countries this role is termed *rehabilitation engineer* and in others *rehabilitation technician*. In the United Kingdom, clinical and bioengineers are usually registered under the protected title of *clinical scientist*. Furthermore, *orthotists* and *orthopedic engineers* may be involved in the clinical casting process for the production of custom contoured wheelchair seating.

One must be aware of this inconsistency in terms when communicating with different services or organizations who may have little understanding of engineering applied to the clinical setting.

Principles of Communication and Patient Assessment

Asking Questions

A fundamental part of the engineer's role is to ask questions to find out precisely what has been requested. It is often tempting to want to jump ahead with getting a result or finding a "solution." In the book *The Hitchhiker's Guide to the Galaxy* (Adams, 1979), a computer is asked a question, and eventually produces a result:

> "The answer to the great question... of life, the universe and everything... is... forty-two," said Deep Thought, with infinite majesty and calm. "I checked it very thoroughly," said the computer, "and that quite definitely is the answer. I think the problem, to be quite honest with you, is that you've never actually known what the question is... so once you do know what the question is, you'll know what the answer means." *(Adams, 1979)*

This is particularly true in clinical practice where a reliable method of communication with the patient must be established, but the problem is that communication is often deeply flawed. We sometimes make sweeping assumptions about what we have heard and are thoroughly convinced that what we have said has been crystal clear, but as George Bernard Shaw pointed out, "The single biggest problem with communication is the illusion that it has taken place."

How many times have you later realized that the outcome from a previous conversation was interpreted entirely differently by the other person involved? Translating this into the clinic environment: A question has been proffered and an answer received. Did that elicit the "correct" information? It often becomes clear in the course of an assessment that the question has been interpreted differently from how it was intended, and/or there is more to a situation than the person is reporting. This is not usually because there is anything to hide, rather that the question was

asked in a way that was not accessible to the recipient. If you are to get to the root of a problem then you need to be patient, persistent, and perceptive. You need to develop the ability to perceive when the response you receive does not contain all the information you require. You will then need to ask the question in a different way, or come back to it later in the appointment. Consider the construction of a robust survey questionnaire: some questions will be asked more than once, but will be worded differently. To sum up, take the time to understand, then be sure to be understood.

Taking the Time to Make a Thorough Assessment

Clinic schedules are often packed too full, forcing the assessor to cut short clinical assessments. Unfortunately, this tends to lead to subsequent problems as insufficient data were gathered at the start, in other words "save now, pay later." This produces poor outcomes for the patient, possibly producing more problems than they had before, and certainly wasting time (and by inference, money) for both them and the professional/service. Far better to allow more time upfront and to proceed with something that is more likely to succeed.

In developing a solution to the problems presented, there are usually conflicting factors, such as the need for a high level of postural support in a wheelchair combined with the requirement for manual self-propulsion. Time will be needed to determine priorities and come to an agreement as to how to proceed. Added consideration should be given to the person having difficulties with communication, and/or who needs to use a communication aid, where additional time will, in most cases, almost certainly be required.

Involving the Person

Regardless of their age and ability, the patient should remain central in the assessment process, their requirements and desires being the focus. It may not be possible to provide everything that is desired, or even required, but this must still be acknowledged if the person is not to feel that their wishes are being ignored. The same applies to the family and carers who play a crucial part in the person's life. In the case of equipment provision, their support is often vital to a successful outcome. One must be careful, however, that family and carers do not control the outcome where the person is able to advocate for themselves. They (almost always) have the best interests of the person at heart, but it is worth making a point of speaking to the person, using their name, and even saying "I'd like to know what John thinks," leaving enough time for John to collect himself and provide a response. An additional challenge is where the patient has a mild intellectual disability, and you believe that the carer is right, while the person continues to make their point to the contrary. There are no rules about how to respond in these situations, but the assessor must slowly and sensitively work their way through to a conclusion, taking care to acknowledge what the person is saying. It may be necessary to involve a further person close to and respected by the patient.

Further points include maintaining eye contact and avoiding physical contact without first establishing a rapport. Be aware that some disabled people are particularly prone to fatigue and may struggle with a long appointment or one that is timed later in the day. Is there a carer sitting quietly in the background? Ask them if there is anything they want to contribute. Get alongside people when you have the chance; develop opportunities for people to talk to you. With care, humor can be used to establish a rapport, but avoid over familiarity. Don't feel the need to fill what might seem an uncomfortable lull in conversation as that might be just the opportunity someone will take to speak.

Using a Clinical Methodology

Whatever branch of this field we happen to work in, a methodology is required if we are to achieve consistency in approach and outcome. Returning to *The Hitchhiker's Guide to the Galaxy*, Arthur Dent (the main character), in attempting to find in modern times the location of the cave in which he had lived temporarily on prehistoric earth, wrote a computer program to carry out the calculations, and:

> ... decided not to mind the fact that with the extraordinary jumble of rules of thumb, wild approximations and arcane guesswork he was using he would be lucky to hit the right galaxy, he just went ahead and got the result. He would call it the right result. Who would know? As it happened, through the myriad and unfathomable chances of fate, he got it exactly right, though he of course would never know that (Adams, 1984).

Let us not fall into the trap of hoping for lucky guesses, or cutting short our procedures which, arguably, could amount to the same thing, as for every one of these there will be a hundred (or more) that are wrong. Even if your guess, or corner cutting, turns out right you will have no idea how you got there, and obtaining the result a second time will be troublesome. Clearly, a clinical methodology is required. This will vary according to clinical area but should be scrutinized for sensitivity to the question in hand. The most basic form of methodology follows:

1. Clarify what has been requested by the referrer and the patient/carer.
2. Compile a detailed list of requirements and problems.
3. Take the relevant medical, functional, social, environmental, physical, and psychological details.
4. Define the constraints, i.e., those problems that cannot readily be overcome or that are fixed.
5. Evaluate all the data collected.

6. Produce a list of aims and objectives in conjunction with the patient/carer, acknowledging and resolving any conflicting priorities.
7. Develop conceptual solutions.
8. Finalize the outcome, detailing the specification of any equipment.
9. Form a plan of action.

Clearly there are many substages within this process but this at least provides an overview. Clinical assessment may be thought of as akin to the assembly of a complicated jigsaw puzzle where all the pieces must be fitted together in a specific way and which are not all visible at the beginning of the process. As pieces are assembled, the picture becomes clearer and it is more obvious where other pieces might fit.

Prescription

In the case of equipment provision it is difficult to prescribe without first having sight and touch of the equipment. As a result, it will be necessary to invite companies to demonstrate their products, or to attend a suitable exhibition. Be analytically critical, ask difficult questions, and, as the saying goes, if it looks/sounds too good to be true, it probably is. It is also worth remembering that while the clinician is the expert in the needs of their particular patient, the company representative is (usually) more knowledgeable about the product and will bring useful experience from having worked alongside a wide range of people.

There will, of course, be restrictions on funding which will vary according to the source. It will be necessary, on occasion, to form an argument for something out of the ordinary. This should be made significantly easier by having completed the assessment in the manner defined above. However, be sure to under-promise and over-deliver.

Finally, it can be helpful to ask oneself the following questions (adapted from Pope, 2006):

- Is the piece of equipment needed?
- Is it wanted?
- Do the patient and carers know how to use it?
- Can the patient and/or the carers manage it?
- Does it fit in with their lifestyle?
- What is the tradeoff?

Fitting the Equipment to the Person

It sounds obvious, but equipment should be made to fit the person; it would be reasonable to expect this to be the case. However, in services that are under pressure, either financially or from a lack of time or staff, the tendency is for the person to be made to fit the equipment, the assessor possibly managing to convince themselves of the opposite. There is a balance to be struck: if the person uses the equipment seldom, a precise fit may be less critical. If use is likely to be regular and prolonged, tailoring to the individual is imperative. We use the same decision-making processes in everyday life.

Equipment for Children

Consideration must be given to the needs of children who require equipment designed to promote their development. This may be to develop the physical skills required to control other equipment, or to be provided with equipment that will grow with them. It is important that children are engaged with the assessment process wherever possible. They have an opinion and it is important that this is heard.

Overall Aim of Provision

It is all too easy to become so embroiled in the provision of equipment that sight is lost of the original purpose of intervention.

As such, it is important that reference be made to the aims and objectives at regular intervals during the process. The art is in balancing these often conflicting requirements. Honest and open discussion with the patient is critical to achieve a satisfactory outcome.

Potential for Learning

Sadly, it is often the case that equipment is denied because the person cannot demonstrate that they can control it adequately. However, one is not born with the ability to drink without spillage, or to read poetry without first learning nursery rhymes. In other words, clinical assessment and provision must allow for the *potential* of a person to learn a new skill. If you are a car driver, recall the moment you first attempted to wrestle with and coordinate the pedals, steering wheel, mirror, and lights. Did you remember to put your indicator on at every turn? You built up this skill over time and it started to become second nature as your brain laid down new neural pathways so you could avoid having to relearn every move.

This applies to people with disabilities in just the same way. Where someone has an intellectual disability, the process may take a little longer. People can be fearful of using equipment, perhaps because they are worried they will damage something, or perhaps because they lack self-belief. It is your job to facilitate (not to pressure) them in learning a new skill.

Consent to Assessment and Treatment

One must be cognizant and respectful of the wishes of an individual. "Well, of course," you may say. However, this presents challenges where the person has difficulties with communicating, has an intellectual disability, or has impaired cognition. If a person does not wish to be assessed or to be treated, and cannot be persuaded otherwise, that is their decision and should be respected, wherever possible.

One should tread very carefully in casually taking action in the person's "best interests"—you are taking a judgment on their behalf.

Increasingly in the United Kingdom, professionals such as occupational therapists, physicians, and social workers are being given formal, Master's-level training to enable them to act as "best interests assessors" to carry out deprivation of liberty assessments for people who are not able to be sectioned under the Mental Health Act. Best interests assessments are descended from the Mental Capacity Act 2005 and put in place deprivation of liberty safeguards (DOLS) that protect people who have been formally diagnosed as lacking capacity from being unlawfully restrained/constrained. At present, this applies only to adults living in a residential setting or who have been admitted to the hospital.

A common example is where an elderly person with dementia lives in a care home, requiring total assistance with all aspects of care, and who must be constantly supervised as they have a history of walking out into the main road without checking for traffic. The care home locks the doors to prevent this from happening. The best interests assessor will investigate whether the care home is depriving the person of their liberty and breaching their human rights by effecting complete and total control of their life and, if so, whether or not it is being done in their best interests. If the assessment concludes that it is in the person's best interests they will grant a DOLS authorization.

"Challenging" Behavior

Some people, notably those with profound intellectual disabilities, an acquired or traumatic brain injury, or dementia may exhibit what is termed "challenging behavior." This is behavior that is not culturally acceptable and/or puts the person or others at risk, and includes aggression, self-harm, destructiveness, and disruptiveness (NHS Choices, 2012).

People displaying these symptoms are less able to participate in daily life, which is likely to include conforming to a clinical assessment process. Such behaviors are likely to be heightened by anxiety brought about by the presence of strangers and unfamiliar surroundings. Pain can also be a contributory factor, as can boredom (NHS Choices, 2012). One must be mindful to work with the family, and professional and care staff to minimize the effects of such behavior and the distress of the person.

Outcome Measures

Having completed your clinical assessment and made your intervention, how do you know that it met its requirements and will continue to meet them? This leads us to enter what some might call the "murky" world of outcome measurement, of which there are myriad tools. It can be difficult to identify one that is appropriate to the question being asked, particularly whether it is sufficiently sensitive to detect changes in the specific area of interest (Laver Fawcett, 2007). As an example, if you desired to measure changes in sitting ability as a result of an active rehabilitation program, a tool providing the options of the person being able to "sit without support" or "requiring additional support to sit" would not detect small, incremental changes in ability as rehabilitation progressed.

Having identified a suitable tool, it is then necessary to ask whether it has been tested for validity, that is, does it measure what it is said to measure. Taking the survey questionnaire as an example, it is simple enough to produce a set of survey questions, particularly with some of the tools available on the Internet, but the design of questionnaires having validity is a very complicated subject and far easier to get wrong than right. Have you ever struggled to answer a question on a survey either because the wording could be interpreted in different ways or because you were not able to select

from the options presented? Have you ever felt a question to be leading?

It is not possible to cover within this book the range of outcome measures that might be suitable in the field of rehabilitation engineering and assistive technology. Interested readers are directed to their organization's research department, library, or, with care, online resources.

Documentation

This is a critical factor for all clinical, technical, and scientific work. It is important that healthcare decisions are clearly described for future reference, be that a colleague taking over the care of an individual, or yourself in being able to recall precisely what you decided and why when the patient returns to see you a few months/years in the future. When we have multiple patients on our caseloads it is all too easy to forget critical details at a future date, even if they appear with absolute clarity in your mind at the time of assessment.

Registered professionals are, of course, required by statute to keep adequate records. This is not only to promote good practice and quality outcomes, but also to protect their employer against any potential future legal case mounted against them. It should also be remembered that, in the United Kingdom at least, patients have the right to request access to their own notes.

Patient Groups and Their Characteristics

There are thousands of medical diagnoses used to explain a person's clinical presentation. As a result, it is often necessary to carry out research into a particular condition, looking for certain characteristics that will inform the assessment and provision process. These may be grouped as follows:

- Prognosis: Is the condition terminal, rapidly deteriorating, slowly deteriorating, fairly stable, or temporary; i.e., is the person expected to recover? Is there a specific life expectancy?

- Rehabilitation: Is it possible that the person will recover some or all of the function they have lost? Does your provision need to reflect this, both in terms of the potential for adjustment and the time taken to supply?
- Physical impairment: How is the person likely to be affected? Are there specific difficulties, e.g., limited joint mobility/pain, poor balance?
- Intellectual disability: Are there aspects of the condition that limit learning or understanding, or the ability to make judgments or form opinions that are based on logical reasoning?
- Specific risk factors: Are there aspects of the condition that are critical to understand, e.g., potential for bone fracture for those with osteoporosis?

The following conditions are among those more commonly seen by clinical and biomedical engineers:

- Acquired brain injury (ABI)
- Amputation (see the "Prosthetics" section in Chapter 26 for further details)
- Arthritis (osteoarthritis (OA) or rheumatoid arthritis (RA))
- Cerebral palsy (CP)
- Cerebrovascular accident (CVA) or "stroke"
- Dementia (including Alzheimer's disease)
- Multiple sclerosis (MS)
- Muscular dystrophy (MD)
- Parkinson's disease (PD)
- Spinal cord injury (SCI)
- Spinal muscular atrophy (SMA)
- Traumatic brain injury (TBI)

There is a wealth of information available on the Internet but the usual precautions apply: look for a known organization and/or check information across two or three sites, particularly in the case of rare conditions.

Rehabilitation

Rehabilitation implies recovery, or part thereof. There are many centers specializing in rehabilitation and which are typically, but not exclusively, for people having had a stroke, acquired/traumatic brain injury, or spinal cord injury. There is a window of opportunity in which it is possible for improvements to the person's condition to be made through a formalized and intense treatment process. This usually involves a multidisciplinary team including physicians, physiotherapists, occupational therapists, speech and language therapists, and nurses. Engineers can play an important role, particularly where provision of equipment is required.

Change in the body can happen very quickly, particularly in respect of plastic adaptation of the musculoskeletal system. Muscle starts to waste and to shorten surprisingly quickly, leading to contracture and joint range limitation. These changes can be extremely difficult, and sometimes almost impossible, to overcome. Posture management, of which physiotherapy is a core component, plays a crucial role in maintaining joint ranges and body symmetry.

On the other hand, the plasticity of the nervous system, known as neuroplasticity, (i.e., the ability of the brain to lay down new or revised neurological pathways), is remarkable. A substantial amount of recovery can take place with an intense treatment program as skills are relearnt and practised, embedding movement patterns within the brain.

Figure 17.1 illustrates a range of outcomes from rehabilitation, with time shown along the x-axis. After the acute phase, a number of long-term outcomes are possible. Following the end of the program of rehabilitation, further significant increases in function and ability are less likely, the focus often being on preventing deterioration.

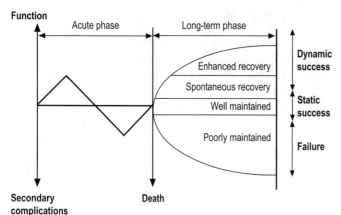

FIGURE 17.1   Ranges of outcomes. *Source: Pope (1988).*

There are many resources available on the subject of rehabilitation. The reader is directed to any good bookstore for a formal textbook, or to relevant journals for the most up-to-date research findings.

MEDICAL ENGINEERING DESIGN

Clinical engineering is about finding a solution to the specified need of a client. Usually this will be through the use of an existing product or application of an existing process or procedure. There will be times, however, when an existing product is not appropriate: it may need modifying or a new piece of technology may need designing and producing. The discipline of "design" applied to medical engineering is wide, but in the context of clinical engineering it is about solving the stated need of a client/patient by designing a new (or modified) piece of technology.

There are many models and methodologies described in the literature for the design process (Pugh, 1990; Pahl and Beitz, 1988; Ulrich and Eppinger, 2011; Design Council, 2013; BSI, 1997), as well as more general project management (Prince2[1]).

[1]<www.prince2.com>

Some are more appropriate for a product design environment while others are more suitable for the design of a "one-off" item. In general terms they all contain the following aspects:

- Understand the problem
- Identify and evaluate concepts
- Embody and develop the design
- Build or manufacture the device
- Verify/validate/hand over

For medical engineering design the identifying features are that the work (1) starts with an identified need, (2) is by definition "user centered," and (3) ends with a device delivered to the client. This is different from product design which starts with an idea and ends with a product. A description of such a design process is given by Orpwood (1990).

To conform as closely as possible with one of the well-documented design methodologies, the method described as follows is based on Pahl and Beitz (1988). Using their headings, the phases, with their main deliverables, are:

- Clarification of the task → specification
- Concept design → concept
- Embodiment design → preliminary layout/ definitive layout
- Detail design → documentation
- Final deliverable = solution

Note: As this book does not attempt to repeat what is already well covered in the engineering literature, particularly in the area of engineering design, only the aspects of the model that are distinct for clinical engineering are highlighted; this book does not attempt to cover the aspects of engineering that should be covered in a good engineering degree syllabus (e.g., tolerancing, technical drawing, stress analysis, electronic circuit design, etc.).

Clarification of the Task

The input to this phase is the specified task, also described as the project brief in the Prince2[2] project management method, or the problem definition (Orpwood, 1990). It will initially be described in terms of the user (in rehabilitation technology, the patient or end-user is not the only person using the technology: it may include a carer, formal or voluntary, or therapist, as well as the end user) and the functional need to be met.

At the outset this will be in qualitative terms. The engineer must develop this to a more detailed and, where possible, quantitative specification. Some aspects will be known in quantitative terms from the start (e.g., weight of the user), while others will be determined in this and later phases of the project. Some parts of the specification will be defined as demands (what the product *must* meet) and wishes (what the product *should* meet). A detailed specification is vital to allow evaluation of the design at all stages (evaluating concepts, design review, verification, and validation).

Other aspects involved in clarification of the task include an understanding of the environment the device must fit into: both the physical environment and the regulatory framework. There will be financial constraints for the project, in terms of an overall budget, or a

cost/benefit judgment, or the cost of other solutions. There must obviously be a clear understanding of the clinical (physiological, etc.) aspects of the problem to be addressed.

Concept Design → Concept

An engineering device has a primary function that can be divided into a number of subfunctions. The first stage of concept design is to identify the essential problems, establish the overall functions and the subfunctions, and then to find a number of solution principles to fulfill the subfunctions. For a typical assistive or rehabilitation device the user interface will be a critical part of the design. It can be useful to separate the user interface aspects from the supporting engineering features (Orpwood, 1990). Technical aspects will often (but not always) be well-known technology; it is the user interface aspects that are likely to be the most challenging. As an example, a patient with impaired sitting ability and balance requires a perching seat/stool for their professional work which results in it needing to be transportable. The design considerations are:

- Folding/dismantling mechanism
- Must not slip or become unstable
- Structural: Strength, stiffness
- Position/angle of main seat
- Other body supports
  - Back
  - Thoracic
  - Feet
  - Knees
- Material/shape/padding of seat
- Aesthetics (needs to look professional)

It is the skill of an engineer to identify a wide range of solution principles. At this stage it is less important whether they are in detail feasible. Quantity is more important than quality. A brainstorming or mind-mapping technique (whether carried out individually or

[2]<www.prince2.com>

with a team) is valuable in generating a large number of potential solutions. There are many sources of inspiration, both technical and aesthetic. For example, other assistive technology products, literature review, mainstream and consumer products, nature, and so on.

The feasibility of these solutions is often investigated through the use of prototypes. A prototype may not necessarily be physical; it could be a sketch or CAD drawing to illustrate a point to the client, or a computer model for stress analysis. To be effective, prototypes need to answer a question and need to be as simple as possible to get the required answer. It is important to try to get as much iteration done at this stage before funds are committed to the final device.

It should be noted at this point that for every test with an end user, an adequate risk assessment should be in place, although that assessment must bear in mind the constraints within which the evaluation is taking place.

It is important that evaluations consider the reaction of the user (be this the patient, carer, or healthcare professional). At one extreme it is important not to unrealistically raise expectations, but at the same time the appearance of the prototype should not produce negative reactions. In this respect, simple materials such as cardboard are a very useful tool as they are clearly not the finished device but can communicate the function and scale of the final device. In all such interactions a good working relationship with the user is essential.

Having identified a range of solution principles, these are then evaluated (with reference to the specification) and the most promising combined to produce a small number of concept variations. At the start of this phase the functional requirements were separated. Now the task is to put them back together. Using the same features to solve more than one functional demand produces the most economical and efficient solution. These concepts are then evaluated against the specification. Those that

do not meet the "demands" of the specification must be either discarded or modified. This will leave a number of concepts that are evaluated against the criteria described in the specification "wishes."

Although an engineer's judgment will often lead to a preferred concept, it is valuable to use a quantitative selection method in which the concepts are scored against the product specification. The simplest is to score a concept as $+1/0/-1$ against a datum concept for each specification point. More sophisticated methods use a wider scoring range and introduce weighting for the importance of each specification point. To avoid or reduce bias, the weighting and scoring should be carried out, or at least reviewed, by someone other than the project engineer.

Embodiment Design → Preliminary Layout/Definitive Layout

In the embodiment phase, a concept that is still loosely defined is refined first to a preliminary layout and finally to a definitive layout. This is an iterative stage as the functions are fitted into an overall solution.

Certain features will define the size of the solution. For most rehabilitation devices the overall size will be based around the human scale and the interface support features. Also to be considered are size-determining features such as power and strength requirements and any known standard components. It is now possible to start laying out the overall spatial arrangements.

Pahl and Beitz (1988) make no reference to the need to consider the aesthetics of a design. However, the words of William Morris (Morris, no date) are pertinent "Have nothing in your house that you do not know to be useful, or believe to be beautiful." This is particularly important for equipment that relates so closely to the user. Pullin (2009) discusses these issues in detail in his book *Design Meets Disability*. The question to ask is whether the design gives a sense of value: Does the user feel valued by using this piece of equipment? Appropriate aesthetics can be incorporated into a device without an unacceptable cost penalty; they are not a final detail but must be iteratively included throughout the embodiment phase, and sometimes even earlier in the concept generation phase.

Starting with the main functions that define the embodiment, the design is developed, including the auxiliary functions. Materials are chosen and the form of the individual components defined. Preliminary analysis of stresses and performance are carried out. As a preliminary design is developed it must be checked against the product specification. Design review may be carried out informally or more formally (e.g., BS EN 61160; BSI, 2005).

Design freeze is an important discipline (Pietzsch et al., 2009). This has already happened once in choosing which concept to take forward to embodiment design. It must also be implemented after the preliminary layout has been selected.

When the preliminary layout has been fixed, the details are optimized and weak points are improved or eliminated. Checks for errors must be made on both detail points and the overall design. This gives the definitive layout which must again be checked against the product specification.

Detail Design → Documentation

In this stage the parts are defined in detail and documents prepared for manufacture. The most important document is obviously the manufacturing drawing. This will include accurate material specification, tolerances, and guidance for manufacture and assembly. Parts lists will be prepared for manufacture and bought in parts. As documentation is developed, the (European) requirements of the technical file for regulatory purposes will be met; see the next section "Regulations and Risk Management" for further details. This will include the identification of relevant standards to allow appropriate testing to take place. Documentation will not only include manufacturing instructions, but also risk assessment and user instructions.

Final Deliverable = Solution

The final device must undergo verification and validation. Verification asks the question: Does it meet the quantitative specification? Validation is concerned with whether the device works for the user, in his or her environment. This is why it is so important to have a detailed definition of the requirements of the user(s). Making assumptions about intended use/function is very likely to lead to the development of a device which is, at best, ineffective or worse, useless.

In an ideal world the device would work perfectly the first time. The reality of rehabilitation devices is that there may be some need for modification or fine tuning. At earlier stages, intelligent use of prototypes should minimize the risk of this happening. To de-risk the solution, the design should include the facility for adjustment and fine tuning without the need for major changes. A device will often be tested with the client in the "bare metal" before being disassembled, finished, and reassembled.

REGULATIONS AND RISK MANAGEMENT

The Medical Devices Directive

Where rehabilitation equipment is placed on the open market there are nearly always regulatory requirements that must be met. In Europe, the relevant legislation is the Medical

Devices Directive (Council Directive 93/42/EEC of 14 June 1993 concerning medical devices; EC, 1993), referred to as the MDD. Most equipment used in this field is classed as a medical device. The following part of the MDD definition applies:

> The term "medical device" means any instrument, apparatus, appliance, software, material or other article, whether used alone or in combination, including the software intended by its manufacturer to be used specifically for diagnostic and/or therapeutic purposes and necessary for its proper application, intended by the manufacturer to be used for human beings for the purpose of:
>
> - diagnosis, prevention, monitoring, treatment or alleviation of disease,
> - diagnosis, monitoring, treatment, alleviation of or compensation for an injury or handicap,
> - investigation, replacement or modification of the anatomy or of a physiological process,
>
> and which does not achieve its principal intended action in or on the human body by pharmacological, immunological or metabolic means, but which may be assisted in its function by such means. *(EC, 2007)*

It is important to note that within the 2007 amendment (EC, 2007) there is a helpful addition to previous advice in respect of software used for medical purposes:

> It is necessary to clarify that software in its own right, when specifically intended by the manufacturer to be used for one or more of the medical purposes set out in the definition of a medical device, is a medical device. Software for general purposes when used in a healthcare setting is not a medical device.

This is particularly relevant for assistive technology where software is used increasingly in a clinical context for improving functional ability.

Purpose

The purpose of the MDD is to allow manufacturers to trade throughout Europe without having to comply with multiple national legislation. It is based on standards of quality and safety accepted across the member countries. Other countries have similar legislation, for example, the U.S. Food and Drug Administration (FDA) requirements. It is critical that a device complies with the requirements of whatever country in which the device is to be marketed.

Categorization of Devices

Medical devices are categorized according to the level of risk, often in terms of whether a device is invasive or inputs energy to a patient. Most rehabilitation equipment will come within the lowest level of risk, but its manufacture and provision must be properly regulated. Some equipment; for example, in functional electrical stimulation, will be classified as slightly higher risk, in this case because the use of surface electrodes transmits energy to the body. To demonstrate compliance with the MDD a more robust system must be in place. There are rules that govern risk categorization and these are detailed within the directive.

CE Marking

It is necessary to place a CE mark on a device before it is placed on the market. This demonstrates that the manufacturer has considered the requirements of the MDD and believes their product to comply. In the case of custom-made devices the CE mark should not be used. These are defined as follows:

> A "custom-made device" means any device specifically made in accordance with a duly qualified medical practitioner's written prescription which gives, under his responsibility, specific design characteristics and is intended for the sole use of a particular patient. Mass-produced devices which need to be adapted to meet the specific requirements of the medical practitioner or any other professional user shall not be considered to be custom-made devices. (EC, 2007)

Regardless of whether the device is custom made, the manufacturer must still ensure that the device meets the essential requirements of the MDD. It should be noted that the term "manufacturer" applies not only to commercial organizations but to any person or organization manufacturing medical devices.

Requirements

The essential requirements of the MDD define and describe how the device should be designed and manufactured. There is a set of general requirements and then a more detailed list of design and construction requirements under the following headings:

- Chemical, physical, and biological properties
- Infection and microbial contamination
- Construction and environmental properties
- Devices with a measuring function
- Protection against radiation
- Requirements for medical devices connected to or equipped with an energy source
- Information supplied by the manufacturer

Full information is available in the directive. These requirements are aligned with both good engineering practice and common sense, giving no surprises to a competent engineer. A helpful summary is provided:

> The devices must be designed and manufactured in such a way that, when used under the conditions and for the purposes intended, they will not compromise the clinical condition or the safety of patients, or the safety and health of users or, where applicable, other persons, provided that any risks which may be associated with their intended use constitute acceptable risks when weighed against the benefits to the patient and are compatible with a high level of protection of health and safety. This shall include:
>
> - reducing, as far as possible, the risk of use error due to the ergonomic features of the device and the environment in which the device is intended to be used (design for patient safety), and

- consideration of the technical knowledge, experience, education and training and where applicable the medical and physical conditions of intended users (design for lay, professional, disabled or other users). *(EC, 2007)*

Regulatory Process

While it is not appropriate to go into any level of detail here, the most important parts of the regulatory process are:

- Risk assessment: An assessment of whether the level of risk posed by the intervention is acceptable within the context of the benefit to the patient, which must be justified, and how the level of risk will be minimized and controlled.
- Conformance with standards/requirements: These may be specified as national or international standards or a list of the essential requirements. Conformance with "harmonized standards" infers compliance with the essential requirements in specific respects.
- A technical file: This will fully describe the design and manufacture of the product, and include the risk assessment and standards conformance data. It will also include information provided to the client, such as user instructions and labeling.
- Quality assurance for manufacture: This may be an accepted national or international standard such as BS EN ISO 9001, Quality Management Systems (ISO, 2008), or BS EN ISO 13485, Medical devices — Quality management systems — Requirements for regulatory purposes (ISO, 2012a), or an in-house standard. In the case of devices in higher-risk categories, quality must be audited by a designated body.

Risk Management

On any day, any one person carries out a number of risk assessments, such as crossing a

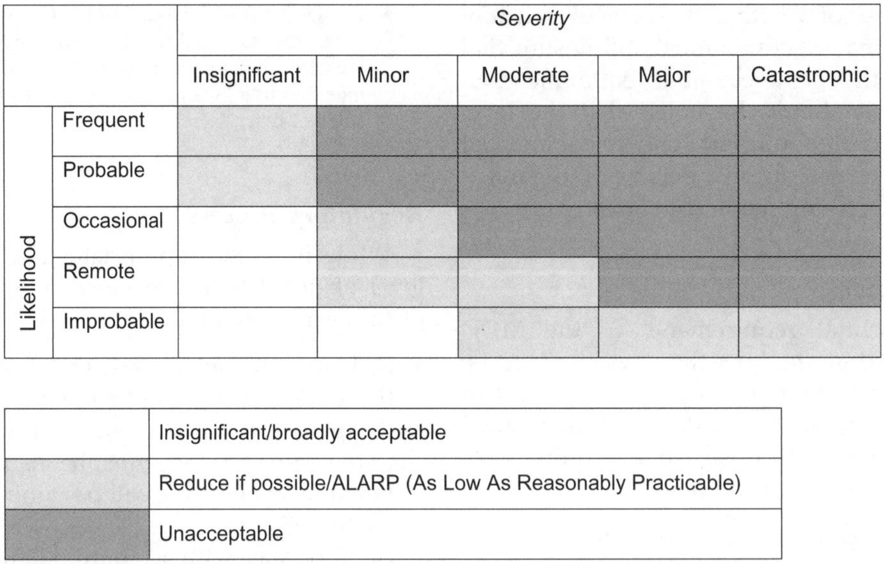

		Severity				
		Insignificant	Minor	Moderate	Major	Catastrophic
Likelihood	Frequent					
	Probable					
	Occasional					
	Remote					
	Improbable					

	Insignificant/broadly acceptable
	Reduce if possible/ALARP (As Low As Reasonably Practicable)
	Unacceptable

FIGURE 17.2   A 5 × 5 risk assessment matrix. *Source: Derived from BS EN ISO 14971, Medical devices − Application of risk management to medical devices (ISO, 2012b).*

busy road, driving a car, or pouring boiling water. We are used to taking risks: we would go nowhere and do nothing if we were to attempt to avoid them. What we are less used to doing is documenting the process; we have little requirement for this when the main person affected is ourselves (although this is not always the case, of course). In respect of rehabilitation engineering and assistive technology it is necessary to manage risks and document the process so that safe and efficient solutions to problems are provided, and that this can be demonstrated at any point in the future.

Definitions and Process

Risk can be considered in three parts: (1) the potential for harm to occur (i.e., a hazard), (2) the nature and severity of that harm, and (3) the likelihood of it occurring. *Management* can mean many things but perhaps in this context it is being in control, having an awareness of the end goal, ordering and applying logic, monitoring, and problem solving. BS EN ISO 14971, Application of risk management to

medical devices, describes risk management as "... a framework within which experience, insight and judgment are applied systematically to manage risks" (ISO, 2012b).

Once the risk has been defined, consideration should be given as to whether it can be avoided, either altogether or in part; a different method or product may achieve the same aim. Next, one must determine whether the risk is acceptable. This requires a knowledge of the potential harm and likelihood of this harm occurring (see next subsection). Unacceptable or high risks must be reduced by putting in place specific control measures. A risk/benefit analysis will be needed to justify the acceptance of risk. This is a complex area to manage and, ultimately, comes down to objective clinical judgment and practical control measures. Some interventions, diagnostics, and pieces of equipment carry an inherent risk with use, making removal of risk impractical.

As an example, the parent of a child with quadriplegic cerebral palsy has requested that a powered wheelchair be provided since

Likelihood rating	Score	Guideline definitions
Frequent	5	Expected to occur in most circumstances/almost certain 1:1–1:10 Once per day/week/use
Probable	4	Likely to occur/will probably occur, but not persistently 1:10–1:100 Once per week/month/100 uses
Occasional	3	May occur occasionally 1:100–1:1000 Once per month/year/1,000 uses
Remote	2	Unlikely to occur/not expected but possible 1:1000–1:10,000 Once per year/decade/10,000 uses
Improbable	1	May occur only in very exceptional circumstances/rare 1:10,000–1:100,000 or more Once per decade/100,000 uses

FIGURE 17.3　Guideline likelihood ratings for risk management.

manual self-propulsion has become difficult. Such a device has the potential to cause significant harm to the child, other people, and the environment, were the child to lack or lose control of the chair. It is anticipated that this could happen often, at least initially. However, there is the potential for clinical benefit if the child is able to learn to drive: greater independence and participation. Following initial assessment, the service agrees to supply the chair and, to reduce the risks, stipulates that it can be used only indoors initially and always with adult supervision. A training program is organized. The service puts in place a review for six weeks' time to establish if the child has learnt the requisite driving skills for use without adult supervision and driving outdoors.

Rating Likelihood and Severity

Most risk assessment policies use a matrix to determine whether a risk is acceptable. An example is shown in Figure 17.2 where the shading provides an outcome. For example, if the perceived harm is likely to occur regularly (probable) and is serious (major), the risk is deemed unacceptable and must be avoided or reduced to an acceptable level. If the harm is negligible (insignificant) and unlikely to occur

Severity rating	Score	Guideline definitions
Catastrophic	5	Death
Major	4	Major permanent harm/disability Extensive injury
Moderate	3	Semi-permanent harm (up to 1 year) Significant injury Medical treatment required
Minor	2	Non-permanent harm (up to 1 month) First Aid treatment required
Insignificant	1	Inconvenience or temporary discomfort

FIGURE 17.4 Guideline severity ratings for risk management.

regularly (remote), no additional action need be taken. Many risks fall into the middle bracket and it is here that clinical benefit must be shown to outweigh the inherent risks. Note that:

- The shaded areas should be adjusted to suit the application.
- Any matrix combination is possible, e.g., 3×3, 4×6, to ensure that the risk assessment is specific to context.
- This table displays three risk outcomes: it is appropriate to have more, depending on the application.

The previously mentioned standard (BS EN ISO 14971; ISO, 2012b) provides further examples and explanation in the appendices.

You will quickly realize that categorizing harm and its likelihood can be extremely difficult to achieve quantitatively, accurately, and objectively. Figures 17.3 and 17.4 provide further guidance on choosing appropriate levels. There is still a problem, of course, because determining likelihood numerically requires a precise knowledge of past events. If this information is not available (often/usually the case

in rehabilitation engineering and assistive technology) then one can only produce an estimate based on experience and knowledge of similar incidents, or lack of incidents. An estimate is worth making, however, because it demonstrates that the risk has been considered, were this ever to be called into question.

Fault Detection

In some applications it will be appropriate to consider the likelihood of a fault being detected before it causes harm. Before a component of a device fails, the problem may be detected; for example, a rattle that gets progressively louder. Other times, the chances of detection will be almost none; for example, sudden failure of an electrical switch. In the case of safety-critical components/operations it is safest to assume that the risk will not be detected.

Methodologies for Identifying and Analyzing Failures/Faults

Failure mode and effects analysis (FMEA) is a qualitative tool used to identify and evaluate the effects of a specific fault or failure mode at

Clinical
Communication difficulties
Muscle weakness
Limited range of movement
Involuntary movements
Cognition
Poor skin condition
Development/deterioration of pressure ulceration
Management of future changes in condition (requirement for review)
Use of wheelchairs as vehicle seats
Use
User error
Reasonably foreseeable misuse
Complexity of use/operation related to the ability of the person/carer
Manual handling
Context, e.g., delicate equipment in a busy school environment
Technical
Interfaces and compatibility of equipment from multiple manufacturers
Durability
Wheelchair stability
Service/review interval

FIGURE 17.5 Example areas for consideration in risk management.

a component or subassembly level. Human error is considered, which makes it particularly suited to this field. In contrast to an FMEA, a fault tree analysis (FTA) takes an undesirable event and works backwards to identify potential failure modes. This has the advantage of allowing the process to be evaluated, as opposed to looking at the failure in isolation. The hazards identified in an FMEA can be used within an FTA.

There is extensive literature available on these and other methodologies within many engineering textbooks to which the interested reader is directed.

Areas for Consideration in a Clinical Context

Figure 17.5 lists some of the more common areas for consideration in rehabilitation engineering and assistive technology; it is by no means exhaustive and is not in any particular order. The full set of hazards applicable will be derived from the clinical assessment and the context of use.

References

Adams, D., 1979. The Hitchhiker's Guide to the Galaxy. William Heinemann, pp. 128–129.

Adams, D., 1984. So Long, and Thanks for All the Fish. William Heinemann, p. 521.

British Standards Institution (BSI), 1997. Design Management Systems - Part 2: Guide to Managing the Design of Manufactured Products BS 7000-2:1997. British Standards Institution.

British Standards Institution (BSI), 2005. Design Review BS EN 61160:2005. British Standards Institution.

Design Council, 2013. The design process. <www.design-council.org.uk/designprocess>.

European Commission (EC), 1993. Medical Devices Directive: Council Directive 93/42/EEC of 14 June 1993 concerning medical devices.

European Commission (EC), 2007. Medical Devices Directive; Directive 2007/47/EC of the European Parliament and of the council of 5 September 2007 amending Council Directive 93/42/EEC concerning medical devices.

International Organization for Standardization (ISO), 2008. BS EN ISO 9001 Quality Management Systems. International Organization for Standardization.

International Organization for Standardization (ISO), 2012. BS EN ISO 13485 Medical devices – Quality Management Systems – Requirements for Regulatory Purposes. International Organization for Standardization.

International Organization for Standardization (ISO), 2012. BS EN ISO 14971 Medical Devices – Application of Risk Management to Medical Devices. International Organization for Standardization.

Laver Fawcett, A., 2007. Principles of Assessment and Outcome Measurement for Occupational Therapists and Physiotherapists. John Wiley & Sons Ltd.

Morris, W., (no date). British craftsman, early Socialist, Designer and Poet, whose designs generated the Arts and Crafts Movement in England. (1834–1896).

NHS Choices, 2012. Dealing with challenging behaviour. <www.nhs.uk/CarersDirect/guide/practicalsupport/Pages/Challenging-behaviour.aspx>.

Orpwood, R.D., 1990. Design methodology for aids for the disabled. J. Med. Eng. Technol. 14 (1), 2–10.

Pahl, G., Beitz, W., 1988. Engineering Design – A Systematic Approach. In: Wallace, K. (Ed.), The Design Council. Springer-Verlag.

Pietzsch, J.B., Shluzas, L.A., Paté-Cornell, M.E., Yock, P.G., Lineham, J.H., 2009. Stage gate process for the development of medical devices. J. Med. Device. 3, 0201004-1, ASME.

Pope, P.M., 1988. A model for evaluation of input in relation to outcome in severely brain damaged patients. Physiotherapy. 74 (12), 647–650.

Pope, P.M., 2006. A physical management programme for people with severe disability. In: Lecture Notes: Posture Management for Adults and Children with Complex Disabilities. Oxford.

Pugh, S., 1990. Total Design – Integrated Methods for Successful Product Engineering. Prentice Hall.

Pullin, G.M., 2009. Design Meets Disability. MIT Press.

Ulrich, K., Eppinger, S., 2011. Product Design and Development. McGraw Hill.

Further Reading

There is a wealth of literature available on the general principles of risk management. In addition, the following two documents provide specific information.

Medicines and Healthcare Products Regulatory Agency (MHRA) 2004. Guidance on the Stability of Wheelchairs. Medicines and Healthcare Products Regulatory Agency.

Institute of Physics and Engineering in Medicine (IPEM) 2008. Risk Management and its Application to Medical Device Management Report 95 Institute of Physics and Engineering in Medicine.

Local employer risk assessment tools may provide useful information and guidance, but are unlikely to be suited to all applications.

18

Functional Electrical Stimulation

Duncan Wood and Ian Swain

Salisbury NHS Foundation Trust

INTRODUCTION

Functional electrical stimulation (FES) is a means of producing useful movement in paralyzed muscle. Electrical stimulation is not a new technique: it dates back to the Ancient Greeks who used rubbed amber and torpedo fish to produce a number of physiological responses, primarily to cause muscle contractions. Its development followed that of advances in physics by Volta and Faraday during the eighteenth and nineteenth centuries which led to more reliable and controllable sources of electricity and advances in our understanding of neurophysiology as a result of the work of Galvani and Duchenne during that same period. Following these advances, various groups showed that denervated muscle only responded to stimulation by connecting and disconnecting a direct current source and not to alternating or Faradic current. However, in the case of an innervated muscle (i.e., where there was an intact motor neuron), a contraction did occur with Faradic current. This opened up the possibility of using electrical stimulation to restore some level of muscle contraction, with the goal to aid function, and remains the focus for this chapter.

Since those early days, electrical stimulation has been used in many medical conditions to restore function and movement. These include assisting cardiac function through the use of pacemakers, controlling bladder, bowel, and sexual function in spinal cord injured persons, and improving hearing ability through implanted devices in the cochlear. Some applications are now widely used in the clinical environment, including the applications listed above, whereas others, such as restoring visual feedback, are still in the research stage with limited clinical trials. One other major application is in the use of electrical stimulation to restore limb movement and hence improve function and activities of daily living. Though this application is certainly applicable clinically, in some cases without the direct need for clinical engineering support, it is still an emerging field in rehabilitation. This chapter describes one of the more successful applications, drop foot correction, concentrating more on the thought processes undertaken during the initial design as well as giving a basic overview of how electrical stimulation works.

DEFINING FES

As stated, FES is a means of producing useful movement in paralyzed muscle. This differs from perhaps the wider definition for electrical stimulation (sometimes called neuromuscular electrical stimulation) and that in itself presents some of the main clinical engineering challenges. For example, if it is possible to produce a muscle contraction in an innervated muscle by the simple technique of applying a Faradic current, what methods can be used to control both that contraction and the resulting limb movement? Both aspects therefore need to be considered: the physiological principles underlying FES and the design concepts relating to the desired application to make it functional and clinically relevant.

PHYSIOLOGICAL PRINCIPLES OF FES

In FES, small electrical impulses are applied to the nerves that supply the affected muscles using either self-adhesive electrodes placed on the skin or implanted electrodes on the nerve or muscle close to the motor point. The electrical current generates an electric field between the pair of electrodes (Figure 18.1), and, with the right conditions, may induce a nerve impulse that is propagated along the nerve to the muscle, causing the muscle to contract in a manner very similar to natural contraction. Though there are obvious differences between these two delivery techniques, for this section they can be considered in the same way.

As a result, each nerve in the vicinity of this electric field may be excited. What is meant by the phrase "the right conditions" is that a nerve is required to be in an excitable state and that the level of stimulation intensity needs to be

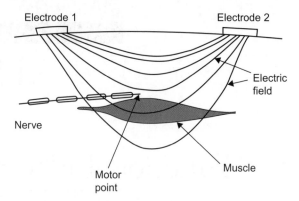

FIGURE 18.1 Electric field between a pair of stimulation electrodes on the skin surface. The closeness of the electric field lines nearer the electrodes indicates the higher current density in that region.

sufficient to cause excitation. It needs to be remembered that just as in normal nerve excitation, the all-or-none principle is maintained. The characteristics of the electrical pulses are therefore important: the amplitude, the pulsewidth, and frequency of electrical pulses. This perhaps over-simplifies the nature of these pulses, but is fundamental to how a train of pulses can generate a required movement useful for function.

The amplitude and pulsewidth can be regarded as being synonymous with the stimulation intensity. Below a certain level of intensity there is no muscle response because the level is insufficient to cause any nerve excitation. The point where nerves begin to become excited is called the threshold of stimulation. As the intensity increases, the response curve follows the classic S-shape, with a steep linear slope over much of the middle part of the curve, leading to a plateau, where the resulting muscle response does not increase even with an increase in intensity (Figure 18.2). This increase in muscle response over the S-shaped curve is caused by more motor units being recruited, partly due to the electric field penetrating deeper (i.e., more nerves are in its vicinity) and partly because the intensity is now above the excitability threshold of more

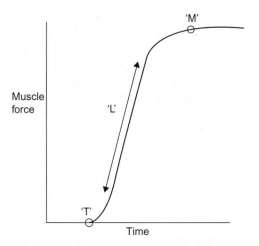

FIGURE 18.2 Effect of stimulation intensity on muscle response. The threshold of stimulation is indicated by "T," the region where the stimulation response plateaus is indicated at levels above "M," and the steep linear slope is indicated by the region "L" (adapted from Baker et al. 1993).

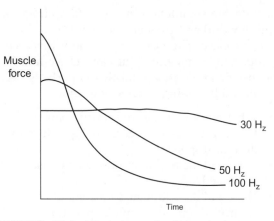

FIGURE 18.4 Effect from stimulation frequency on muscle fatigue, showing that with a sustained contraction at an increased frequency the response reduces more quickly (adapted from Baker et al. 1993).

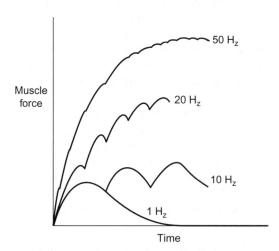

FIGURE 18.3 Effect from stimulation frequency on muscle response, illustrating the higher force generated with a higher frequency and the level of graded control (or smoothness in response) at the different frequencies (adapted from Baker et al. 1993).

frequency (i.e., reducing the inter-pulse interval) also causes an increase in force produced from the muscle, but not from exciting more motor units. The cause here is from summating contractions from the same nerve (or nerves) being excited, since increasing the frequency does not allow the muscle to return to its state of rest between pulses (Figure 18.3). At a certain frequency, these contractions summate to the point where they become fused; this is termed *tetany*. For most functional activities, a fused, or smooth, contraction is required, demanding a higher frequency, however a higher frequency shortens the rest time between pulses and therefore has the disadvantage of increased muscle fatigue (Figure 18.4). A compromise between a sustained smooth contraction and muscle fatigue must therefore be considered when selecting the stimulation frequency.

DESIGNING A PRACTICAL FES SYSTEM

We have now examined the characteristics of the train of electrical impulses and how

nerves. The plateau is at the point where no additional nerves can be excited.

The effect from the frequency of the stimulation pulses is slightly different. Increasing the setting the stimulation intensity (amplitude or pulsewidth) and frequency can affect the muscle response. The next stage is how this technique can be integrated into an FES system for a specific application, and this is perhaps best achieved by posing a series of questions:

- What outcome is desired for the patient?
- Which muscles and nerves will produce the desired response?
- How will the required muscle response be generated?
- How will the stimulated movement be controlled for the specific application?

It would be wrong, though, to assume that FES systems are necessarily simple in their design. As inferred earlier, FES has been used for different outcomes in mobility, such as grasping, standing, and walking and with many patient groups. In these cases, individual patients often present with individual movement patterns and deficits that need addressing, even though there are some generic characteristics. Their different needs and expectations, along with the recent advances in technology, have therefore led to different solutions being applied to FES systems. These have included:

- Using FES with multiple muscle groups, such as in the study by Kim et al. (2012), which demonstrated improvements in spatiotemporal parameters of gait when stimulating gluteus medius in the stance phase, alongside stimulating tibialis anterior in the swing phase. In the work that has come out of Cleveland in the United States, one example demonstrates an 8-channel implanted system to assist persons with spinal cord injuries to exercise, stand, and maneuver (Rohde et al., 2012).
- Using skin surface electrode arrays to accurately detect optimal electrode positions (e.g., Heller et al., 2013),

demonstrates the principle of an automated setup which could lead to FES becoming more viable for patients who, at present, have difficulty in setting up current commercial stimulators.
- Applying stimulation through implanted systems, two examples being the BION device from the Alfred Mann Institute in the United States which is injected in or near muscles (Popovic et al., 2007) and the STIMuSTEP device from Finetech Medical in the United Kingdom which uses electrodes inserted into the epineurium of a nerve (Kottink et al., 2007), as well as the previously mentioned work from Cleveland.
- Detecting physiological signals to control FES devices, such as cutaneous nerve signals to induce grasp force (Inmann and Haugland, 2012) and intramuscular (or surface) muscle activity (EMG) to detect the intention to step (Dutta et al., 2009).
- Applying new sensor detection methods to more accurately detect specific events in a movement pattern; for example, gyro and accelerometer measurements for gait events (Park et al., 2012).
- Employing different control methodologies to determine stimulation envelopes, such as the work of Kordjazi and Kobravi (2012) which used measured activity from a "healthy" muscle to train an artificial neural network to predict the activation pattern for a disabled muscle, in this case for drop foot correction.

This introduces the reader to the complexity of FES systems, but for simplicity and clarity of thought in discussing the steps involved in the concept design of an FES system, only one case study will be considered here, as mentioned above. That case is of using FES to correct for dropped foot following neurological damage, such as from a stroke, and is probably the most widely used application for FES in the United Kingdom. Its beauty is that it is relatively simple in nature, but has the potential of prompting further discussion when answering the previously posed questions.

THE DESIRED OUTCOME FOR THE PATIENT

Dropped foot is the inability to lift the foot as the leg swings forward when walking. It is caused by weakness in the muscles that lift the foot, the dorsiflexors such as tibialis anterior, and/or excessive activity (typically spasticity) in the antagonist muscles, such as the gastrocnemius. It is very common in stroke patients and can result in increased trips and falls, and reduced mobility, leading to loss of confidence, reduced social participation, and loss of independence. The problem can be addressed by using a passive ankle splint or ankle foot orthosis (AFO) and though these can be very effective, some people believe that a more active means, such as from using FES, can have additional advantages, such as promoting a more normal movement pattern following a stroke. Patients also often walk with excessive inversion that leads to poor foot placement

into the stance phase of gait and can often mean the patient being unstable as they bear weight through their affected side (the ankle sometimes "turns over"), as well as weight transference being limited since the foot becomes flat on the floor too early into stance, with the tibia having not advanced forward. Other more proximal problems and compensatory strategies also play a part in the gait of a person who has had a stroke, but these will be ignored at the current time for simplicity.

An FES system therefore needs to provide floor clearance during the swing phase and improved stability into stance. The first could be achieved by eliciting dorsiflexion, whereas the second could be addressed through ensuring a positive heel strike with eversion at initial contact and through the loading response phase to aid weight transference. This can be shown pictorially with reference to a gait cycle, as shown in Figure 18.5. It is perhaps worth noting here, however, that for some people a well-tuned AFO could provide more effective control during the stance phase, but would limit dorsiflexion during swing. (Further information on the gait cycle may be found in Chapter 25, "Clinical Gait Analysis.")

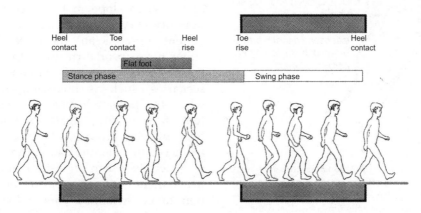

FIGURE 18.5 Representation of the gait cycle and stimulation phases for drop foot correction. A typical gait cycle from an unimpaired person is illustrated, with the stimulation phases represented by blocks from toe rise through to toe contact for floor clearance and tibial progression, respectively.

CORRECT MUSCLES AND NERVES TO PRODUCE THE DESIRED RESPONSE

In this application, dorsiflexion with eversion is required (again, the more proximal muscles have been ignored in this case study). This is produced by activating the tibialis anterior muscle, but since this also produces inversion its response needs to be somewhat compensated for by activating the peronei muscles for eversion. Pairs of electrodes could be placed over both muscle groups, but by understanding the physiology and also how unimpaired walking is achieved, another solution can be achieved by considering not the muscle groups and motor units, but the nerve supplying those two muscle groups, that is, the common peroneal nerve. By using just two

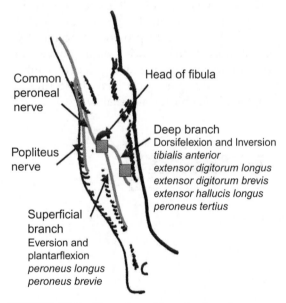

FIGURE 18.6 Electrode positions for drop foot correction, illustrating the anatomy of the common peroneal nerve with both the deep and superficial branches shown. One electrode is placed over the fibula head, and the other is placed more anteriorly and distally over the bulk of tibialis anterior (from Odstock Medical Ltd. Clinician's Instruction Manual 2012).

electrodes, one over the motor point of tibialis anterior and the second over the fibula head, where the common peroneal nerve typically branches into its deep and superficial branches (with motor units at tibialis anterior and peronei muscles, respectively), careful electrode positioning can produce the desired response (Figure 18.6). The added advantage of stimulating at the nerve is that it can also sometimes result in a flexion withdrawal reflex by stimulating the 1a afferent fibres causing hip flexion with abduction and external rotation, knee flexion, and dorsiflexion and eversion, and hence improved responses at the more proximal joints.

GENERATING THE REQUIRED MUSCLE RESPONSE

It has already been explained how a series of electrical stimulation pulses can produce a muscle response. However, before discussing how these can be controlled for this application, it may be worthwhile considering how to "set" the three basic parameters, that is, amplitude, pulsewidth, and frequency. Stimulation intensity (amplitude-pulsewidth) is generally set to be at the middle portion of the steep slope in the muscle response—intensity curve (Figure 18.3). This allows the patient (or clinician) the dynamic range to either turn up or turn down the intensity as the day progresses to accommodate different scenarios; such as increasing the intensity when the patient begins to fatigue and the foot starts to drop again, or when walking over uneven ground and more foot lift is required. In this application the stance phase is generally when the stimulation is "off," hence there is no requirement for the contraction to be sustained over a long time and hence fatigue occurring. The frequency can therefore be set slightly higher than for other FES applications, with the added advantage

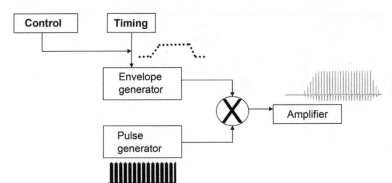

FIGURE 18.7 Basic design of an FES system illustrating the envelope and pulse generator. Two main components are shown: a pulse and an envelope generator. Though it is possible for both components to be placed under some level of control (with or without feedback), in this case only the timing and control aspects of the envelope generator are shown.

of increased speed of muscle response and possibly high enough to more likely elicit the withdrawal reflex. For stimulation of more postural muscles, where stimulation is continuous, it may be necessary to reduce the frequency.

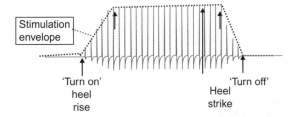

FIGURE 18.8 Representation of the FES envelope. The simple transition points are indicated by the arrows, alongside the solution selected for their practical implementation in a dropped foot system.

HOW TO CONTROL THE STIMULATED MOVEMENT FOR THE SPECIFIC APPLICATION

The FES system consists of a simple pulse generator which is amplified to deliver the stimulation pulses to the body to cause a contraction. However, for this application the stimulation is required to be "on" at certain times of the gait cycle and "off" at others. This perhaps provides the most challenging, and at the same time interesting, aspects of designing a practical FES system: How will the muscle responses be controlled for the required application? A simple way of looking at this is to think of the FES system being the pulse generator plus amplifier, alongside an envelope generator, as shown in Figure 18.7.

Looking now just at the envelope, Figure 18.8, the transition points need to be determined. For the simple envelope described, these transition points are at "turn on" and "turn off" (i.e., at the start and end of the ramp periods). Earlier it was illustrated

that the stimulation needs to be on during the swing phase, with some extension into early stance (see Figure 18.5). For this relatively simple case study, gait events at the foot (i.e., heel/toe strike/rise) could be considered to be used for these two trigger points, that is, turn on and turn off. One solution may be to use toe rise (to turn the stimulation on at the start of swing) and toe strike (to turn the stimulation off toward mid-stance). This would work, but it does require the stimulation to turn on instantly at toe rise to ensure sufficient lift at early swing. Bringing this trigger point further forward in the gait cycle would now take it to the previous gait event (i.e., heel rise), ensuring that the foot is sufficiently prepared prior to the start of swing. It also allows for there to be an up ramp period which is essential to minimize any responses from stretching the

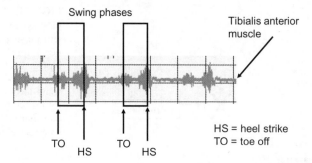

FIGURE 18.9   Surface EMG recording from tibialis anterior muscle in unimpaired walking. Two bursts of muscle activity are observed: the first during the first part of the swing phase for floor clearance, and the second toward the end of swing, but extending into early stance for eccentric activity to control the lowering of the foot to the ground for weight transference.

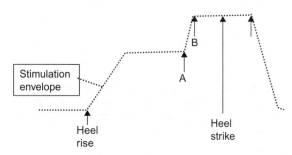

FIGURE 18.10   A proposed new FES envelope to reflect the real activity in tibialis anterior. This has an envelope similar to Figure 18.8, but with an increased stimulation intensity at the time relating to the second burst of unimpaired EMG activity. The challenge, then, is to accurately determine the appropriate point for A and B (*note*: there is no floor contact since these points are during swing).

antagonist muscle too quickly and eliciting a spastic response and plantarflexion. Such a system now has its "on" trigger at heel rise and its "off" trigger at toe strike. This would require switches under both the heel and the toe (or first metatarsal head), so a simpler solution could be to use just one switch under the heel, with the "off" trigger now at heel strike and just simply extending the stimulation into stance for a fixed time, hence turning off by mid-stance. This is the approach taken by the team at the National Clinical FES Centre,

Salisbury, United Kingdom with its Odstock Dropped Foot Stimulator system (www. odstockmedical.com).

This envelope is relatively straightforward, but, even for this simple application, there are some strong grounds to make it more complex. First, it could be argued that the "off" trigger should reflect the actual time when the tibia has progressed forward into stance. Using a second switch at the toe goes some way to achieving this, however a sensor on the tibia could make that more accurate by determining its actual position relative to the vertical, or by investigating the acceleration and deceleration components of its movement. Second, in the scenario described, a simple up-hold-down envelope is used when, in fact, the tibialis anterior muscle in unimpaired physiological gait is seen to have two bursts of activity, as shown in Figure 18.9. The first burst is replicated with the dropped foot system for floor clearance and the second burst relates to eccentric activity of that muscle as the tibia advances forward, as replicated in the extension phase of the stimulation envelope. The issue here is that the second burst is typically stronger than the first and so perhaps the envelope should be more like that shown in Figure 18.10. The challenge, then, is how to control the two levels of the envelope and also when to make the transition from the first to the second burst. Again, this would need to be quite accurately timed to ensure that the gait is not to be impeded and therefore would require additional movement or positional sensors. Examples of this work can be found in the papers by Chen et al. (2010) and O'Halloran et al. (2004). This may make the system more complicated, however, possibly less reliable, and certainly more expensive and hence less likely to be used routinely in a healthcare system such as the NHS.

CONCLUSIONS

FES has been around for many years, but has tended to stay in the research laboratories of academic and clinical institutions. There have been some successes in using FES more clinically, but these have been slow, partly reflecting the limited resources applied to this emerging field. However, as a technique, it does have considerable potential to restore limb movement and as such needs to be taken further.

In this chapter, the basic principles of FES have been considered alongside the specification and design of an FES system for a specific application, that is, dropped foot. It is imperative for there to be cooperation between engineering, scientific, medical, and therapy staff in the clinical environment to specify, design, test, and evaluate such products for patients. Clinical engineers need to be at the forefront of that to ensure that the design requirements and specification are correct, and not just contributing to the engineering design and construction of the product. The testing and evaluation is equally important for all staff to be involved with so that the ergonomics and usability of the product, for both clinicians and patients, can be appropriate and that the functional outcomes from patients using the products can be assessed objectively.

For further information related to the medical applications of electrical stimulation visit the websites of the International FES Society (http://ifess.org/) and the International Neuromodulation Society (www.neuromodulation.com).

For a more detailed description of how nerves are excited, the reader is referred to standard text books on (neuro) physiology.

References

Chen, M., Wang, Q.B., Lou, X.X., Xu, K., Zheng, X.X., 2010. A foot drop correcting FES envelope design method using tibialis anterior EMG during healthy gait with a new walking speed control strategy. Conf. Proc. IEEE Eng. Med. Biol. Soc. 4906–4909.

Dutta, A., Kobetic, R., Triolo, R., 2009. Development of an implanted intramuscular EMG-triggered FES system for ambulation after incomplete spinal cord injury. Conf. Proc. IEEE Eng. Med. Biol. Soc. 6793–6797.

Heller, B.W., Clarke, A.J., Good, T.R., Healey, T.J., Nair, S., Pratt, E.J., Reeves, M.L., van der Meulen, J.M., Barker, A.T., 2013. Automated setup of FES for drop foot using a novel 64 channel prototype stimulator and electrode array: results from a gait lab based study. Med. Eng. Phys. 35 (1), 74–81.

Inmann, A., Haugland, M., 2012. Regulation of FES-induced grasp force based on cutaneous nerve signals: experiments and modelling. Med. Eng. Phys. 34 (1), 467–1255.

Kim, J.H., Chung, Y., Kim, Y., Hwang, S., 2012. FES applied to gluteus medius and tibialis anterior corresponding gait cycle for stroke. Gait Posture. 36 (1), 65–67.

Kordjazi, N., Kobravi, H.R., 2012. Control of tibialis anterior FES envelope for unilateral drop foot gait correction using NARX neural network. Conf. Proc. IEEE. Eng. Med. Bio. Soc. 1880–1883.

Kottink, A.I., Hermans, H.J., Nene, A.V., Tenniglo, M.J., van der Aa, H.E., Buschman, H.P., Ijzerman, M.J., 2007. A randomised controlled trial of an implantable 2-channel peroneal nerve stimulator on walking speed and activity in post-stroke hemiplegia. Arch. Phys. Med. Rehabil. 88 (8), 971–978.

O'Halloran, T., Haugland, M., Lyons, G.M., Sinkjaer, T., 2004. An investigation of the effect of modifying stimulation profile shape on the loading response phase of gait, during FES-corrected drop foot: stimulation profile and loading response. Neuromodulation. 7 (2), 113–125.

Park, S., Ryu, K., Kim, J., Son, J., Kim, Y., 2012. Verification of accuracy and validity of gait phase detection system using motion sensors for applying walking assistive FES. Comput. Methods Biomec. Biomed. Eng. 15 (11), 1129–1135.

Popovic, D., Baker, L.L., Loeb, G.E., 2007. Recruitment and comfort of BION implanted electrical stimulation: implications for FES applications. IEEE Trans. Neural Syst. Rehabil. Eng. 15 (4), 577–586.

Rohde, L.M., Bonder, B.R., Triolo, R.J., 2012. Exploratory study of perceived quality of life with implanted standing neuroprostheses. J. Rehabil. Res. Dev. 49 (2), 265–278.

Further Reading

Baker, L.L., McNeal, D.R., Benton, L.A., Bowman, B.R., Waters, R.L., 1993. Neuromuscular Electrical Stimulation – A Practical Guide. Published by the Rancho Los Amigos Research and Education Institute.

Rushton, D.N., 2003. Functional electrical stimulation and rehabilitation – an hypothesis. Med. Eng. Phys. 25 (1), 75–78.

Swain, I.D., Taylor, P.N., 2004. The clinical use of functional electrical stimulation in neurological rehabilitation. In: Franklyn, J. (Ed.), Horizons in Medicine 16 – Updates on major clinical advances. Pub. Royal College of Physicians, London, pp. 315–322.

Functional Electrical Stimulation for drop foot of central neurological origin: Buyer's Guide (CEP10010), Market Review (CEP10011) and Economic Report (CEP10012). An independent evaluation by the NHS – visit the Cedar website (www.cedar.wales.nhs.uk/home) for access.

For a more detailed description of how nerves are excited, the reader is referred to standard text books on (neuro) physiology.

Posture Management

David Long
Oxford University Hospitals NHS Trust

INTRODUCTION

The content of this chapter has been drawn principally from the subject matter of the posture management courses provided by physiotherapists Wendy Murphy and Pat Postill and the author (clinical scientist David Long), which were developed originally by physiotherapist Pauline Pope and educational advisor Janet Wells. The underpinning theories and principles are documented in the book *Severe and Complex Neurological Disability* (Pope, 2007), published by Elsevier. The work of Dr. Linda Marks, a consultant in rehabilitation medicine, physiotherapist Linda Walker, and clinical scientist Phil Swann has also been influential.

The chapter is aimed primarily at nonambulant individuals having significant limitations to movement, although the principles may be applied more broadly. Please note that it should be considered an outline of posture management, this being a highly complex subject area. A substantial amount of clinical practice with experienced colleagues is necessary to attain proficiency.

POSTURE

One often hears of good postures and bad postures, but posture can be a difficult concept to define. What might be a good posture for typing an email would be a bad one for relaxing, or a good posture adopted within a military parade a bad one for talking casually with friends. Pope (2007) writes that posture is "... the attitude or configuration of the body." Expanding on this, she suggests that it is the ability to organize and stabilize the body segments relative to one another, and then to be able to offload one segment without losing overall stability. Able-bodied people take for granted the ability to lean forward or to one side to reach for an object out of their immediate grasp, but for the physically disabled person this can be enormously challenging, if not impossible. This brings us to functional activity which is a critical aspect of posture management because, ultimately, we arrange ourselves according to the task in hand. Consider how your posture changes as you carry out different activities. One makes these changes to optimize functional performance and to conserve energy. We organize, balance, and stabilize our body segments relative to the supporting surface and are able to respond to changing requirements and external forces.

Able-bodied people are able to move in and out of position at will and in response to stimuli, such as tiredness or discomfort/pain. Importantly, what might be considered a good posture for a particular activity might also be a damaging one if it is sustained over an extended period of time. Those less able to move are at high risk of damage to the body system (i.e., plastic change), which can take place in two ways:

- Orthopedic: The development of soft tissue contracture (muscles, tendons, and ligaments) and change in bony shape, leading to secondary complications such as impaired internal organ function and pressure ulceration.
- Neurological: The laying down of altered neural pathways through repetitive asymmetrical, atypical, unusual, or anomalous movements, having a negative effect (Scrutton, 1991).

The ability to organize one's posture is learnt: it is not something with which we are born. We gradually adapt our primitive reflexes to functional movement, learning how to hold our head up, roll over, sit up, crawl, stand, walk, and run. The steps in this process have been researched extensively by Chailey Heritage Clinical Services which published the Chailey Scales of Postural Ability for sitting, standing, and lying (Pountney et al., 2004).

EFFECTS OF GRAVITY

The effects of gravity are profoundly important for posture. They can be used positively to secure body segments in a particular position and can have significant negative effects in causing the body to buckle and bend.

FIGURE 19.1   Human sandwich.

Gravity is (more or less) a constant force, that is to say one cannot escape it. There is, however, a common, misplaced conception that gravity has no impact on posture in lying. It has just as much effect, of course, the difference being that the body is orientated to it in a different plane from sitting or standing.

A helpful visualization of posture was described by Whitman (1924): "a constant struggle against the force of gravity." This was further developed by Hare (1987) who presented the idea of the "human sandwich" (Figure 19.1), where we are the filling being held in place and compressed between the force of gravity and the surface of the Earth (or the supporting surface). When able-bodied people become tired of standing, they sit to conserve energy; when they become tired of sitting, they lie down. In this position, minimum energy is required, thus resting the musculoskeletal system. This is a profoundly important concept in relation to people with a physical disability.

THE ROLE OF THE SUPPORTING SURFACE

The supporting surface has a vital part to play in posture control. Take, for example, sitting erect on a stool. One must be able to balance on a small area of support and sustain relatively high pressures because the load is being distributed over a small area. One tires of such a position relatively quickly and so additional support is recruited, perhaps by moving the stool near a wall against which one's back can be leant. The legs may also be crossed to secure the position of the foot in contact with the ground/support surface to help secure the position of the pelvis. These changes in position not only offload some of the weight from the ischial tuberosities but also widen the base of support, making the posture more stable and energy efficient. As further tiredness creeps in, one may wish to sit in an armchair, which offers a more tilted position, allowing gravity to help secure a position. Some people choose to sit with their legs flexed and placed to one side, the heels positioned close to or even underneath the pelvis, perhaps leaning on the arm of a couch. It is becoming clear, then, that functional postures are not necessarily symmetrical.

While we require stability to function, it should be remembered that in offloading body segments we create partial instability. It is critical that in providing a piece of equipment for posture management we do not provide so much stability that function is impaired; there has to be a balance between the two. One must ascertain the level of support required set against the functional requirements.

24-HOUR POSTURAL MANAGEMENT

In providing a piece of equipment for posture management it is necessary to consider the full 24-hour period of each day (Gericke, 2006; Pope, 2007; Pountney et al., 2004). There is often great focus on providing seating for use in the daytime, whether this be an armchair, classroom chair, or wheelchair, but an area often neglected is positioning requirements at night. It is not difficult to conceive that the amount of time spent in bed can consume anywhere between a quarter and two-thirds of the 24-hour period. Even the smaller of the two figures is a significant proportion of time and this is important because postures adopted at night can have a profound effect on postures adopted during the day (Goldsmith and Goldsmith, 1998). Tissue adaptation caused by poor positioning in bed affects positioning in sitting; discomfort in lying may transfer to sitting; lack of sleep caused by discomfort may reduce sitting tolerance in the day; pressure management issues in bed may create difficulties in sitting.

Getting the right balance of support, comfort, and ease of use is a difficult result to achieve, but this should not stop the clinician involved in one area of posture management from addressing the other areas, which may require making onward referral to other professionals or services.

BIOMECHANICS AS APPLIED TO POSTURAL MANAGEMENT

This section does not cover basic mechanical concepts, nor does it provide any degree of depth in what is a very complex subject. The concepts of biomechanics most critical to the implementation of 24-hour postural management are discussed. There are many texts available for both mechanics and biomechanics for those who wish to pursue the subject further.

Shearing Forces

Where two masses pass each other in opposite directions, shear forces are generated at

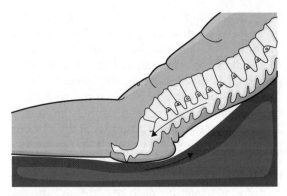

FIGURE 19.2   Patient sliding in bed.

the interface. These comprise a mixture of tensile and compressive forces which also develop heat. An example of this is in the development of tissue ulceration, typified by the hospitalized patient being propped up in a semisitting position with the legs out straight and the head of the bed raised (Figure 19.2), but is applicable to many seated postures. The ischial tuberosities and sacrum travel forward because the center of mass of the person falls behind (outside) their base of support and because a slope has been created down which the trunk can slide. However, motion of the soft tissues between the pelvis and supporting surface is impeded by friction at the surface of the skin.

Development of pressure ulceration is discussed in greater detail in Chapter 20 "Pressure Ulceration."

Stability and Instability

These concepts are central to posture management. We require stability to conserve energy and to improve certain functions, but we also require instability to facilitate other functions. Static stability has three basic attributes:

1. The center of mass must fall within the base of support.

2. The wider the base of support, the greater the stability.
3. The lower the center of mass, the greater the stability.

These principles are carried through the whole of this chapter, both in terms of securing a position and in facilitating functional movement.

Inertia

Inertia has to do with the effort required to overcome a static or stable position. It is more effortful to stand up from a low couch than it is from a dining chair: the center of mass must be moved further forward and raised through a greater height. It translates, then, that a tilted seating system will create greater stability than an upright wheelchair. There are many other clinical examples.

Stress and Strain; Elasticity and Plasticity

We know that an elastic response to loading means that a material will return to its original shape and length when the load is removed. We also know that too much loading will give a plastic response when the elastic limit is passed, causing permanent deformation. These properties are depicted by a stress/strain curve, as shown in Figure 19.3.

What we are generally more concerned with in posture management is sustained loading/positioning over time, often measured in hours and days in a rehabilitation setting post-stroke, spinal cord injury, or head injury, and in months and years for anyone with a long-term condition with reduced or impaired movement. In materials science the term *creep* refers to the situation where an increase in strain can be detected while the applied load remains constant over time, as shown in Figure 19.4. This leads ultimately to failure in static materials, but in biological materials we can observe plastic adaptation to structure, both in bone and soft tissue, which presents clinically as fixed postural

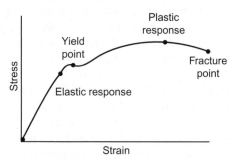

FIGURE 19.3 Sample stress/strain curve.

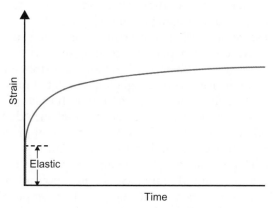

FIGURE 19.4 Sample strain/time curve for a constant applied load.

asymmetry, such as scoliosis and kyphosis, and as joint contracture (i.e., a limitation in movement). Notice in the graph that there is an immediate jump in strain as the load is applied, which corresponds to the elastic response of the material, but that after this the rate of increase is reduced.

Moments

Moments help us to describe the actions and results of forces applied about a pivot point or fulcrum; they are the product of force and distance (Figure 19.5).

In the human body we have long levers in the arms and legs. The spine is multijointed and many moments act simultaneously. As a result, supporting the body and/or facilitating movement requires an understanding and application of the theory of moments. If the leg of a seated person tends to fall out to the side of the chair, a corrective force can be applied to control the position of that leg. The force is most effective when placed distally (i.e., at the knee), and so may be reduced in magnitude compared to application at mid-thigh. One may go a step further and support the thigh along its length so that the load is distributed over a larger area, allowing the applied force to be reduced further still.

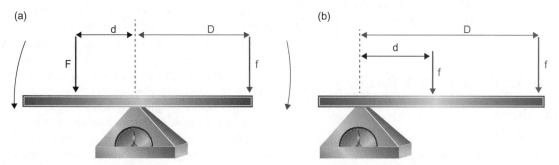

FIGURE 19.5 Moments: (a) the smaller force is able to balance the larger force because it is placed further away from the fulcrum; (b) the same force produces less mechanical leverage if applied closer to the fulcrum.

Day-to-Day Clinical Applications

These biomechanical principles are readily applied to many other clinical scenarios. To some, this is intuitive, perhaps most of all to the engineer, but the ability to explain what is happening and communicate this in a meaningful way to the patient is very important. Why is a support needed in a particular place? Why is a particular orientation of sitting/standing/lying required? The person and/or their relatives or carers must be convinced of the value of the advice they are being given, a clear description of the reasoning behind decisions greatly aiding this process.

INTRODUCTION TO THE PHYSICAL ASSESSMENT

General principles of patient assessment for rehabilitation engineering and assistive technology were covered Chapter 17 "Introduction: Medical Engineering Design, Regulations, and Risk Management." To develop a program of physical management over a 24-hour period, one must take great care to establish all relevant physical abilities and limitations. This is not a quick process and with some patients will be physically demanding of the assessor, as well as of the patient. It is important for the patient to feel comfortable with being physically handled/ touched. What may be routine for the professional may feel extremely intrusive to the person concerned. In some instances it will be appropriate to request the presence of a chaperone.

BODY CONFIGURATION

Introduction

The process starts with an assessment of the presenting body configuration, usually in sitting. It is important that this is a typical presentation: there is always a temptation for the person or their carers to correct what they deem to be a "poor" posture. It is necessary to see the person as they typically sit to establish any postural tendencies and patterns of movement; for example, leaning to the side, sliding forward in the seat, or head dropped forward/ to the side. It is advisable to take photographs, one from the front and one from each side, keeping the camera on the level of the person to avoid perspective distortion.

The international standard ISO 16840-1:2006 "Wheelchair seating — Part 1: Vocabulary, reference axis convention and measures for body segments, posture and postural support surfaces" provides a method for defining body configuration in sitting. This standard is designed to "... specify standardized geometric terms and definitions for describing and quantifying a person's anthropometric measures and seated posture, as well as the spatial orientation and dimensions of a person's seating support surfaces" (ISO, 2006). The method is extremely comprehensive, providing a full data set of measures. While it clearly has a very useful place in a research setting, the level of detail included makes it more difficult to apply in clinical practice, although Waugh and Crane (2013) have developed an accompanying document that provides clinical guidance on its application. There are elements that may clearly be applied successfully, although the terminology is inconsistent with that used widely in healthcare settings, notably by physiotherapists (or physical therapists), acknowledging that this terminology itself is not always applied consistently.

More significantly, it is not possible to apply the standard to the practice of 24-hour posture management because it does not provide a means to describe lying or standing postures. In addition, the detail of spinal shapes, notably those of a complex nature, are also difficult to describe using the prescribed terminology. The standard does, on the other hand, illustrate clearly the complexities of describing body orientation and shape and has, undoubtedly, moved forward the science underpinning our understanding of body configuration.

In this chapter we use the more conventional terms and reference this standard where applicable to illustrate its potential use.

Pelvic Orientation

The first measurement is of pelvic symmetry or asymmetry and is described in three planes of motion: obliquity, rotation, and tilt. It is important that these are understood clearly as they form the basis of postural support, which then allows the more complex picture of body segment organization to be established (Pope, 2007; Frischhut et al., 2000).

To define the position of the pelvis, it is first necessary to find two alike bony landmarks, typically the anterior superior iliac spines (ASISs). These are easily visible in an anatomy text and on people having little excess tissue, but may be difficult to find on some patients. Where the pelvis is so oblique that one ASIS is tucked up inside the lower portion of the rib cage, one cannot usually reach the ASIS and so some other landmark must be used. In these cases it is possible to palpate for the posterior superior iliac spines, iliac crests, or the ischial tuberosities. It is also possible to palpate for the greater trochanters of the hips but this may lead to a false reading where the hip is asymmetric or dislocated, so should be avoided.

- Pelvic obliquity, measured relative to the horizontal, is identified where one ASIS is higher than the other (Figure 19.6(a)). It is referred to in ISO16840-1 as the *frontal pelvic angle*. Ideally, an angular measurement would be recorded, but it can be difficult to align a goniometer with the ASISs and reliable visual estimation of small angles is difficult to achieve. An alternative is to make a visual estimation of the linear difference in height of one ASIS compared to the other. However, the significance of

this will vary according to the overall distance between the ASISs, most notably between small children and adults. In spite of this limitation, a linear measurement allows comparison between postures and between clinic appointments, having greater reliability where it is taken by the same assessor. Note that the ischial tuberosities are closer together than the ASISs, so the differences in their height will be less.

- Pelvic rotation, referred to in ISO16840-1 as the *transverse pelvic angle*, describes the situation where one ASIS is forward of the other (Figure 19.6(b)), i.e., one is more forward of an imaginary or real symmetrical flat seat back. Again, an angular measurement would be ideal but a visual linear estimation is easier to attain, acknowledging that it has the same limitations as described above.

- Where a line drawn between the ASIS and PSIS (posterior superior iliac spine) is horizontal, as viewed in the sagittal plane, the pelvis is said to be in neutral tilt (Figure 19.6(c)). Pelvic tilt is referred to in ISO16840-1 as the *sagittal pelvic angle*. If this line tips down at the front (relative to the horizontal) the pelvis is in anterior tilt (Figure 19.6(c)) and will usually be accompanied by an enhanced (hyper) lumbar lordosis (see the following subsection "Spinal Alignment"). If the line tips down at the back, the pelvis is in posterior tilt (Figure 19.6(c)) and will usually be accompanied by a flattened lumbar spine. Remember that in normal anatomy one would expect to find a gentle inward curve (lordosis) in the lumbar region of the spine. An angle would be helpful to record but without instrumentation this is difficult to estimate. Physiotherapists use the terminology of " + " to indicate a slight increase from normal/neutral, "+ +" and "+ + +" to indicate increasing amounts.

(a)

(b)

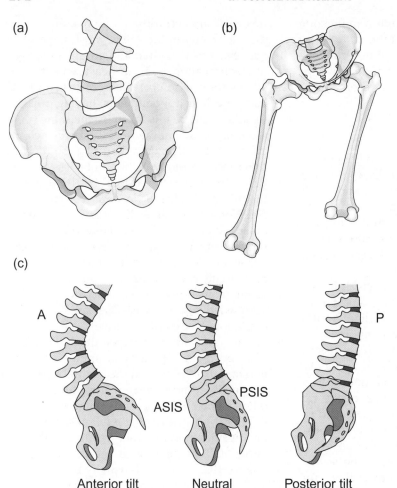

(c)

A

ASIS

PSIS

P

Anterior tilt   Neutral   Posterior tilt

FIGURE 19.6 (a) Pelvic obliquity: viewed from the front, the left side of the pelvis is raised. (b) Pelvic rotation: viewed from above the pelvis is rotated forward on the right side, toward the left. (c) Pelvic tilt: anterior, neutral, and posterior orientations.

Alignment of the Hips

Having noted the alignment of the pelvis, the symmetry or otherwise of the hips may then be determined, but it is important to remember that you are now noting position relative to the *pelvis* and not to any other plane. In this way, the *actual* position of the body segment is described, that is, it is a relative measurement, not an absolute. A grossly adducted hip, for example, would not only indicate a high chance of subluxation or dislocation of the hip, but would also require accommodation to

avoid inducing pelvic rotation. An absolute measurement may mask the severity of the adduction where there is pelvic rotation forward on the opposite side, were the measurement system not to be fully understood.

- Abduction/adduction is measured relative to pelvic rotation in sitting, i.e., the amount of true abduction is increased where the pelvis is forward on the same side as the measured hip, and is decreased where the pelvis is forward on the opposite side.
- Internal/external rotation is measured relative to pelvic obliquity in sitting, i.e., the

true amount of external rotation is less than the apparent measurement where the pelvis is oblique up on the opposite side, and is more where the pelvis is raised on the same side.

Again, the " + " and " + +" terminology can be a useful means of denoting the severity of asymmetry and in making comparisons between hips. An angular measurement could also be used.

Position of Feet

Next, the position and orientation of the feet is described. Are they resting on the supports or falling off the back, front, or sides? Do they become trapped between the footplates? Are they restrained by strapping and, if so, why, and is this all the time? Are the feet angled down at the front (plantarflexed), at the back (dorsiflexed), supinated (turned in, as if to clap the soles together), pronated (turned out), varus (toes closer together than heels), or valgus (toes further apart than heels)? Are shoes worn?

Orientation of the Shoulder Girdle

Having completed the assessment of the pelvis and lower limbs, we start to think about the trunk, but before this it is necessary to analyze the orientation of the shoulder girdle. Obliquity and rotation is described in the same way and with the same reference planes as the pelvis, but one must be mindful that the shoulder girdle is not a fixed structure like the pelvis. The shoulders can independently protract (move anteriorly from neutral), retract (move posteriorly from neutral), elevate, and drop, often with minimal impact on the shape of the spine.

Spinal Alignment

Analyzing this can be challenging as the spine is multijointed and highly flexible,

allowing very complex shapes to emerge. The term *scoliosis* is used to describe a sideways curve, as viewed in the frontal plane, and is referred to as either convex or concave. It is not important which of these terms is used, only that one is used consistently. Figure 19.7(a) shows a simple scoliotic curve, running from sacrum to occiput, which could be described as either convex left or concave right.

ISO16840-1 would refer to the frontal sternal angle to describe such a curve, although this becomes more challenging in Figure 19.7(b) which shows a more complex shape having two component curves: convex right in the thoracolumbar region and convex left in the upper thoracic region.

Scolioses give rise to and are caused by rotation in the spine and this in turn causes asymmetry in the ribs, usually referred to as a rib "prominence" or "fullness." The term "rib hump" has negative connotations and should be avoided, similarly the term "deformity." A typical prominence is shown in Figure 19.8, in this example on the right side. Also note that the pelvis is significantly oblique up on the left, which has resulted in the person sitting on and taking weight not only through the right ischial tuberosity but also the right greater trochanter.

Posterior prominences are very important as they can throw postural alignment if not adequately accommodated. Anterior prominences are less critical for support surfaces unless the person lies prone at night, but they do indicate significant complications with the spine and, most likely, internal organ function. Anterior prominences usually appear on the opposite side to the posterior prominence.

In the sagittal plane, an outward curvature of the spine is referred to as a kyphosis and an inward curve a lordosis, as previously mentioned. Figure 19.9 demonstrates a significant upper thoracic kyphosis which has resulted in the person only being able to look into their lap, with obvious functional and social

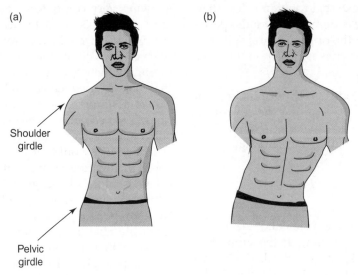

FIGURE 19.7   (a) Simple scoliotic curve. (b) Complex scoliotic curve.

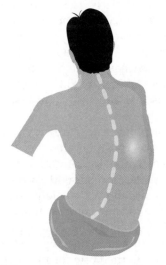

FIGURE 19.8   Prominent ribs posteriorly on the right.

FIGURE 19.9   Implications of an enhanced thoracic kyphosis.

implications. Also note that the lumbar spine has flattened, that the pelvis is in posterior tilt, and that the hip is more extended than flexed.

It is difficult to define precise measurements that can be taken to describe spinal asymmetry.

Instead, descriptive language is required, such as subtle, enhanced, pronounced, marked, or significant. A photograph will assist in the description.

When making an assessment it is necessary to remember that the spine has a natural lumbar lordosis and thoracic kyphosis which can vary significantly between people. The apex of the kyphosis is more posterior than the most posterior point of the sacrum. This means that unless support surfaces are shaped to stabilize the sacrum, posterior tilt will result, with associated kyphosis, problems with head control, and a tendency to slide forward in the seat.

While the neck forms part of the spine, one tends to refer to either neck extension, meaning lordosis, or flexion, meaning kyphosis. One also describes movement in the neck in terms of lateral flexion and by axial rotation to the left/right.

Position and Movement of the Arms

The significance of the arms is easily neglected but they play a vital role. Where they are functional, it is important to understand their movement. Where they are dysfunctional they tend to pull forward and down on the shoulder girdle due to their weight, which is often significant. When describing position, consideration should be given as to whether the shoulders are abducted or adducted, internally or externally rotated, extended or elevated. Following this, it should be established if the elbows and wrists are flexed or extended as this clearly has implications for function. It is necessary to make a detailed assessment of arm function where functional movement is required. People with higher level spinal cord injuries in particular use a variety of "trick" movements to facilitate function and independence that will sometimes limit the amount of postural support that can be provided.

Weight Bearing

It is helpful to define in detail the areas of the body bearing weight, which can be surprisingly few. This will provide additional information to the description of posture developed from the assessment outlined. Bony prominences are particularly vulnerable, notably the ischial tuberosities, greater trochanters, coccyx, sacrum, lateral/medial knee condyles, lateral/ medial malleolus, heels, lateral aspects of feet, apex/rib prominence of scoliosis/kyphosis, scapulae, elbows, and the back of the head.

Patterns of Movement and Neurological Phenomena

Throughout the assessment, observe any movement the person is making, be this volitional or involuntary. The latter, otherwise known as neurological phenomena, includes hypertonia, hypotonia, spasticity, spasm, clonus, ataxia, athetosis, chorea, rigidity, and any reflexes such as the asymmetric tonic neck reflex (ATNR) or startle. Each can have a profound effect on posture and make day-to-day management extremely difficult. It is sometimes necessary to refer the person to their physician for a medical review prior to or in conjunction with the development of a posture management program.

There are some familiar patterns of postural asymmetry, as described by Porter et al. (2007) and as often clinically observed; for example, windsweeping of the hips to one side with an associated scoliosis convex to the contralateral side. An adducted hip tends to be internally rotated and an abducted hip to be externally rotated. In supine lying, flexed hips and knees will usually cause the legs to fall to one side whereupon the pelvis will be pulled forward on the contralateral side. These are not rules, and many variations will be observed in clinical practice.

Lying Position

Having completed the assessment of body configuration in sitting, it is then necessary to repeat the process in lying. It is tempting to think if one is assessing for seating that lying is irrelevant. I should like to propose that it is difficult, if not impossible, for a satisfactory assessment of seating to be carried out without having also assessed posture in lying. In sitting, gravity is having a profound impact on posture, most noticeably on the spine, but in lying this effect is reduced. Supine, it is possible to explore just how much correction of spinal asymmetry is possible. Having supported the legs and stabilized the pelvis, the trunk can be manipulated to establish how much of the curve is fixed and how much is flexible, that is, whether there is contracture and what internal forces are at work. Having done this, the process is to carry out the same measures of pelvic asymmetry, hip position, shoulder girdle asymmetry, and spinal alignment as with sitting. Only then is it possible to accurately determine the amount of correction that may be achieved in sitting.

Note that when assessing hip position, the pelvic plane against which one is measuring is opposite to that in sitting, that is to say hip abduction and adduction become relative to pelvic obliquity and hip rotation relative to pelvic rotation. This is caused by the legs being in an extended position rather than flexed, that is, their position relative to the pelvis has been transposed through 90°. Where the legs do not straighten fully one must take care in describing hip orientation.

Summary

The body is a complex, organic structure that can give rise to an almost infinite number of physical presentations. The job of the assessor is to pick out the critical elements, most notably to determine what is fixed and what is flexible, and not to dwell on detail that has little impact on the end result. Knowing where to focus can only be learnt with practise.

CRITICAL MEASURES: JOINT RANGE OF MOTION AND PELVIC/ TRUNKAL ASYMMETRY

Introduction

The next step, with the patient lying supine, is to measure joint ranges of motion, most notably the hips, knees, and ankles, and to identify critical limitations in the alignment of the pelvis and trunk. Together, these are referred to as critical measures. They are dealt with in the following subsection, "Critical Measures to Consider for Sitting, Lying, and Standing." These measures are critical because without consideration being given to each one, the postural management strategy may be jeopardized. Prior to their introduction we briefly discuss the reasons why we take measurements and general points on carrying out a physical assessment.

Reasons for Taking Measurements

Why do we measure ranges of joint motion? We are interested in the presence of limitations and the effect of these on posture. Joints undergo plastic change that impairs their ability to move. Pain can also limit range and has many potential sources, often being difficult to control and resolve.

In which joints are we interested? This depends on the clinical question we are trying to answer. In terms of postural management, we are most interested in the ranges of motion of the hips, knees, and ankles. The feet, arms, and hands are also important and we are also interested, of course, in any limitations of movement in the (multijointed) spine.

How precise do our measurements need to be? Again, this depends on the clinical question. A pediatric physiotherapist working in a school will carry out a physical assessment of a child with the aim of determining the present range available in a given joint. This data will then be compared to past measurements to determine if there has been a change and, if so, whether treatment should be adapted to suit. As a result, the physiotherapist is interested in the extremes of motion available to the child. In posture management, we are often seeking to find out if the joint ranges allow the person to be positioned in either sitting, lying, or standing. As such, we are interested more in what are termed *critical measures* and less in the extremes of motion (Pope, 2007).

General Points on Carrying Out a Physical Assessment

It is possible to carry out an assessment of lying position on a bed mattress, but it is far preferable to use a physiotherapy plinth as there is less movement compared to a mattress and, as a result, less margin for error. A floor mat can also be used as an alternative to a plinth but this is best suited to children as hoisting to the floor can present difficulties, as can manual handling at that level.

If you are unused to measuring joint ranges, work alongside an experienced colleague such as a clinical scientist/engineer, physiotherapist, occupational therapist, or consultant in rehabilitation medicine. In terms of taking a measurement, it is possible to use visual estimation or goniometry. The latter is limited in cases where anatomical landmarks are altered or not in their normal positions, as is often the case with people having complex postural presentations. If your spatial awareness is less developed, however, goniometry might be a helpful way of improving the accuracy of your estimations.

An assistant is always required to stabilize the more proximal body segment, particularly the pelvis. Trying to carry out a physical assessment alone is extremely difficult. If a fellow professional is unavailable, carers can be recruited for such tasks but only with very clear instruction as to what they are to look for. There is less certainty that the assessment will be valid in such circumstances.

When moving the limbs, be sure to provide support such that the joint is not loaded abnormally. Support the resting position with pillows or towels where required. Move around the person to ensure you can support them without injuring yourself, particularly your back. Adults have heavy limbs and when mixed with spasticity can make a physical assessment an extremely demanding process. Use postural supports and manual handling aids, such as slide sheets, to aid your assessment.

Where neurological phenomena persist (e.g., spasticity), try a different approach with holding the limb or joint and return to it later in the assessment if need be. It can be surprising to see full range in a joint that was previously thought to have significant restriction.

Immobile and/or older people are more prone to bone fracture due to osteoporosis. The soft tissues around a joint that has reduced movement will have adapted to the new position. As a result, it is important to move joints slowly and within comfortable range. If the person is able to tell you if they are in pain, ask them to do so (people will often mask pain). If not, keep a check on their face or ask a carer known to the person to advise. Other people call out in the anticipation of pain, which is misleading to the assessor.

It is important to display confidence (without being over-confident or patronizing) when manipulating joints. Plan your next move and don't dwell too long on any one range. It is possible to be too gentle. The person is more likely to relax if they have faith in your ability. Be tactful in what you express verbally to people having profound postural asymmetry.

Critical Measures to Consider for Sitting, Lying, and Standing

This section describes which joints, joint positions, and other critical measures to focus on for posture management in respect of sitting, lying, and standing. It is a guide and will require interpretation for each individual being assessed. Some joint ranges become critical, or more critical, where there are proximal limitations.

Sitting
- Hip flexion (Figure 19.10(a))
  - Measured relative to pelvic tilt
    (a posterior tilt will mask a reduced
    range of hip flexion and an anterior tilt

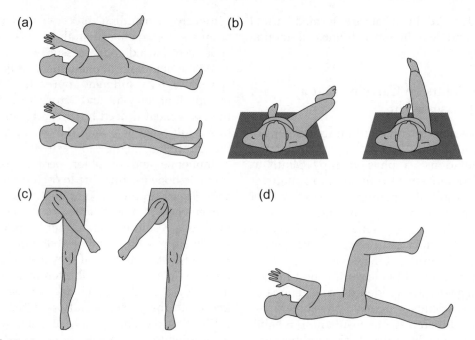

FIGURE 19.10   (a) Hip flexion to extension. (b) Hip abduction to adduction (hip flexed). (c) Hip external to internal rotation (hip flexed). (d) Knee extension with the hip flexed.

will make the range appear less than is available).

- If the hip will not flex to 90°, a normal sitting position cannot be attained; positioning the person in a normal wheelchair or seating system will result in posterior pelvic tilt and sliding forward if the limitation is bilateral, or in the pelvis lifting and rotating forward on one side if the limitation is unilateral (sliding forward may also occur).
- Hip abduction (with the hip flexed as close to 90° as possible; Figure 19.10(b))
  - Measured relative to pelvic rotation, e.g., what appears to be an abducted right hip may in fact be neutral or even adducted where the pelvis is rotated forward on the left.
  - Ideally for sitting, the hips should be able to be positioned in 10° of abduction bilaterally to widen the base of support, thus increasing stability.

- If the hip is fixed in an abducted position but the seat demands a less abducted position, the pelvis will tend to rotate forward on the same side as the abducted hip due to tightness in the abductor muscles.
- Hip adduction (hip flexed as close to 90° as possible; Figure 19.10(b))
  - Measured relative to pelvic rotation, e.g., what appears to be an adducted left hip may in fact be neutral or even abducted where the pelvis is rotated forward on the left.
  - If the hip is fixed in an adducted position but the seat demands neutral, the pelvis will tend to rotate forward on the opposite side to the adducted hip due to tightness in the adductor muscles.
- Hip rotation (hip flexed as close to 90° as possible; Figure 19.10(c))
  - Measured relative to pelvic obliquity, e.g., what appears to be an externally rotated

left hip may in fact be neutral or internally rotated where there is a pelvic obliquity raised on the right.

- If the hip is fixed in external rotation but the lower leg is forced into neutral by the foot support, significant forces will be applied to the knee joint which is likely to lead to pain, and the pelvis may rise on the side of the affected hip.
- If the hip is fixed in internal rotation but the lower leg is forced into neutral, the knee will again be placed under great stress, and the pelvis may rise on the opposite side to the affected hip.

- Knee extension (with the hip flexed as close to 90° as possible; Figure 19.10(d))
  - Measured relative to the line of the femur.
  - The hamstring muscles pass over two joints, the hip and the knee. As such, the position of one joint will affect the available range in the other. In sitting, a flexed hip position is required and this consumes some of the available hamstring length. It is in this position that knee extension must be measured. If there is insufficient range then the conventional position of the feet, i.e., just in front of the knee, will either result in the feet sliding back or, if the feet are fixed in position, in the pelvis being dragged forward in the seat with resultant posterior pelvic tilt, thoracic kyphosis, and difficulties with head control, not to mention the generation of shear forces under the buttocks/thighs which may give rise to skin damage.
  - A hip unable to flex to 90° will demand a greater amount of extension at the knee if feet are not to be tucked back under the front edge of the seat.
  - This measurement is usually taken between the tibia and an imaginary line

projected distally from (and in line with) the femur, being somewhere in the region of 30 to 40° in a healthy subject. It is *not* normal to gain full knee extension with the hip flexed to 90°. The measurement is sometimes taken as the angle between the posterior aspects of the femur and tibia, called the "popliteal angle," i.e., somewhere in the region of 140 to 150° in a healthy subject.

- Ankle dorsiflexion/plantarflexion
  - Measured relative to the tibia.
  - For the foot to sit flat on the supporting surface, the ankle must be able to attain a neutral, or plantigrade, position.
  - The foot is a complex structure, being able to adopt a variety of profoundly asymmetrical shapes. Where this is the case, a simple measurement of dorsi/plantarflexion is not possible, a fuller description of shape being required.

- Pelvic asymmetry
  - This is determined by analyzing the presenting postures in sitting and lying.
  - If the pelvis is unable to attain neutral in obliquity, rotation, or tilt, the sitting base is compromised and must be accommodated to provide postural stability and to avoid high peak pressures in soft tissues.
  - A certain amount of asymmetry may be accommodated in modular seating, but for the more pronounced presentations, a custom contoured support surface is required.

- Spinal asymmetry
  - Again, the previous analysis of presenting postures will be called on at this point.
  - If there is fixed kyphosis and/or scoliosis, it is not possible to sit in the conventional manner. The fixed component of spinal asymmetry must be accommodated and stabilized.

- The potential for complexity of shape in the spine is greater than in the pelvis due to the significantly greater number of joints. As a result, it is more difficult for modular seating to offer appropriate support for an asymmetrical spine. The decision to use a custom contoured support surface will, therefore, come sooner.
- Neck flexion/extension
  - To facilitate respiration, a safe swallow, and a functional and social line of vision, it is necessary to accommodate any limitation in neck flexion or extension. More commonly, it is a loss of range into extension that presents, i.e., chin on or close to chest. Loss of range into flexion usually occurs where there is a marked anterior tilt to the pelvis and the person has adopted a leant-forward position.

Lying

- Hip extension: Can the hip be extended fully to allow supine lying? (Figure 19.10(a))
  - Measured in side lying and relative to pelvic tilt.
  - Where there is a limitation and it is intended that the person will be positioned in side lying, the leg must be supported, otherwise it will tend to fall to one side or the other, pulling the pelvis into rotation and axially rotating the spine, predisposing to scoliosis; a limitation in hip extension (where the knee is able to reach full extension) will tend to pull the pelvis into anterior tilt which can be uncomfortable: again, the legs must be supported under the knees.
  - The ankle must also be supported, usually with a resting night splint, to prevent it remaining in a plantarflexed position for extended periods, which is likely to lead to a shortened Achilles tendon.

- Knee extension with the hip extended: Can the knee be extended fully to allow supine lying?
  - Measured in side lying and relative to the line of the femur.
  - Again, the leg must be supported, otherwise it will tend to fall to one side or the other; limitations in hip and knee extension will tend to cause increased loading under the heel if the leg is not supported, with the potential for pressure ulcer development.
- Hip abduction/adduction
  - Measured relative to pelvic obliquity, i.e., opposite to sitting, as the femur is now transposed through 90° to that of full extension.
  - Where the hip is fixed in a degree of abduction or adduction, any attempt to straighten the leg will cause the pelvis to become oblique and the spine to develop scoliosis.
- Pelvic asymmetry
  - If the pelvis is unable to attain neutral in obliquity, rotation, or tilt, the base of support is compromised and must be accommodated to provide stability.
- Fixed kyphosis
  - An increased kyphosis that is fixed reduces the surface area over which weight is borne, tending to lead to pressure marking; it also creates instability, the trunk effectively being able to roll to either side, potentially giving rise to asymmetry in other body segments.
- Fixed scoliosis
  - There is usually a fullness to the ribs accompanying a lateral curve; a scoliosis tends to have a posterior rib prominence associated with it on the same side as the convexity, and an anterior prominence on the opposite side; any posterior fullness can have the same effect as an enhanced kyphosis (see previous).

- Neck flexion/extension
  - Any limitation in this respect must be accommodated within the supports used for bed positioning.

Standing

- Pelvic asymmetry
  - Fixed obliquity, rotation, or tilt will destabilize posture and may be a contraindication to standing, depending on severity.
- Ability to achieve a plantigrade foot
  - The feet bear all the person's weight in free standing and often a significant proportion in a standing frame. As a result, little fixed asymmetry is tolerated before rendering standing unsuitable.
- Hip abduction/adduction
  - If the hip is fixed in adduction, the base of support is smaller and the feet are at different heights. Without support this will lead to a pelvic obliquity and scoliosis; if one hip is more abducted than the other, the same results will occur.
- Hip/knee extension: Is the hip/knee fixed in a degree of flexion?
  - Standing requires full extension of the hip and knee if it is not to be substantially effortful; where supports are used to block the knees, care must be taken to avoid high shearing forces within the knee joint.
- Severe fixed spinal curvatures cannot be accommodated in a standing frame and are a contraindication in their own right.

MODELING A STABLE AND FUNCTIONAL SEATED POSTURE

Having identified the critical measures for your patient it is now time to sit them over the side of the physiotherapy plinth. It is possible to use a bed for this purpose but the mattress will not provide the stability required to assess pelvic orientation accurately. A further complication occurs where the person has tight hamstrings and the feet need to tuck back under the surface, something which is not possible with most beds and particularly so where side rails are fitted.

All the critical results of your physical assessment are combined at this point to determine a realistic position for sitting. Supports and bolsters are added to accommodate joint range limitations and to assist in aligning and stabilizing the body segments relative to each other and the supporting surface. Limitations in hip flexion are accommodated by building up under the pelvis to let the femurs drop; minimization of any pelvic obliquity is achieved by iteratively adjusting the heights of supports under the ischial tuberosities. Limitations in hip abduction/adduction and rotation are accommodated, as are tight hamstrings (limited knee extension with the hip flexed). The trunk is supported manually and the best possible alignment determined through the application of forces to the skeletal structure. The assessor will kneel behind the person to take their weight and hold the trunk. A second person is required to remain in front of the person to ensure they remain in position on the plinth. The arms should be raised to reduce the drag on the shoulder girdle and spine to open out the chest to enhance respiration (Chan and Heck, 1999). Small stick-on spots may be applied to each spinal vertebra to highlight spinal shape and to facilitate evaluation. Photographs will assist in subsequent clinical reasoning. Figure 19.11 shows how a flexible scoliotic curve can be corrected with application of forces to the apex of the curve on the convex side and to points on the trunk above and below this on the concave side, the weight of the arms being supported.

This part of the process must be adapted to the individual, making each assessment

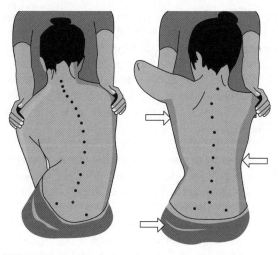

FIGURE 19.11   Correction of a flexible scoliosis.

4. Define the best corrected position in supine lying.
5. Assessment of joint ranges.
6. Identification of all critical measures.
7. Sitting the person over the edge of the bed or plinth.

RECOMMENDATIONS AND RATIONALE FOR POSTURE MANAGEMENT

Setting Objectives

This is the point where all the assembled information must be synthesized into a meaningful and realistic set of objectives, balancing conflicting priorities (including those of cost). These objectives will allow the development of a posture management program being sufficiently detailed to allow precise recommendations to be made. They will list postural abilities and limitations, functional requirements, environmental restrictions, and psychological considerations.

Sitting

Seating is prescribed for wheelchairs, armchairs, classroom chairs, office chairs, toilet seats (commodes), shower chairs, and car seats. The same principles apply regardless of what type of seating is required; it is the application that differs.

The first question is what shape of material is needed, for example, plain slab, contoured, or custom contoured? What properties must it have, for example, for pressure relief, postural support/correction, dynamic properties, and potential for adjustment? What are the functional requirements, for example, transfers, toileting, folding? What size is required?

These initial questions allow conceptual ideas to be developed and from this it is possible to appraise individual pieces of equipment

slightly different from the next. It is important to keep the person safe as they are in a potentially precarious position on the side of the plinth. If practicable, a tilted position should be modeled and is particularly relevant where a kyphosis is flexible.

Throughout this process one must keep in mind the differences between the presenting postures in sitting and lying. That attained in lying will usually provide a best possible reference point. During this stage of the assessment process, that is, sitting over the side of the plinth, one determines a realistic optimal position. What is achieved is an understanding of the type and amount of support required. This information will feed directly into the process of making recommendations for the seating system as a full understanding of the postural requirements will have been obtained.

SUMMARY OF THE PHYSICAL ASSESSMENT PROCESS

1. Describe the presenting posture in sitting.
2. Describe the presenting posture in lying.
3. Consider the differences between the two.

to determine suitability. Very often it is necessary to combine products from different manufacturers. In this case the method of assembly must be considered. It may be necessary or appropriate to liaise with the manufacturers about compatibility. Consider that a different set of products may need selecting. There may be nothing available commercially to meet clinical need and it is here that bespoke manufacturing is indicated. This will require a substantial amount of design work, risk assessment, and technical documentation, although if a service is set up to carry out this type of work routinely, the additional workload will be diminished.

The structure of the support surface can have a significant effect on activities of daily living, particularly the ability of a person to reach beyond their base (Aissaoui et al., 2001). Some surfaces are inherently more stable than others. A very high level of stability can hinder function. The structure can also have an effect on the pressure distributing qualities of the cushion (Apatsidis et al., 2002).

It is beyond the scope of this book to list specific types of seating (and lying or standing equipment), with the exception of custom-contoured seating, a subset that is detailed in Chapter 22 "Wheelchair Prescription." Any form of seating equipment must be appraised for suitability and will usually require trialing with the person. The results of the assessment process should be used as a guide to determine suitability, with particular reference to the list of objectives. Some services have a preferred list of equipment or manufacturers, this making sense from the perspective of holding stock, but this periodically requires revision where better equipment becomes available. Equipment designs and specifications evolve, mostly for the better but sometimes to cut costs. Where this happens, it may be necessary to review other products on the market.

The choice of base support (e.g., wheelchair, armchair, etc.) is critical to the success of the seating system. Some base supports lend themselves better to the fitting of seating than others. The articulations of the base in terms of tilt and recline will be critical, as will its ability to support appropriate peripheral supports such as those for the legs, arms, and head. With wheelchairs, there is often a requirement for a seating system to be transferred between two bases, typically between a powered chair and a manual chair. Depending on the type of seating that is being used and its method of interface, this can present significant technical challenges, and it is sometimes necessary to use a different combination of wheelchair bases to obtain a satisfactory result.

Lying

As has been identified previously, posture in lying is a crucial part of the 24-hour program of management. A supported posture can facilitate sleep for the individual and their spouse, parents, or carers. It can reduce moving and handling requirements, discourage detrimental bony and soft tissue adaptation, facilitate respiration and the management of saliva, cause a reduction in neurological phenomena, and assist in the management of tissue integrity.

The principles are the same as for prescribing seating, with the following additional considerations which are brought about by there being a reduced amount of supervision at night:

- Increased risk of asphyxia where there is an ability of the person to move into a compromised position and where there is also the inability to move out of such a position.
- Increased risk of aspiration/choking/asphyxia if the person is known to vomit.

There are a variety of commercially available sleep systems, each having particular attributes that can be appraised against the requirements of the individual. Additionally, simple positioning aids such as bean bags, which come in a wide variety of shapes and sizes, can be used for less formal systems. It is often preferable to start with very basic, commonly available items to explore the possibilities for support. These are as follows:

- Pillow folded lengthways and wrapped in a bath towel to create a simple support for the legs where the hips and knees do not extend fully; a central "funnel" can be created to separate the knees.
- Pillows placed *under* an additional bed sheet orientated at 90° to normal, the trailing ends of which are tucked back under the pillow and slightly under the person, thereby stabilizing the position of the support.
- Folded towels placed under a pillow to create a nest for the head where it tends to turn to the side.
- Any combination or variation on the above to support other body segments as necessary.

These simple supports are sometimes appropriate for long-term use. Most sleep systems support supine lying but it is possible with some to support side lying or, more commonly, semiside lying. This is required typically where there is a posterior rib prominence caused by scoliosis, the person being unable to attain a stable supine position. Severe kyphosis may also be an indicator for side lying, again due to instability.

Very great care should be taken in making any recommendation for bed positioning due to the reduced levels of supervision, as mentioned above. Sleep systems, in being supportive, can also cause problems with temperature regulation, although some manufacturers claim to have materials that will assist in maintaining a normal body temperature. Where a person uses a urinal bottle independently at night, the level of support able to be provided will be significantly reduced. It would be unacceptable to demand that bottle use be ceased as this would cause a loss of dignity and the requirement for alternative means of bladder management.

It is worth noting that support at night is usually contraindicated where the person moves around significantly. It is possible that the introduction of supports may help to stabilize their position, but where this does not happen there is often little that can be introduced without compromising safety.

It should also be noted that a further contraindication can be where the introduction of a sleep system places a higher burden of care on carers or family members such that their sleep is disturbed regularly through the night. In all likelihood the equipment will not be used and the regime not followed.

Standing

This can be broken down into two broad areas, as follows:

1. Standing wheelchairs: There are a small number of wheelchairs that translate from a seated posture to one of standing. They are mechanically/electromechanically complex and allow increased independence and participation for many people. Significant forces are taken through the knees which are blocked to stabilize the joint so that standing is achieved. It is difficult to fit custom-contoured seating to such chairs as the articulation of the seat does not match the biomechanical joint centers of the person, and also because body shape changes between sitting and standing.
2. Standing frames: These are typically used in conjunction with physiotherapists in schools and rehabilitation units, and sometimes in private homes, clinical engineers rarely becoming involved. The same principles of assessment apply. Standers are either prone, upright, or supine and come in a wide

variety of models, shapes, and sizes. They are static devices and so functional tasks should be given to the person while they have their period of therapy.

The joint range limitations for standing must, of course, be considered, with particular attention paid to the symmetry and loading of the foot, knee, and hip. Marked spinal asymmetry will be an additional contraindication. It can be difficult to maneuver a person into a standing frame, sometimes making the procedure unsafe.

Complementary Interventions

Alongside postural support for sitting, lying, and standing, there are a number of other interventions that will require consideration. Onward referral to other services is often required, for example, for physiotherapy, orthotics, speech and language therapy, surgical opinion, medical review, or nursing review.

It may be appropriate to initiate a physiotherapy stretching regime and to arrange periods of time where "counter" postures are adopted, that is, those which oppose postures adopted for the majority of the day. This may involve lying over a wedge in the prone position, or using side lying to oppose the normal position of a flexible scoliosis. Orthotics are often used as part of a physical management program. Further details are available in the relevant chapter of this section. The involvement of speech and language therapists can be sought where there are particular difficulties with swallowing.

A surgical opinion can be useful in relation to management of an orthopedic condition, particularly the hips and spine. Where there is pain that cannot be controlled by position, or where there is spinal instability that cannot, or is only partially, controllable through postural support, the view of an orthopedic surgeon may be sought. They may not recommend surgery straightaway, there being an (anecdotal) tendency nowadays toward conservative management, at least in the United Kingdom, but they will be able to take a surgical overview and refer on to other parts of the medical profession as appropriate.

A medical review can be helpful to ensure that the drug regime continues to be appropriate. It can be difficult to predict the effect of a drug on an individual. Some people respond well to a certain medicine where others do not. When more than one drug is being taken, the effect becomes even more difficult to predict.

Where there is spasticity and where muscles have shortened, typically in the hips, knees, and ankles in relation to postural management, consideration may be given to the use of botulinum toxin injections. This temporarily blocks nerve transmission at the neuromuscular junction of specific muscles, that is, it has a targeted rather than a global effect, this being the case with oral administration of, typically, baclofen. It has a short-term effect, usually around three months, during which time an active physiotherapy program of stretching is required to make the most of the available time that joint motion is not resisted by spasticity. It should be noted that botulinum toxin will have no effect on contracted muscle. Many people have repeated doses at three-month intervals.

An increasingly common method of managing severe spasticity in the lower limbs is the use of intrathecal baclofen (ITB). Oral administration requires a higher dose to have an effect on severe spasticity but one of the most significant side effects of baclofen is drowsiness. An intrathecal dose, being targeted, can be significantly smaller. It is administered into the spinal tract by a catheter connected to a pump which is surgically inserted into the abdomen. There is a reservoir for the drug which must be refilled periodically. ITB can be extremely effective, making postural management significantly easier and, as a result, allowing the person to function far more effectively.

Pressure sores are usually managed by nurses who will have a variety of techniques at their disposal. In the case of sores that do not respond to conservative treatment, it is sometimes necessary to carry out plastic surgery to repair the skin. This is not straightforward and the area affected will not return to normal state. The area from where the skin graft was taken must also then be allowed to heal. Where there are pronounced postural asymmetries that are part of the cause of the pressure sore, typically bony prominences such as the ischial tuberosities, clinical engineers have an important role to play in the use of pressure mapping equipment to determine peak loads and pressure distribution, and in the provision of bespoke equipment shaped to the individual. Further information is available in Chapter 20 "Pressure Ulceration."

Overview

It is important to consider the effects of a postural management program that is too intense, that places too great a burden on the carers or family, and that lacks scientific evidence to support its use (Gough, 2009). One should take care in using words like "prevent" in the context of managing a scoliosis. This is likely to lead to false hope by the person and their family. It would be more realistic to use words like "discourage" or "to slow down the development of," and so on. If the end result is that deterioration *is* prevented, no one will be unhappy. In some circumstances, an improvement in condition can be brought about. This requires a very active physiotherapeutic regime and can be extremely advantageous to the individual. However, as mentioned, placing too great a burden on carers and family members

can be counterproductive and it should be remembered that we cannot always say with confidence that our recommendations will have the desired effect.

It is a valiant aim to attempt to reduce discomfort, but the truth is that this can be very difficult for a person with altered neurology and posture (Crane et al., 2004). Comfort relates very strongly to functional attainment and is impacted by many physiological, environmental, and psychosocial factors. In the same way that a pair of shoes do not immediately induce discomfort, so it is with postural support that the person must be allowed time to become familiar with the new support surfaces.

In certain circumstances it is necessary to initiate a multidisciplinary review (case conference) where there are a number of interlinked strands needing to be addressed. Trying to address one area in isolation may be unproductive or could even make the situation worse.

SECONDARY COMPLICATIONS

Many medical conditions are static, such as cerebral palsy or head injury, but the medical and functional condition of people having such diagnoses often deteriorates. This is caused by what are termed secondary complications, that is, those difficulties experienced that are secondary to the main diagnosis. Of course, people with progressive conditions, such as multiple sclerosis or muscular dystrophy, are also affected by secondary complications.

As an example, postural instability tends to lead to unequal loading of tissues which, in turn, can lead to pressure ulceration, treatment for which includes extended periods of bed rest. Where an open pressure sore exists there is a risk of infection to the surrounding soft tissue and, eventually, to the underlying bone itself. Prolonged immobility can, in extreme cases, lead to profound shortening of muscle and critically reduced ranges of joint movement, which renders some people unable to adopt a sitting position.

Altered spinal postures have an impact on internal organ function. Organs become compressed, stretched, or even displaced. This leads to impaired respiratory and gastrointestinal function. According to Stewart (1991), "Seating imposes significant effect on the cardiovascular, respiratory, abdominal, renal and neurological systems." Swallowing can become profoundly impaired, leading to food and drink being aspirated into the lungs. Where the person's ability to cough is impaired, this food and drink may cause a chest infection, potentially requiring hospitalization.

Immobility causes bones to become osteoporotic. At the simplest level it is the dynamic loading/unloading cycles of normal movement that keep bone structures healthy. Osteoporotic bones are more prone to fracture, which may further impact the person's quality of life. The clinician manipulating joints should be mindful of the condition.

SUMMARY

Postural management is a highly complex but rewarding clinical area in which to work. It presents multiple, simultaneous challenges to the clinician and engineer, requiring considerable thought and the development of sound clinical reasoning. It has a significant impact not only on the person's health, but on their functional ability and participation in society. It is a building block for electronic assistive technology since without postural management, operation of such equipment can be extremely difficult, if not impossible.

References

Aissaoui, R., Boucher, C., Bourbonnais, D., Lacoste, M., Dansereau, J., 2001. Effect of seat cushion on dynamic stability in sitting during a reaching task in wheelchair users with paraplegia. Arch. Phys. Med. Rehabil. 82, 274–281.

Apatsidis, D.P., Solomonidis, S.E., Michael, S.M., 2002. Pressure distribution at the seating interface of custom molded wheelchair seats: effect of various materials. Arch. Phys. Med. Rehabil. 83, 1151–1156.

Chan, A., Heck, C., 1999. The effects of tilting the seating position of a wheelchair on respiration, posture, fatigue, voice volume and exertion outcomes in individuals with advanced multiple sclerosis. J. Rehabil. Outcomes Meas. 3 (4), 1–14.

Crane, B., Holm, M.B., Hobson, D., Cooper, R.A., Reed, M., Stadelmeier, S., 2004. Development of a consumer-driven wheelchair seating discomfort assessment tool (WcS-DAT). Int. J. Rehabil. Res. 27 (1), 85–90.

Frischhut, B., Krismer, M., Stoeckl, B., Landauer, F., Auckenthaler, T., 2000. Pelvic tilt in neuromuscular disorders. J. Pediatr. Orthop. B. 9, 221–228.

Gericke, T., 2006. Postural management for children with cerebral palsy: consensus statement Developmental. Med. Child Neurol. 48, 244–244.

Goldsmith, E., Goldsmith, J., 1998. Physical management. In: Lacey, P., Ouvry, C. (Eds.), People with Profound and Multiple Learning Disabilities: A Collaborative Approach to Meeting Complex Needs. David Fulton Publishers, pp. 15–28.

Gough, M., 2009. Continuous postural management and the prevention of deformity in children with cerebral palsy: an appraisal. Dev. Med. Child Neurol. 51, 105–110.

Hare, N., 1987. The Human Sandwich Factor Congress Presentation, Chartered Society of Physiotherapy, Oxford.

International Organization for Standardization (ISO), 2006. ISO 16840-1:2006 Wheelchair seating – Part 1: Vocabulary, reference axis convention and measures for body segments, posture and postural support surfaces International Organization for Standardization, p. vi.

Pope, P.M., 2007. Severe and Complex Neurological Disability. Elsevier.

Porter, D., Michael, S.M., Kirkwood, C., 2007. Patterns of postural deformity in non-ambulant people with cerebral palsy: what is the relationship between the direction of scoliosis, direction of pelvic obliquity, direction of windswept hip deformity and side of hip dislocation? Clin. Rehabil. 21, 1087–1096.

Pountney, T.E., Mulcahy, C.M., Clarke, S.M., Green, E.M., 2004. The Chailey Approach to Postural Management: An Explanation of The Theoretical Aspects of Posture Management and Their Practical Application Through Treatment and Equipment. Chailey Heritage Clinical Services.

Scrutton, D., 1991. The causes of developmental deformity and their implication for seating. Dev. Med. Child Neurol. 15, 199–202.

Stewart, C., 1991. Physiological considerations in seating. Prosthet. Orthot. Int. 15, 193–198.

Waugh, K., Crane, B., 2013. A clinical application guide to standardized wheelchair seating measures of the body and seating support surfaces Assistive Technology Partners, University of Colorado.

Whitman, A., 1924. Postural deformities in children. N. Y. State J. Med. 24, 871–874, in Pope P.M (2007) Severe and Complex Neurological Disability, Elsevier.

Pressure Ulceration

Dan Bader

University of Southampton

SKIN AND SOFT TISSUES

The skin represents the largest organ of the body with a surface area of 1.8 m^2 for a standard person. It is divided into three separate layers: the epidermis, the dermis, and the subdermis or hypodermis. The outermost layer (epidermis) is approximately 75 to 150 μm thick, although it is considerably thicker in the palms and plantar aspects of the feet. It is divided into five strata, the deepest of which, the stratum basale, is the site in which cell division occurs to form the main epidermal cells, the keratinocytes. As these cells migrate outward they increase in size, change to a flattened morphology, and their organelles start to change and degrade. The most superficial layer, the stratum corneum, consists of 15 to 20 layers of dead anucleated cells that are hexagonal thin flat squames. At this stage the cells represent terminally differentiated keratinocytes, termed *corneocytes*.

The epidermal–dermal junction provides a physical barrier for cells and large molecules, and forms a strong molecular attachment enhanced by parts of the epidermis penetrating the outermost dermis resulting in large cones and rete ridges, or papillae.

The human dermis constitutes the major thickness of human skin, contributing between 10 and 20% of the total body weight. It contains many structural features including blood and lymph vessels, nerve endings, and skin appendages, such as hair follicles, sebaceous glands, and sweat glands. The predominant cell type in the dermis, the fibroblast, is responsible for synthesizing a moderately dense extracellular matrix of solid constituents, typically:

- Collagen fibers: Approximately 75% of the fat free dry weight and 18 to 30% of the dermal volume. The relatively inextensible collagen fibers bundles form an irregular network that runs almost parallel to the epidermal surface. Indeed, the collagen orientation is distinctive for each body area, a feature known as Langer's lines, which is routinely used during surgical procedures.
- Elastin fibers: Approximately 4% of the fat free dry weight and 1% of the dermal volume. The extensible elastin fibers,

interwoven among the collagen bundles, is important in restoring the fibrous array to its original dimensions and organization when an external load is removed.

- A supporting matrix of amorphous ground substance, composed of long chain glycosaminoglycans (GAGs), which attract and bind a high proportion of water, forming a biological hydrogel.

The dermis is divided into two arbitrary layers: (1) The outermost thin papillary dermis which contains smaller and more loosely distributed collagen and elastin fibers associated with considerable ground substance, and (2) the reticular dermis, representing the majority of the dermal thickness, which contains dense larger collagen and elastin fibers interspersed with small amounts of ground substance.

The subcutaneous fat, or hypodermis, is a fibro-fatty layer which is loosely connected to the dermis. Its thickness varies with anatomical site, age, gender, race, endocrine, and nutritional status of the individual. It acts as an insulating layer and protective cushion, constituting about 10% of the body weight. Subjacent to this layer can be a muscle layer, which overlies either bony prominences or internal tissues and organs.

The complex skin structure interfaces with the external environment where it is exposed to a range of insults, which may be mechanical, physical, biological, and chemical in nature. Of its many functional roles, the highly organized skin structures are designed to permit gas/fluid transport across its surface and, critically, maintain the internal body homeostasis.

In a normal state, skin exhibits viscoelastic behavior similar to other biological tissues, such as muscles, articular cartilage, blood, and lymphatic vessels. Accordingly, skin demonstrates creep, stress relaxation, and load-rate dependent properties. The most common form

of loading reported in both in-vivo and in-vitro studies involves either uniaxial or biaxial tension. The resulting force–extension curves from the uniaxial tests are nonlinear in form and vary in magnitude depending on whether the direction is parallel or perpendicular to the preferred orientation of the underlying collagen fibers—the characteristic anisotropic behavior of skin. By contrast, there are relatively few studies in which the compressive behavior of skin and underlying tissues has been investigated. One exception revealed an in-vivo viscoelastic response which varied considerably with both subject age and tissue site (Bader and Bowker, 1983). Indeed structural changes in the soft tissues associated with age and disease will inevitably compromise this critical mechanism and endanger internal body organs.

PRESSURE ULCERS (PUs)

The condition known as *pressure ulcers* or *decubitus* has represented a problem since time immemorial; indeed, there is evidence of its presence in the Egyptian section of the British Museum (Rowling, 1961). It was recognized in the nineteenth century by Florence Nightingale that there is a relationship between effective nursing care and the incidence of pressure ulcers. It has been defined as a localized injury to skin and/or underlying tissue, usually over a bony prominence, as a result of pressure, or pressure in combination with shear (National Pressure Ulcer Advisory Panel (NPUAP) or the European Pressure Ulcer Advisory Panel (EPUAP) Guidelines, 2009[1]).

PUs are generally categorized in terms of the extent of the associated tissue damage. Thus PUs confined to the epidermal tissues are

[1]*www.epuap.org/guidelines/Final_Quick_Treatment.pdf*

referred to as Grade (or Stage) I ulcers, and are often indistinguishable from incontinence-associated dermatitis (IAD) or moisture lesions. Grade II ulcers affect deeper dermal tissue. Both of these PU grades should, with time and effective management, lead to successful healing. By contrast, damage affecting subcutaneous tissues is classified as Grade III and IV PUs, which may account for between 11 and 31% of the total (Vanderwee et al., 2007), and may require some form of surgical intervention to close the wound. Another form of extensive damage initiating in vulnerable muscle tissues close to bone and progressing upward undetected toward the skin has been recently identified and termed *deep tissue injury* (DTI). The prognosis of such a DTI is highly variable and might even prove fatal. Indeed the high-profile actor Christopher Reeve developed a deep PU following a spinal cord injury, which later led to an untimely death due to septicemia.

PUs can occur in any situation where people are subjected to sustained mechanical loads, and are particularly common in those who are bedridden supported on mattresses or confined to sitting in chairs for much of their waking day. This situation will only be exacerbated at loaded tissue sites also exposed to a hostile microclimate involving elevated temperatures and humidity, such as the plantar aspects of the foot, the residual amputee stump-socket interface of amputees, or at tissues where high forces are transmitted to enable orthotic control. Some people may also experience bladder dysfunction leading to incontinence, whose management is often associated with urinary and fecal infection. Common sites of soft tissue damage invariably involve locations adjacent to bony prominences, as indicated in Figure 20.1, such as the sacrum, ischial tuberosity, heels, and the back of the head.

Subjects who are insensate and/or immobile are at particularly high risk of developing soft

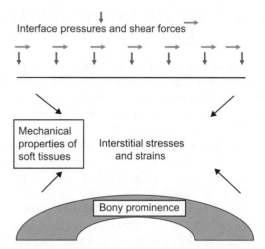

FIGURE 20.1 External load transmission into soft tissue areas adjacent to bony prominences.

tissue breakdown overlying bony prominences. Additionally, the skin tissues of some individuals are compromised by intrinsic factors and, as a consequence, exhibit mechanical properties that are not able to tolerate even normal mechanical loading conditions. PUs have been traditionally associated with the elderly population, particularly those who are malnourished and dehydrated with additional medical complications. However, PUs affect a wider age range with those potentially at risk including neonates nursed in incubators, young patients in intensive care pediatric units, and, commonly, the spinal cord injured population, for which PUs may occur throughout their lifetimes. With respect to body mass index (BMI), it is generally accepted that subjects with extreme values of BMI (<19.0 kg/m^2 and >30.0 kg/m^2) may be predisposed to PUs (Kottner et al., 2011). However, this criteria does not exclude the development of PUs in individuals within the normal BMI range who exhibit intrinsic characteristics that are known to predispose to tissue damage, such as immobility, lack of sensation, and metabolic disorders.

CARE QUALITY AND FINANCIAL IMPLICATIONS

Pressure ulcers represent a massive burden to individual sufferers and their carers worldwide, and health services and the community at large across each nation. Traditionally, the perceived importance of PUs, particularly by medical doctors, has been under-estimated and PUs have been considered to be a direct result of poor nursing care. In the last two decades, however, a number of factors have raised awareness worldwide, associated with the ever-aging population in many countries. As an example, in 1997, results from a financial audit in the Netherlands indicated that the treatment of PUs represented the fourth costliest of all medical conditions to be managed within the Dutch Health System. This financial assessment led to considerable activity in the Netherlands, where prevalence measurements have been performed since 1998 in a range of healthcare settings. Most recent data have indicated a significant reduction in prevalence values, most probably due to the heightened awareness and education of carers; as an example, in general hospitals, the values have decreased from 23 to 10.5% over a 14-year period. By contrast there has been no change from the 13% prevalence in academic hospitals, with 75% of these ulcers being classified as hospital-acquired PUs. This issue is particularly relevant when considering the recent laws passed in the United States, which state that if a patient was not considered to be at risk of acquiring a PU on entering a hospital but subsequently acquired one during their stay, no financial support would be given to the hospital from insurance companies for any of the treatment costs of the patient.

Apart from the inevitable human suffering and the high risk of complications, treatment costs are significant in cash-strapped health services in many countries. Indeed average costs of treating PUs in the United Kingdom range from £1214 (Grade I) to £14,108 (Grade IV) (Dealey et al., 2012). Costs increase with ulcer severity because the time to heal is longer and the incidence of complications is higher in more severe cases. Also, 2010 estimates suggest a financial burden of £4 billion per annum for treating PUs and related conditions and they are now recognized as associated with patient safety in the United Kingdom. In the Netherlands, Schuurman et al. (2009) reported that up to 1.4% of the total budget for healthcare expenditure in the Netherlands is spent on hospital-acquired PU care. Inclusion of the costs in nursing homes, rehabilitation centers, and home care will increase these numbers considerably. The totality of this data provides compelling evidence for the need for effective preventative strategies to reduce the unacceptable high incidence of PUs.

Epidemiological research using subjective clinical assessment of risk factors highlights general characteristics of patients at risk, but these factors are generally indirect measures of the causative factors. Current "macro" measures of immobility (moves a lot, moves a little, etc.) do not provide sufficiently sensitive data to identify those at risk of developing a PU, and of those, approximately 90% will not develop a PU (Nixon et al., 2006). Health-related quality of life literature highlights that PUs impact greatly on physical, social, and psychological domains of the patient resulting from one or more of the following: distressing symptoms including pain, exudates, and odor; increased care burden; prolonged rehabilitation; requirement for bed rest; hospitalization; prolonged work-related sickness (Gorecki et al., 2009).

Current Knowledge on the Pathogenesis of Pressure Ulcers

The etiopathogenesis of PUs has long been considered to involve the obstruction of blood vessels within loaded soft tissues leading to pressure-induced ischemia. This mechanism will result in a limited delivery of vital nutrients, such as oxygen, to the cell niche. The resulting cell death would impede any remodeling processes and lead to the accumulation of soft tissue breakdown. Other mechanisms had been proposed but their examination had largely been limited to hypothetical concepts (Reddy et al., 1981; Krouskop, 1983) or to the use of invasive techniques, such as radioisotope tracers (Miller and Seale, 1985). In the last decade these mechanisms, each of which is influenced by mechanical-induced soft tissue breakdown, have been revisited. It is now recognized that, in addition to pressure-induced ischemia, PUs can result from other mechanisms with compelling research evidence (Bouten et al., 2003), namely:

- Impaired interstitial and lymphatic flow: This will result in an accumulation of toxic intercellular waste products, which are both damaging to the cells and can influence the local cellular environment, e.g., reduced levels of local pH.
- Ischemia-reperfusion injury associated with load removal: This results in the reperfusion of blood and transport of other nutrients, which may result in an over-production and release of oxygen-derived free radicals, which have been implicated in the damage of many tissues and organs (Pierce et al., 2000; Unal et al., 2001).
- Cell and tissue deformation: Tissue deformation triggers a variety of effects which may be involved in early cell damage, such as local membrane stresses leading to buckling and rupture of the membrane. This loss of membrane integrity will lead to altered transport of biomolecules and ions, volume changes, and modifications of cytoskeletal organization, all of which can affect cell

viability and limit the remodeling capacity of the tissues.

To examine these mechanisms, a hierarchical approach can be adopted involving a range of cell-based studies (Gawlitta et al., 2007), tissue and animal models, and human studies. As an example, in vivo animal models were developed in which real-time damage from mechanical loading could be visualized using magnetic resonance (MR) imaging. Typical results highlighted the nature of the loading regimens that caused changes in tissue structure, as reflected in MR parameters, such as T2-weighted index. Additionally, the acuity of these MR images enabled the production of computational models representing body segments. The resulting finite element (FE) analysis predicted the internal mechanical state of the tissues, which could then be compared to the changes in tissue structure (Ceelen et al., 2008). This continuing series of animal-based studies has revealed that:

- Soft tissue damage due to large deformations occurs at an earlier loading period than pressure-induced ischemia (Stekelenburg et al., 2007).
- Loading periods of as short as 10 minutes can cause small, but significant levels of muscle damage (Loerakker et al., 2010).
- There is considerable variability in the response associated with individual animal models.
- A reperfusion period can lead to muscle damage additional to that produced by ischemia alone (Loerakker et al., 2011).

A similar combined MR imaging-modeling approach was also adopted to examine a group of able-bodied volunteers (Gefen et al., 2006; Linder-Ganz et al., 2007) and spinal cord injured (SCI) subjects in a sitting position (Linder-Ganz et al., 2008). This research yielded threshold values of internal strain above which damage might be predicted. At the present time, research involving MRI represents the "gold standard" for imaging the soft tissue composite overlying bony prominences. However, it is a complex and expensive modality and, as such, cannot be considered for routine use of assessing risk of developing PUs in a clinical setting.

Extrinsic Factors

Effective prevention of PUs is only practically possible if the initiation of tissue breakdown can be detected at an early stage and rigorous intervention strategies are introduced. From a bioengineering perspective there are a number of goals related to PU prevention that could be achieved in the research and clinical setting:

- Development of an integrated system to monitor conditions at the loaded body-support interface
- Prediction of the interface/interstitial conditions which may lead to tissue breakdown
- Assessment of novel materials and advanced support systems
- Establishment of objective screening techniques, which are reliable and robust
- Identification of those subjects particularly at risk of developing PUs

The following sections focus on the first two goals. With respect to the interface conditions, there are a number of factors that can be measured. These include pressure, shear, temperature, humidity, and time, at the loaded patient support interface. Pressure and shear have been long known to be relevant in load-induced ischemia (Figure 20.1). Despite the limited number of studies in the literature, there is a growing realization of the importance of the microclimate at the loaded patient support interface. For example, a 1°C rise of temperature increases the metabolic demands of the skin tissues by approximately 13%.

Thus, tissue demands an increased delivery of vital nutrients, typically oxygen, although the critical blood supply will inevitably be compromised by the localized mechanical environment within the soft tissues. In addition, an increase in skin temperature is likely to lead to an increase in moisture released in the form of sweat at the loaded interface. This will increase skin susceptibility to damage and increase frictional forces at the interface.

In a much-quoted retrospective study, Reswick and Rogers (1976) confirmed that time of exposure as well as magnitude of pressure is critical in determining the risk of developing a PU. A sustained pressure of low magnitude can result in an ulcer, as well as a high pressure existing for shorter time periods, as reflected in a hyperbolic risk curve. However, more recent work cited previously (Stekelenburg et al., 2007; Ceelen et al., 2010; Linder-Ganz et al., 2007) has indicated that the curve of internal mechanical state, as a function of the loading period, has to be adapted to a sigmoidal form because of the damage mechanisms associated with deformation at short loading periods.

MECHANICAL LOADING AT THE PATIENT-SUPPORT INTERFACE

Knowledge of the nature of the stresses occurring at the interface is essential in the assessment of the potential damage to soft tissues. Thus, body tissues can support high levels of hydrostatic pressure, with equal components in all directions, with no resulting tissue distortion. This may be illustrated in the case of deep sea divers, who are regularly exposed to hydrostatic pressures in excess of 750 mmHg (100 kPa) for prolonged periods with no deleterious effects to the soft tissues. However, if the pressures are applied nonuniformly then localized tissue distortion and associated damage can result. This can arise if the pressures are applied locally or if pressures are applied in association with external shear forces (Hobson, 1992). In the former case, maximum pressure gradients will often be associated with periphery of the locally compressed area. This establishes the need for the measurement of interface pressures.

INTERFACE PRESSURE MAPPING

Regardless of other factors involved in the etiology of PUs, external pressures must be present when either the body forces are transmitted through support surfaces or, alternatively, when an orthosis applies correctional forces through soft tissues. Accordingly, for many years, bioengineers have focused their attention on developing accurate and reliable pressure measuring systems. They have incorporated sensors based on a number of physical principles involving pneumatics, electropneumatics, force-sensitive resistance materials, and capacitance methods. A number of these systems have become commercially available to map interface pressure distribution, using a range from single cells to arrays incorporating in excess of 1000 discrete measurements.

Pressure measurement systems are generally used in either a research or clinical setting. In the former, the systems can be used in the laboratory to evaluate the relative performance of different pressure relief/redistribution systems under controlled conditions. To compare products, two alternative approaches have been adopted. Indenters have been designed to simulate the loading patterns at the patient support interface. Traditionally they have involved simple domed indenters (Bain, 1998), although more sophisticated anthropomorphic mannequins have been developed with an internal skeleton covered by simulated soft tissues (Bain, et al., 1999). Meaningful data can also be obtained from measurements on human subjects, provided that the experimental

TABLE 20.1 Critical Evaluation of Interface Pressure Measurements

Potential of interface pressure monitoring
• Well-established clinical measure to compare support surfaces for individual subjects
• Ideal for feedback for individuals to indicate support postures and areas of high pressure
• Useful lab-based tool for comparative evaluation of new materials and support surfaces

Limitations of interface pressure monitoring
• Analysis of large data sets (Bogie et al., 2008)
• Relevance to interstitial pressures?
• Relevance to site of initial tissue breakdown?
• There is no reliable indicator of safe pressure, or band of pressures, in association with time, which would be appropriate for all patients at risk

Conclusions
• Pressure measurements alone are not sufficient to alert the clinician to potential areas of tissue breakdown
• It is important to examine the effects of interface pressure on tissue viability/status

protocols are standardized, hence minimizing the inherent variability. Alternatively, interface pressure monitoring systems can be used, typically in a seating clinic, as an adjunct to subjective risk assessment (e.g., Braden or Waterlow scales), as well as providing an aid to clinical prescription of an appropriate support surface. Such measurements can also be used as a biofeedback tool to the patient providing evidence of postural factors associated with pelvic obliquity, tilt, and rotation and the efficacy of pressure relief regimes. A critical analysis of these systems is provided in Table 20.1. Irrespective of how the tests are performed, regular calibration must be undertaken for all pressure measuring systems.

It is generally accepted that interface pressure measurements are subject to great variability. There are differences between anatomical sites, between individuals, and even when the sensor is kept on a single anatomical site on a given individual, there are differences due to small

TABLE 20.2   Sources of Variability Associated with Interface Pressure Monitoring

Source of Variability	Examples
Measurement system	Spatial and temporal resolution of sensor Methods of recording, displaying and interpreting data
Positioning	Change in footrest height Changes due to leaning forward/side lying Tilt maneuvers, including change of head-of-bed angle Subject posture
Anatomical location	Local body curvature Proximity to bony prominences
Inter-subject	Subject morphology Body mass index

changes in posture (Table 20.2). There are also differences due to clothing/covering materials at the interface, the type of measurement system used, and the interpretation of the data. This is particularly relevant when using systems that provide a vast data set to be processed into a pressure index which can be used in a research and/or clinical setting (Bogie et al., 2008). Common parameters include peak pressure, peak pressure gradient, average pressures, and symmetry index (i.e., comparing values on two sides of the body). However, it is inevitable that no one parameter can provide a ubiquitous index of pressure distribution that is applicable to all subjects at risk of developing PUs.

As pressure is force per unit area it is obvious that the shape of a subject will have an effect on the interface pressure distribution. Both shape and form of a subject in a load-bearing site will depend on the bony skeleton; the quantity, tone, and shape of the musculature; the amount of subcutaneous fat; as well as the resilience of the intervening skin. Accordingly, there is a complex interaction between body morphology and interface pressure and there are many cases for which

subjects with very similar body types can exhibit quite different interface pressures. This was exemplified by a series of studies examining the performance of mattresses with able-bodied volunteers commissioned by the U.K. Department of Health and several healthcare companies, as described by Swain and Bader (2002). As an example, one subject (height of 1.57 m and a BMI of 23.6 kg/m^2) recorded a mean interface pressure under the sacrum when semirecumbent in bed of $61.9 + 9.2$ mmHg $(8.3 + 1.2$ kPa), whereas another subject (1.57 m, BMI 23.1 kg/m^2) yielded an interface pressure of $86.4 + 12.2$ mmHg $(11.5 + 1.6$ kPa). This effect is even more marked when the heels are considered as there is great variation between individuals' anatomy: On the standard King's Fund mattress we have seen variations from a mean of $107 + 22.6$ mmHg $(14.3 + 3.1$ kPa) to a mean of $231 + 46.6$ mmHg $(30.8 + 6.2$ kPa). Clearly, the contact area of heels with a small radius of curvature will be limited in size on a relatively nonconforming surface and thus any slight difference in contact area will have a major effect on the interface pressure. Errors in measurement will also increase as the placement of the sensor is far more critical if the contact area is small. Therefore, the range of interface pressures on various anatomical sites vary widely with the subject's underlying anatomy, and in our experience it is not possible to predict the interface pressures from the individual body type.

It is evident from the above factors that the monitoring of interface pressures alone will not prove sufficient to alert the clinician to potential areas of tissue breakdown. This has motivated a number of investigators to utilize additional measurement techniques, which can provide early objective indicators of compromise to the health of soft tissues in the loaded state. These include physical sensors, typically involving transcutaneous gas tensions and laser Doppler fluxmetry and biosensing systems, involving the analysis of blood and urine markers, sweat metabolites, and inflammatory biomarkers. In some cases these have been combined to evaluate the effects of different loading regimens on able-bodied (Knight et al., 2001) and SCI patients (Bogie and Bader, 2005), as well as to evaluate the effectiveness of specialized support surfaces (Goossens and Rithalia, 2008). For a more comprehensive summary of some of these multisensor approaches the reader is directed to a review article (Bader and Oomens, 2005).

INTERNAL MECHANICAL STATE OF LOADED SOFT TISSUES

The externally applied pressures will inevitably deform the soft tissue composite over the bony prominences establishing an internal mechanical state. However, the relationship between interface or external pressures, which can be measured, and the resulting mechanical state (i.e., the internal interstitial stresses/strains), is necessarily complex in nature (Figure 20.1). It is important to note that shear can be present externally at the skin surface, due to subject sliding and transfer across a support surface, as well as internally within the soft tissue structures due to pressures applied normal to the skin surface.

Experimentally there are a few studies, using invasive methods, that have examined the load transfer across the interface. As an example, Sangeorzan et al. (1989) noted that the values of interface versus interstitial pressures were not equivalent and were highly dependent on the nature and the mechanical properties of the intervening soft tissues. Thus the thickness, tone, and mechanical integrity of subcutaneous tissues and the proximity of bony prominences will influence this relationship (Figure 20.1). An investigation of subjects during surgical procedure examined the response of tissues adjacent to

the lateral aspect of the proximal thigh. Results indicated that skin interface pressures were dissipated within the depth of the tissues resulting in reduced internal stresses (Bader and White, 1998). Indeed, linear models of the data suggested interstitial stresses range between 29 and 40% of the applied interface pressures. This highlights the protective nature of tissues to attenuate the effects of sustained pressure.

An alternative approach to analyze load transmission across the patient interface involves developing mathematical solutions to the problem, as described in a number of publications. A summary of the typical findings is provided in Table 20.3. It reveals considerable variation in the ratio values relating interface pressures and predicted interstitial pressures. A description of this mathematical approach is beyond the scope of this chapter, but it is worth noting that the more recent models incorporating nonlinear and hyperelastic behavior are more realistic in terms of the mechanical response of the soft tissues. Additionally, appropriate values of mechanical parameters for all layers (skin, fat, and muscle) within the human soft tissue composite is essential if the models are to predict critical internal threshold levels (Then et al., 2009). Early MRI studies indicated that changes in structural stiffness values could be demonstrated in healthy tissues adjacent to ischial tuberosities when compared to atrophied buttock tissues of an age-matched paraplegic with flaccid paralysis (Reger et al., 1990). The tissues were observed to be more distorted under load, suggesting an increased risk of tissue trauma. More recently, the use of elastography combined with either MR or ultrasound imaging has been proposed as a reliable means of estimating soft tissue properties under compressive loading in the research environment (Deprez et al., 2011). Although such techniques have been introduced to evaluate structural inclusions in breast and liver tissues, they are

TABLE 20.3 The Relationship Between Interface Pressures and Interstitial Pressures, as Determined by a Selection of Experimental and Computational Approaches

Study	Model System	Values	Interface Pressures: Interstitial Pressure Ratio
Bader and White, 1998	Loaded greater trochanter of surgical patients		0.28–0.57
Lee et al., 1984	Pressure sensors implanted in a pig model		3–5
Ragan et al., 2002	Axisymmetric 3D (FE) model of buttock	37 kPa/10 kPa	3.7
Oomens et al., 2003	3D FE model, variable properties of muscle, fat, and skin	120 kPa/50 kPa	2.4
Gefen et al., 2005	3D FE model	4 MPa/15 kPa	266
Sun et al., 2005	FE model based on nonsitting MRI	76 kPa/21 kPa	3.5

still to be used to detect the onset of PUs in a clinical setting.

SUPPORT SURFACES AND PRESSURE RELIEF REGIMENS

Support surfaces at the patient interface should provide a safe, stable, and comfortable means of transmitting loads to the body. Their design should for the most part provide a fairly uniform pressure distribution over a significant contact area. In addition, the choice of materials should account for other physical, functional, and aesthetic factors, including durability, permeability to water vapor, and heat dissipation. Additionally, materials should be biocompatible with the host tissues, not evoking any skin irritations or allergic reactions. Compliant viscoelastic materials form the basis of many support surfaces. As an example for seating materials, a combination of foam, gel, and air supports are regularly used at a thickness that will minimize the risk of "bottoming out" without compromising other postural effects associated with arm and footrest supports.

If the pressure is relieved periodically, pressures can be tolerated for longer periods. This forms the basis of pressure relief and redistribution regimens, which are performed by regular turning, lift-off from the support surface, and alternating pressure air mattresses and cushions. The nature of the tissue recovery is determined by the resilience of the specific tissues and the tissue structures, including the blood and lymph vessels. The viscoelastic behavior of the soft tissues will influence the recovery, which will depend on the rate and time of loading as well as its magnitude. Short-term loading generally produces elastic deformation with minimal creep and rapid elastic recovery, whereas long-term loading results in marked creep and requires significant time for complete tissue recovery (Bader, 1990).

Subjects who are considered to be at high risk of developing PUs in hospitals are often provided either low-air-loss or alternating pressure air mattresses. Nonetheless, the relative merits of these high technological interventions, as assessed in a Cochrane systematic review (McInnes et al., 2012), still remain unclear. Clearly, more robust trials are indicated for which individual characteristics are taken into account. Future developments could include establishing a range of optimal design features for support surfaces—the concept of a "personalized support surface" to match the physiological response of the individual.

SUMMARY

- Soft tissues represent an important interface with the outside world.
- When the integrity of soft tissues is compromised, prolonged external loading can lead to the development of pressure ulcers which, in extreme conditions, can be life-threatening.
- Pressure ulcers affect a wide range of individuals in both hospital and community settings, leading to personal suffering and considerable financial burden to health services.
- There are a number of mechanisms associated with the etiology of pressure ulcers, each of which influence the viability at both a cellular and tissue levels.
- A range of bioengineering strategies can be adopted to address this problem and minimize the incidence of pressure ulcers.
- Interface pressure measurements alone are not sufficient to define a damage threshold for pressure ulcer prevention.
- The transmission of load across the body-support interface establishes internal mechanical conditions that are, in part, determined by the mechanical properties of the soft tissues.
- There is a need for the optimal design of support surfaces coupled with effective pressure relief/redistribution strategies to match the individual risk of pressure ulcers developing.

References

Bader, D.L., 1990. The recovery characteristics of soft tissues following repeated loading. J. Rehab. Res. Dev. 27 (2), 141–150.

Bader, D.L., Bowker, P., 1983. Mechanical characteristics of skin and underlying tissues *in vivo*. Biomaterials 4, 305–308.

Bader, D.L., Oomens, C.W.J., 2005. Recent advances in pressure ulcer research. In: Romanelli, M., et al. (Eds.), Science and Practice of Pressure Ulcer Management. Springer-Verlag, Berlin, pp. 11–26.

Bader, D.L., White, S.H., 1998. The viability of soft tissues in elderly subjects undergoing hip surgery. Age and Ageing. 27, 217–221.

Bain, D.S., 1998. Testing the effectiveness of patient support systems: the importance of indenter geometry. J. Tissue Viability. 8 (1), 15–17.

Bain, D.S., Scales, J.T., Nicholson, G.P., 1999. A new method of assessing the mechanical properties of patient support systems (PPS) using a phantom. A preliminary report. Med. Eng. Phys. 21 (5), 293–301.

Bogie, K.M., Bader, D.L., 2005. Susceptibility of spinal-cord injured individuals to pressure ulcers. In: Bader, D.L., Bouten, C.V.C., Oomens, C.W.J., Colin, D. (Eds.), Pressure Ulcer Research: Current and Future Perspectives. Springer-Verlag, pp. 73–88.

Bogie, K.M., Wang, X., Fei, B., Sun, J., 2008. New technique for real-time interface pressure analysis: Getting more out of large image data sets. J Rehabil. Res. Dev. 45 (4), 523–536.

Bouten, C.V.C., Oomens, C.W.J., Baaijens, F.P.T., Bader, D.L., 2003. The aetiology of pressure sores: skin deep or muscle bound? Arch. Phys. Med. Rehabil. 84, 616–619.

Ceelen, K.K., Gawlitta, D., Bader, D.L., Oomens, C.W.J., 2010. Numerical analysis of ischaemia- and compression-induced injury in tissue-engineered skeletal muscle constructs. Ann. Biomed. Eng. 38 (3), 570–582.

Ceelen, K.K., Stekelenburg, A., Loerakker, S., Strijkers, G.J., Bader, D.L., Nicolay, K., et al., 2008. Compression-induced damage and internal tissue strains are related. J. Biomech. 41 (16), 3399–3404.

Dealey, C., Posnett, J., Walker, A., 2012. The cost of pressure ulcers in the United Kingdom. J Wound Care. 21 (6), 261–262.

Deprez, J.F., Brusseau, E., Fromageau, J., Cloutier, G., Basset, O., 2011. On the potential of ultrasound elastography for pressure ulcer early detection. Med. Phys. 38 (4), 1943–1950.

Gawlitta, D., Li, W., Oomens, C.W.J., Bader, D.L., Baaijens, F.P.T., Bouten, C.V.C., 2007. Temporal differences in the influence of ischemic factors and deformation on the metabolism of engineered skeletal muscle. J Appl. Physiol. 103 (2), 464–473.

Gefen, A., Gefen, N., Linder-Ganz, E., Margulies, S., 2005. In vivo muscle stiffening under bone compression promotes deep pressure sores. J Biomech Eng. 127 (3), 512–524.

Goossens, R.H., Rithalia, S.V., 2008. Physiological response of the heel tissue on pressure relief between three alternating pressure air mattresses. J Tissue Viability. 17 (1), 10–14.

Gorecki, C., Brown, J.M., Nelson, E.A., Briggs, M., Schoonhoven, L., Dealey, C., et al., 2009. Impact of pressure ulcers on quality of life in older patients: a systematic review. J Am Geriatr Soc. 57 (7), 1175–1183.

Hobson, D.A., 1992. Comparative effects of posture on pressure and shear at the body seat interface. J. Rehabil. Res. Dev. 29, 21–31.

Knight, S.L., Taylor, R.P., Polliack, A.A., Bader, D.L., 2001. Establishing predictive indicators for the status of soft tissues. J. Appl. Physiol. 90, 2231–2237.

Kottner, J., Gefen, A., Lahmann, N., 2011. Weight and pressure ulcer occurrence: a secondary data analysis. Int. J. Nurs. Stud. 48 (11), 1339–1348.

Krouskop, T.A., 1983. A synthesis of the factors that contribute to pressure sore formation. Med. Hypothesis. 11, 255–267.

Lee, K.M., Madsen, B.L., Barth, P.W., Ksander, G.A., Angell, J.B., Vistnes, L.M., 1984. An in-depth look at pressure sores using monolithic silicon pressure sensors. Plast. Reconstr. Surg. 74, 745–756.

Linder-Ganz, E., Shabshinb, N., Itzchakb, Y., Gefen, A., 2007. Assessment of mechanical conditions in subdermal tissues during sitting: A combined experimental-MRI and finite element approach. J. Biomech. 40, 1443–1454.

Linder-Ganz, E., Shabshinb, N., Itzchakb, Y., Yizhar, Z., Siev-Ner, I., Gefen, A., 2008. Strains and stresses in subdermal tissues of the buttocks are greater in paraplegics than in healthy during sitting. J. Biomech. 41 (3), 567–580.

Loerakker, S., Manders, E., Strijkers, G.J., Nicolay, K., Baaijens, F.P.T., Bader, D.L., et al., 2011. The effects of deformation, ischemia, and reperfusion on the development of muscle damage during prolonged loading. J. Appl. Physiol. 111 (4), 1168–1177.

Loerakker, S., Stekelenburg, A., Strijkers, G.J., Rijpkema, J. J., Baaijens, F.P.T., Bader, D.L., et al., 2010. Temporal effects of mechanical loading on deformation induced damage in skeletal muscle. Ann. Biomed. Eng. 38 (8), 2577–2587.

McInnes, E., Jammali-Blasi, A., Bell-Syer, S., Dumville, J., Cullum, N., 2012. Preventing pressure ulcers – Are pressure redistribution surfaces effective? Int. J. Nurs. Studies. 49 (3), 345–359.

Miller, G.E., Seale, J.L., 1985. The mechanics of terminal lymph flow. J Biomech. Eng. 107 (4), 376–380.

Nixon, J., Cranny, G., Iglesias, C., Nelson, E.A., Hawkins, K., Phillips, A., et al., 2006. Randomised, controlled trial of alternating pressure mattresses compared with alternating pressure overlays for the prevention of pressure ulcers: PRESSURE (pressure relieving support surfaces) trial. BMJ. 332, 1413–1415.

Oomens, C.W.J., Bressers, O.F.J.T., Bosboom, E.M.H., Bouten, C.V.C., Bader, D.L., 2003. Can loaded interface characteristics influence strain distributions in muscle adjacent to bony prominences? Comput. Methods Biomech. Biomed. Eng. 6, 171–180.

Pierce, S.M., Skalak, T.C., Rodheaver, G.T., 2000. Ischaemic-reperfusion injury in chronic pressure ulcer formation: A skin model in the rat. Wound Repair Regen. 8, 68–76.

Ragan, R., Kernozek, T.W., Bidar, M., Matheson, J.W., 2002. Seat-interface pressures on various thicknesses of foam wheelchair cushions: a finite modelling approach. Arch. Phys. Med. Rehabil. 83, 872–875.

Reddy, N.P., Palmieri, V., Cochran, G.V., 1981. Subcutaneous interstitial fluid pressure during external loading. Am. J. Physiol. 240 (5), R327–R329.

Reger, S.I., McGovern, T.F., Chung, K.C., 1990. Biomechanics of tissue distortion and stiffness by magnetic resonance imaging. In: Bader, D.L. (Ed.), Pressure Sores – Clinical Practice and Scientific Approach. Macmillan, Basingstoke, pp. 177–190.

Reswick, J.B., Rogers, J.E., 1976. Experience at Rancho Los Amigos Hospital with devices and techniques to prevent pressure sores. In: Kenedi, R.M., Cowden, J.M., Scales, J.T. (Eds.), Bedsore Biomechanics. Macmillan, London, pp. 301–310.

Rowling, J.T., 1961. Pathological Changes in Mummies. Proc. R. Soc. Med. 54, 409–415.

Sangeorzan, B.J., Harrington, R.M., Wyss, C.R., Czerniecki, J.M., Matsen, F.A., 1989. Circulation and mechanical response of skin to loading. J. Orthopaedic Res. 7, 425–431.

Schuurman, J.P., Schoonhoven, L., Defloor, T., van Engelshoven, I., van Ramshorst, B., Buskens, E., 2009. Economic evaluation of pressure ulcer care: A cost minimization analysis of preventive strategies. Nurs. Econ. 27, 390–400.

Stekelenburg, A., Strijkers, G.J., Parusel, H., Bader, D.L., Nicolay, K., Oomens, C.W., 2007. Role of ischemia and deformation in the onset of compression-induced deep tissue injury: MRI-based studies in a rat model. J. Appl. Physiol. 102 (5), 2002–2011.

Swain, I.D., Bader, D.L., 2002. The measurement of interface pressure and its role in soft tissue breakdown. J. Tissue Viability. 12, 132–146.

Sun, Q., Lin, F., Al-Saeede, S., Ruberte, L., Nam, E., Hendrix, R., et al., 2005. Finite element modeling of human buttock-thigh tissue in a seated posture. Summer Bioengineering Conference, June 22–26, Vail, Colorado.

Then, C., Menger, J., Vogl, T.J., Hübner, F., Silber, G., 2009. Mechanical gluteal soft tissue material parameter validation under complex tissue loading. Technol. Health Care 17, 393–401.

Unal, S., Ozmen, S., Demir, Y., Yavuzer, R., LatIfoğlu, O., Atabay, K., et al., 2001. The effect of gradually increased blood flow in ischaemia-reperfusion injury. Ann. Plast. Surg. 47, 412–416.

Vanderwee, K., Clark, M., Dealey, C., Gunningberg, L., Defloor, T., 2007. Pressure ulcer prevalence in Europe: a pilot study. J. Evaluation Clin. Pract. 13, 227–235.

Introduction to Mobility and Wheelchair Assessment

David Long* and Mike Hillman[†]

*Oxford University Hospitals NHS Trust, [†]University of Bath

INTRODUCTION TO MOBILITY

Historical Context

Within the field of rehabilitation engineering, mobility devices are normally thought of in terms of wheelchairs, other devices often coming within the remit of occupational and physio/physical therapists. Clinical engineers, however, can have an important role in understanding the correct use of such devices, their prescription, modification, and design, where commercial equipment does not meet a specific need.

It is difficult to say precisely when the wheelchair was invented, but it seems clear that it has existed since at least the nineteenth century. Early examples were literally ordinary chairs to which wheels were attached. Designs progressed through the early part of the twentieth century and use became more prominent in the rehabilitation of veterans returning injured from the two world wars and subsequent conflicts. Use of more advanced materials was adopted through the second half of the twentieth century and the knowledge of the concept of posture as related to the wheelchair began to take off.

A number of factors have led to the more widespread use of wheelchairs in society:

- The emergence of specialist spinal injury units leading to a more advanced understanding of wheelchair propulsion, postural biomechanics, and pressure ulcer prevention, coupled with advances in materials technology and the application of this to functioning from a wheelchair
- The integration into society of disabled people previously living in large, specialist hospitals and institutions
- An aging population
- Advances in medical science that have resulted in people surviving traumatic accidents from which previously they would not, often resulting in profound and complex levels of disability

This last point may give rise to difficult moral questions about the sanctity of life, quality of life, prolonged suffering of individuals and their families, the potential of having to grieve for the loss of a loved one while they are still medically alive, the individual's wishes stated pre-trauma, whether these may have changed post-trauma, and their capacity to consent to treatment or the withdrawal of treatment (life support machine).

Most wheelchairs are provided for and used by an individual. However, it is common in some settings, such as residential homes, to find a "pool" of wheelchairs that are available for use by a number of people. This makes sense where not everyone needs a wheelchair all the time and where a standard chair can be used by many people. Hospitals have portering chairs, used purely within the hospital context to move people from one department to another. They are usually more sturdy than standard wheelchairs, but the resultant weight penalty is less of an issue as the chair does not leave the premises.

In many countries, wheelchairs are provided by the state, for those who have need. They are also often purchased privately, sometimes because a particular model is not provided by the state. In the United Kingdom, wheelchair users are able to top up funding of their equipment to allow themselves more choice (known as the voucher scheme). Charitable organizations also provide wheelchairs alongside other necessary equipment.

Basic Mobility Devices

At the simplest level is the walking stick, a device used for millennia. On a medical level this has developed into a number of variations and then to the elbow crutch for those who have less weight-bearing ability through the legs. These simple devices provide support, security, and stability. In the case of regular users, an ergonomic handle design can reduce pressure to the hand.

Those with more advanced mobility needs but still being able to weight bear may be supported by a frame (often referred to as a Zimmer). At the simplest this has four rubber ferrules on the feet, but variations have wheels or castors replacing some or all of the ferrules (these devices are often referred to as rollators). Other variations include seats and luggage-carrying capability.

In essence, a wheelchair offers a means of mobility for people who are less able or unable to walk. However, it also has a critical role in facilitating the completion of functional tasks, such as eating and drinking, and using a computer. Added into this mix is the need for the person to be comfortable, which, in turn, leads to the requirement for the wheelchair to provide postural support, particularly for those having higher levels of physical impairment. Very quickly, our simple means of mobility has become a complex, multifaceted device which, in many cases, will need to be designed for the individual; that is, it will be a bespoke, prescribed solution.

There are a vast array of wheelchairs, seating, and accessories available on the market today. Some of these will be described to aid explanation but an exhaustive list would be impossible to compile and out of date by the time of publication. It is more important that the reader knows how to determine what is needed from a wheelchair, then to be able to appraise what is available on the market. Without this knowledge, the approach to provision would be rather "hit and miss," with questionable outcomes for the patient and poor efficiency for service provision. More detail is given in the next chapter on what considerations should be made.

Environmental Issues and Adaptation

Mobility is not just assisted by devices, but also by careful attention to the environment.

There is a general trend toward improving accessibility for wheelchair users and those who find walking difficult. New buildings are required to have a level access entrance, and buses and trains are slowly becoming more accessible. Air travel, however, can still present substantial difficulties for those who are wheelchair dependent.

Environmental features such as grab rails, hand rails, escalators, and elevators improve mobility for those with varying degrees of disability. Dropped curbs are widely used by wheelchair users to cross from the pavement/sidewalk to the road, although these vary in height and their effectiveness can be reduced by a very steeply cambered roadway or drain grate that is difficult to cross. Access ramps of restricted slope allow wheelchair users entry to buildings. Side slopes are particularly difficult for wheelchairs as there is a tendency for the chair to roll down the slope, that is, to turn away from its desired course. This problem is not isolated to manual wheelchairs as (rear wheel drive) powered chairs rely on adhesion/traction of the drive wheel at the top of the slope to provide a braking effect, this being reduced by slippery surfaces.

Surface treatments can help or hinder mobility: some surfaces, such as cobbles, paving, or grass, make it more difficult for wheelchair users, though textured surfaces can help those with visual impairment. Another issue for those with visual impairment is to avoid overhanging architectural features such as "open" stairs. The design of the environment is often the preserve of architects but engineers can understand the advantages and problems of such features, and can offer advice.

Children and Mobility

There are specific mobility needs related to children. Naturally developing children progress from lying to sitting to crawling and then to walking within the space of about a year. Children with disabilities are often encouraged to develop as much as possible by their own abilities rather than through use of a device such as a wheelchair. There is an opinion, however, that use of a mobility device from a young age can improve the chances of a child developing their abilities through play and social interaction, because they are able to move and explore their environment more independently (Nilsson et al., 2011; Durkin, 2009). For this reason there is a strong argument that even very young children with delayed mobility should be given a powered mobility device which they can control independently. Ideally this should not look like a wheelchair: styling as a toy (or similar) may provide added appeal. There are a small number of products on the market that fill this niche.

The Broader Picture of Wheeled Mobility

Physiotherapy departments often use walkers that support a child in a standing position but allow use of the legs for mobility. A wide range of products are available with an array of levels of support. Used in therapy less commonly than in years past are tricycles, which clearly require the child to be able to pedal and balance the trike. In the context of Paralympic sport, however, people having difficulties with balance, but who are able to pedal, can compete in road-cycling events.

Mobility scooters are becoming increasingly common. These are different to wheelchairs in that they have a pivoting seat to allow access from the side, and a tiller to steer the front wheel(s) rather than a joystick. One would not ordinarily expect to be hoisted into a scooter, and the inability to transfer independently would suggest a level of postural inability not able to be accommodated within a scooter.

They are designed primarily for use outdoors but some of the more compact versions are also suitable indoors where large distances have to be covered; for example, in a residential home. There is far less scope for modification than with wheelchairs and so clinical engineers do not tend to have so much involvement.

WHEELCHAIR ASSESSMENT PROCESS

Introduction

As is clear from the previous section, gaining the right wheelchair is not only complex but is also critical in allowing the person to achieve their functional aims. To determine the most appropriate wheelchair it is necessary to carry out a clinical assessment to discover what the person wishes to achieve coupled with an analysis of their impairments and any constraining factors such as the size of their house.

Factors to Be Considered in Each Wheelchair Assessment

The following list should be used as a guide for enquiry when assessing for a wheelchair. It will be the case that priorities alter according to the nature of the referral and the precise circumstances of the person. It should be noted that an assessment form in its own right will not necessarily lead the assessor to an appropriate conclusion. The following is a prompt sheet that requires the assessor to analyze and synthesize the information collected in order to draw a conclusion.

List of Problems and Aims, as Identified by the Patient

Prior to carrying out a full assessment, it is first necessary to elicit from the patient and family and/or carers what it is that they would like. The difference between the reason for referral stated by the referring healthcare professional and the stated aims of the person themselves can be surprising.

Some aims may be outside the scope of the service but should be acknowledged.

Medical

- Primary and secondary diagnoses
- Prognosis (i.e., likely progression of disease and time scale)
- Stage in rehabilitation program, if applicable (i.e., does equipment need to be adjustable?)
- Age of person, particularly concerning child development
- Current state of health, including respiratory status
- Pain and whether it is likely that this can be reduced by altering the wheelchair, e.g., is it relieved when lying down?
- Skin condition and susceptibility to pressure ulcers, including sensation; past history of pressure ulceration
- Neurological phenomena, e.g., spasm, spasticity, movement disorders, persistent reflexes; observation will elicit useful information
- Past or planned surgical procedures, e.g., hip, spine, abdomen
- Hip joint status, particularly for those who have never walked
- Continence and how this is managed; for example, a urinal bottle used to empty the bladder will preclude certain fixed seat shapes

- Relevant medication, e.g., for pain, spasm, epilepsy, bowel function
- Ability to swallow (i.e., implications for head positioning and orientation of posture in space)
- Communication: Can the person reliably give a yes/no response to questions? Is speech impaired? Is a communication aid in use?
- Cognition: This is a complex area that is beyond the scope of this book; it is important that lack of cognition is not assumed due to impaired communication, and that a judgment is made concerning aspects of safety (i.e., can the person demonstrate that they are able to use their wheelchair without coming to harm?)
- Vision: Implications for safety, communication, and use of equipment, e.g., buttons on a powered wheelchair joystick module
- Hearing: Implications for safety and communication
- Height, e.g., impact on a standing transfer or access into a vehicle
- Weight, in relation to the capacity of the chair to carry the person and any items of additional equipment, usually seating

Social/Functional/Environmental

- Current or intended pattern of use of wheelchair (i.e., is it to be used all day, or for brief periods only?; if the latter, a simpler solution may be appropriate)
- Type of mobility: Manual or powered; occupant or attendant propelled
- Occupation/education/ leisure pursuits/ household duties, mostly in relation to the environment, either access to buildings or to work spaces/surfaces
- Ability to carry out self-care tasks and the potential impact of prescribed equipment on these
- Difficulties eating/ drinking: Is there a required postural orientation?

- Method of vehicular transport: Is the wheelchair to be folded and stowed in the luggage compartment, or does the person travel seated in their chair?
- Social situation: Does the person live at home with family, on their own, or in a care setting?; who are the primary carers?
- Other equipment in use, such as other wheelchairs, any form of armchair, office chair, or classroom chair, orthotic splints, prostheses
- Restrictions in the home or work environment, e.g., narrow doorways, tight turns, through floor lifts
- Method of transfer — see below

Transfers

- Standing pivot: The person is able to stand in front of the wheelchair and lower themselves, probably with assistance, into the seat. A pivoting frame (turntable) can be used for assistance.
- Slide from side: The person is unable to carry out a standing transfer but has sufficient strength in their arms to lift their weight across from one surface to another, the armrests having been removed or swung back. A slide board is often used to provide assistance.
- Lift: The person is unable to bear weight either through their feet or hands/arms, and is light enough not to pose a moving and handling risk to the carer. This type of transfer tends to be limited to small children, though parents often continue to lift their child into their teens and adulthood, and spouses frequently carry the person to avoid the limitations to lifestyle imposed by the requirement to use a hoist.
- Hoisted: Where the person cannot be lifted, a hoist and sling are used. Hoists are either mobile, running on the floor, or mounted on a ceiling track system. A wide variety of slings are available.

Physical Measurements

It is necessary to take measurements to determine the appropriate size of wheelchair. Ideally, this should be carried out on a firm surface, such as a physiotherapy plinth, as softer surfaces tend to mask anatomical landmarks. The following measurements should be considered but will not all be necessary on every occasion.

- Hip width: Across the widest point; consider alongside shoulder width
- Seat depth: Back of the knee, or popliteal crease, to the back of the buttock; care should be taken not to measure with the pelvis in posterior tilt (see Chapter 19 "Posture Management") where this is mobile, as the seat depth will be too long, causing postural complications
- Lower leg length: From the popliteal crease to the bottom of the heel
- Back support height: Taken from the underside of the buttock to a point where support is not required, which is a matter for clinical judgment based on the shape of the spine, the use of tilt-in-space, and the requirement for shoulder movement, particularly related to manual propulsion of the wheelchair

- Sacral support height: It is a common failing of wheelchair seating that posterior support to the sacrum is not considered; without such support, the pelvis tends to fall into posterior tilt; the height at which support stops must be determined
- Head height: The height needed to be achieved by a headrest
- Elbow height: To determine the required height of arm support
- Thoracic width: In relation to the shape of the back cushion and positioning of lateral trunk supports
- Shoulder width: Related to seat width and which is required to avoid impingement of shoulders on the back posts of the wheelchair

Postural Assessment

Usually required for a full-time wheelchair user and adapted according to complexity. Full details are available in Chapter 19 "Posture Management," but note here that for those having even moderate postural impairment, failure to address postural positioning at night may impact heavily on equipment used in the day. This is due to:

- Asymmetrical postures repeatedly sustained through the night giving rise to tissue adaptation affecting sitting postures

Postural Assessment

- Discomfort in lying may give rise to discomfort in sitting
- Fatigue caused by lack of sleep may result in reduced sitting ability in the day
- Pressure management issues not addressed at night may have an impact on the effectiveness of wheelchair seating

The service may not have the remit to address issues of positioning at night but onward referral should be made to fellow healthcare professionals where necessary.

Objective Setting

Having completed the information-gathering sections described in the previous prompt sheets, it is necessary to define the objectives for provision, that is, what it is that is needing to be achieved. This provides a basis for the prescription, clarifies the reasons for intervention, is a way of identifying conflicting demands, and can help in the measurement of outcomes.

There is no fixed method but it is important to pull out the critical elements of the assessment. It can be helpful to describe an overall aim, such as in this example:

It is the intention of this prescription to prevent Mr. X sliding down in his wheelchair, for him to be able to sit out for up to four hours without back pain, and for Mrs. X to be able to take Mr. X out to the local shops.

Further detail can then be added, potentially in the form of a list, such as:

- Support and stabilize pelvis in neutral alignment
- Accommodate limited hip flexion on left side
- Stabilize trunk in central alignment
- Support head in neutral position
- Protect skin over ischial tuberosities and sacrum
- Facilitate ease of attendant propulsion

- Facilitate a hoisted transfer
- Facilitate use of urinal bottle

This list can be used to discuss priorities with the person and their carers. In this example we may wish to prescribe a tilt-in-space wheelchair to stabilize posture and allow sitting for a longer period, but this will not be lightweight in terms of pushing to the local shops. However, such a solution might be acceptable if comfort is significantly improved and has been prioritized ahead of mobility. It can also be helpful to use the SMART goal setting methodology, which is widely used in the clinical setting. Here is an example:

- **S**pecific (allow Mrs. X to independently push Mr. X to the local shops twice per week)
- **M**easurable (was Mrs. X able to achieve this aim?)
- **A**ttainable (Mrs. X does not have any significant physical impairments)
- **R**ealistic (suitable equipment is available)
- **T**ime defined (the journey should not take more than 15 minutes each way)

There are many other goal setting/planning methodologies but their suitability to the task in hand should be analyzed prior to use.

Conceptual Ideas

Having gathered the assessment data and defined aims and objectives, an outline of a prescription may be developed. It is often tempting to jump directly to a particular piece of equipment through familiarity, but this may preclude the identification of potentially better solutions. Questions to ask include:

- What sort of wheelchair is appropriate, manual or powered?
- What type of seating should be used: basic foam slab, contoured, custom contoured?

- What properties must the material have: pressure relief, postural support/correction, adjustability?
- What orientations should the wheelchair move through?
- What are the functional requirements?
- In what setting will the equipment be used?

This will generate a performance specification that provides a basis from which to judge potential pieces of equipment as different options are explored.

Prescription

Finally, one is able to make specific recommendations. These should be detailed and include the full remit of what is needed. A clear, clinical rationale supports each recommendation and is developed through clinical reasoning. Assembly techniques must be considered because it is likely that equipment contained in the prescription will come from more than one manufacturer.

References

Durkin, J., 2009. Discovering powered mobility skills with children: 'responsive partners' in learning. Int. J. Ther. Rehabil. 16, 331–342.

Nilsson, L., Eklund, M., Nyberg, P., Thulesius, H., 2011. Driving to learn in a powered wheelchair: the process of learning joystick use in people with profound cognitive difficulties. Am. J. Occup. Ther. 65 (6), 652–660.

Wheelchair Prescription

David Long

Oxford University Hospitals NHS Trust

GETTING INTO THE DETAILS

Having covered in the previous chapter the outline of the assessment process, we now examine some of the more detailed designs, sections, and components of a wheelchair.

Wheelchair Frame (or Chassis)

Wheelchair frame types generally fall into one of two categories: folding or fixed (Figure 22.1). The former usually has a cross brace arrangement which allows the wheelchair to be flattened to aid stowage in a car. It also allows the chair to flex, one side frame relative to the other, tending to keep all four wheels in contact with the ground over rougher surfaces. The additional tubing components, however, add weight to the chair and the moving parts can work loose over time.

Fixed frame chairs do not fold flat but usually have a backrest that folds down onto the seat. They are generally used for more active wheelchair users because they are lighter and stiffer which uses less energy. With the wheels removed (there is a quick release mechanism) and the backrest folded, some car drivers can pull the wheelchair frame between themselves and the steering wheel, placing it on either the passenger or rear seat. The rear wheels of the chair are stowed separately inside the car.

As they are fixed, these chairs do not conform so well to less even surfaces but this can be less of an issue since they tend to be set up with more weight over the rear axle to allow the user to perform "wheelies," a critical aspect of negotiating curbs and rough terrain.

Most powered wheelchairs have a fixed frame, only a small number being designed to be folded and lifted into a car trunk/boot. More commonly, the user will travel seated in their wheelchair as this avoids them having to transfer and avoids the manual handling issues associated with lifting a heavy chair into a car, even if it does break down into a number of smaller components. Wheelchair hoists installed in the car are available where the person is able to transfer to a normal car seat.

Wheels (Manual Chairs)

Usually measured in imperial units, the rear wheels of an occupant propelled wheelchair

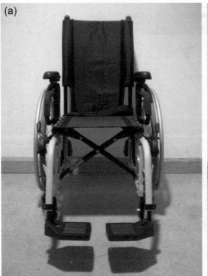 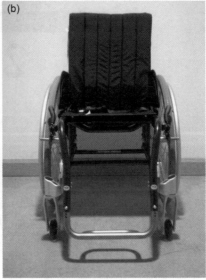

FIGURE 22.1   (a) Folding frame wheelchair (note the cross brace), (b) fixed frame wheelchair.

are either 22″, 24″, or, less commonly, 26″. The smaller size is used with children and the largest size by adults wanting higher "gearing." Tires can be fitted with pneumatic tubes or solid inserts. The latter will clearly avoid punctures but some people prefer the ride of a pneumatic tire which can provide improved comfort. Solid inserts also add weight.

Attendant-propelled wheelchairs usually have 12″ (or slightly larger) rear wheels with solid tires, mostly because they reduce the overall price of the chair and may help with reducing maintenance costs compared to larger wheels fitted with pneumatic tires. They are easier to stow in a car boot but many attendants prefer to have the larger diameter wheel as it makes the wheelchair much easier to push: the larger diameter wheel overcomes obstacles with significantly more ease than the smaller wheel.

Push Rims (Manual Chairs)

A push rim is the additional rim, distanced from the wheel rim, used to grip and turn the wheel. Different materials and surface coverings are available to suit individual need and preference. The distance of the rim from the wheel can be critical, particularly for those with reduced hand function. Also available are "capstan" rims, which have a series of projections against which someone with reduced hand function or inability to grip can push.

Castors

The smaller wheels, generally at the front of the chair, are known as castors. They should ideally be used as stabilizers and not bearers of significant weight. Why? Because the more weight that is applied, the more difficulty the castor has in rotating about its stem, having a profound impact on the maneuverability of the wheelchair. The axle having the largest wheels would, in an ideal world, bear all of the weight, but in reality we know that because this would be unstable there is a need for some form of stabilization. This is discussed in more detail in the section "Wheelchair Stability."

Brakes

There are two groups of braking mechanism on a wheelchair: those that act on the tire, which are either on or off, and those that act on a hub. The former are by far the most common. The latter can be used to slow the wheelchair when descending a hill and may also have a mechanism to allow them to be used as parking brakes (on/off). Brakes acting on the tire might be considered a crude engineering solution as they use a metal bar pushed into the face of the tire which is locked in place by an over-center mechanism. Hub brakes, operated using a handle and cable, are operated only by an attendant and while they may add a little convenience in not having to stoop on either side of the chair to apply the brake, they also add weight and cost to production.

Wheelchair brakes frequently lose efficiency or fail altogether. This applies to both types. A very small number of wheelchairs are available with bicycle style disc brakes.

Foot Support

Providing support for the feet is a challenge to the wheelchair designer and prescriber alike. Supports must offer sufficient adjustment to cater for varying leg length while at the same time being of sufficient strength to bear the weight of the feet and lower legs. Added to this is the common requirement to swing to the side to allow access for a standing transfer. Footplates are in a vulnerable position at the front of the chair and often become entangled with door frames, drop curbs, and all manner of other obstructions, leading to distortion of the supports and loss of position.

Some wheelchair designs are inclusive of a fixed foot support in the front section of the frame and do not, as a result, swing away. They are clearly less susceptible to damage but do not offer so much scope for adjustment.

Third-party components are often fitted, such as brackets to position support outside the scope provided in the original design, as a means of locating the feet more securely (e.g., molded plates and straps).

Elevating leg rests (ELRs) are used to support the knees in an extended position but there are two problems:

1. They do not always allow a return to a position similar to a normal foot plate, which extends the length of the wheelchair, potentially giving rise to problems with access.
2. Where the hamstrings are tight (see Chapter 19 "Posture Management"), elevated leg rests will tend to pull the pelvis into a posterior tilt, causing the buttocks and legs to slide forward in the seat which in turn creates problems with skin damage. Alternatively, raised leg rests may cause the knees to rise and then, because they lack support in this position, they fall to one side, causing a rotation in the pelvis and trunk.

ELRs can be manually adjusted or powered, which may allow independent adjustment by the occupant. Where there is sufficient joint range, they can be a useful adjunct to the wheelchair user having problems with swelling in their ankles or with pain. Furthermore, having an alternative position can extend the period able to be tolerated sitting in the wheelchair, which may in turn improve independence and participation.

Arm Support

These are most commonly used to support the weight of the arms through the elbows and forearms, with hands tending to fall in the lap. Arm supports are also used as a surface against which to push when rising to stand from a chair. Some are reduced in depth to allow access under desks. Some are adjustable in height to suit individual need. Most are removable, either completely or swing back out of the way to allow the person to transfer sideways out of their chair onto another surface.

Arm support is usually accompanied by some form of clothes guard on the inner side to protect clothing from entanglement and dirtying on the rear wheel. Clothes guards also help to contain the seat cushion and can provide lateral location to the pelvis to aid postural stability. Wheelchairs for more active users often dispense with armrests as they can hinder access to the wheels.

Trays can be used to support the weight of the arms, reducing the drag on the shoulder girdle, and/or to provide a surface on which to carry out functional tasks. They can be wooden or of some form of plastic. They are frequently ill fitting and are prone to failure, mostly because of the number of times they are fitted and removed. They can also be a barrier to social inclusion; for example, in not allowing the person to join others at the meal table, or in not allowing a small child to climb/be placed onto the person's lap.

Weight of the Chair (Manual Chairs)

While it is true in theory that a lighter wheelchair may be easier to push, both for the occupant and attendant, it is entirely possible to make a heavier wheelchair feel lighter if the position of the center of gravity is placed over the axle having the larger wheels. Many chairs have adjustable axle positions but they are frequently left in the default position as the chair left the factory, that is, the most stable position, the wheels being set back as far as they will go.

The weight of the chair is most important to people who need to routinely perform awkward lifts of their wheelchair; for example,

wheelchair users who are independently transferring into the driver's seat of a car.

The weight of the individual will also dictate the weight of the wheelchair. Heavier people require stronger chairs and this adds weight. Again, though, the potential for adjustment of the axle positions is critical.

Methods of Propulsion (Manual Chairs)

By far the most common method of propulsion in manual wheelchairs is with the hands on the wheel push rims. An alternative method of propulsion is with one hand and one foot, as in the case of someone having hemiparesis as a result of a stroke. Less commonly, both feet can be used without assistance from the hands. The wheelchair will then typically be propelled backwards since to gain sufficient grip on the floor and to pull forwards is extremely demanding of the leg muscles and will tend to pull the person down in the seat.

A further derivative is the one-arm drive manual wheelchair. Here, one wheel has a push rim as normal, and the other wheel has none. Instead, a linkage connects one wheel to the other, with a second, smaller rim being presented to the user on the same side as the normal rim. It is then possible to propel with one arm and hand, but a high level of dexterity is required to manipulate the rims either together or individually. It also requires a reasonably high level of cognitive functioning to be able to dissociate movement of the rims and to associate each rim with a particular wheel. The linkage between the wheels either folds in a concertina or is telescopic to allow folding of the chair.

Many attempts have been made over the years to develop lever propulsion, that is, similar to rowing but with the "oars" vertical and with both being pushed forward at the same time, rather than pulled. The levers may act either directly on the tire or may activate the hub via a geared linkage.

Self-Propulsion Biomechanics

The biomechanical action of manual propulsion is a complex subject which cannot be covered in detail within the context of this book. The principal considerations are to bring the rims as close to the hips as possible and for the rear axle to be as far underneath the person as possible while maintaining an acceptable amount of instability (see the section "Wheelchair Stability"). A wholly stable wheelchair, of course, is not efficient to push and will prevent the pulling of a wheelie which allows access up and down steps or curbs and over rough ground. The main considerations in terms of adjustment to the wheelchair are as follows:

- Size of rear wheel
- Type of wheel construction
- Type of tire
- Type of push rim
- Position of rear axle, in terms of fore/aft and up/down, relative to shoulder
- Camber angle of rear wheels; a negative camber (i.e., tops of wheels leaning inward) improves straight-line stability of a wheelchair as a leaning wheel will, when rolling, tend to turn in the direction of the lean

It is critical for a wheelchair to be adjusted to suit the person and his or her intended use (Engström, 2002). There are many adjustments possible and each can have a profound effect, both positive and negative, on the functionality of the wheelchair.

Seat Cushions and Back Supports

The former term usually refers to seating placed under the buttocks and thighs; the latter to support for the trunk. There are hundreds, if not thousands, of seat cushions available on the market. As such, it is

necessary to define a specification for what is desired. This will include considerations of:

- Shape
- Comfort
- Stability
- Tissue integrity
- Ease of interfacing with specific wheelchair
- Function (e.g., type of transfer)
- Weight
- Durability
- Cost
- Local service preferences

Of course, the patient may also have preferences and past experience of different cushions which must also be considered. Additional shaping and support is often available in the form of inserts for the cushion.

Modular Seats

Modular seats include seat and/or back in a package that can be fitted to a range of different bases and can be adjusted for growth. They usually include a variety of seat cushion shapes, options for lateral pelvic and trunk support, and a variety of head, knee, and foot supports.

"Special Seating"

This is a commonly used term without a common definition. It always includes custom-contoured seating (see the "Custom-Contoured Seating" section) but may also include modular seating and "comfort" chairs (as described next). As such, it is a term that is perhaps best avoided.

"Comfort" Chairs

This, again, is a commonly used term, but not consistently. It generally refers to a complete modular wheelchair system (i.e., chassis and seating) with adjustable tilt-in-space and recline, with the option to fit a variety of postural supports. It fills the gap between a standard wheelchair and custom-contoured seating, that is, where someone can be positioned symmetrically but where a posteriorly tilted orientation is required to maintain such a position.

Mostly prescribed for adults, there are a smaller number of children's versions available.

Tilt-in-Space Versus Recline

The former term refers to a system of support that moves the body in one unit and in one orientation in the sagittal plane, where the hips, knees, ankles, and spine maintain their relative positions. Recline, by contrast, refers to motion of the back support, hinged at the base, which opens the hip angle, leaving the legs in the same position.

Reclining wheelchairs were developed to provide an alternative position for people having to sit for extended periods. The problem, of course, is that there is a natural tendency to slide in a seat where the backrest is reclined and the seat is level (try this out yourself) caused by the person's center of mass falling behind their pelvis, that is, outside their base of support. Able-bodied people are able to correct such a position but people using wheelchairs are less able. The result is unsustainable shearing forces in the soft tissues under the ischial tuberosities caused by friction between the skin, clothing, and surface of the seat. This will ultimately lead to the development of tissue damage and pressure ulceration.

Tilt-in-space, on the other hand, reduces these problems because although the person's center of mass still falls outside their base of support, their ischial tuberosities are resting on an inclined surface and do not experience the same degree of damaging, shearing forces (one

is careful not to suggest that shearing forces are absent).

Having suggested that recline alone can lead to harmful forces in soft tissues, if used in conjunction with tilt-in-space it can, in fact, provide improved postural alignment and reduced energy expenditure since gravity is used powerfully to secure a position. This is found in a variety of wheelchairs and also in riser-recliner armchairs which often tilt by a small amount as they recline.

The other use for recline is in reducing pressure on the lower abdomen. Immobile people often have difficulties with digestion and bowel movement. Opening the hip angle helps with this, much as sitting back in one's chair after a big dinner is more comfortable. In some cases, recline is required to allow a urinal bottle to be placed effectively between the legs, particularly in the case of an anteriorly tilted pelvis where opening the hip angle draws the penis up and back from its otherwise very low position where access can be difficult.

Further information is available in the paper by Michael et al. (2007) which is a systematic review of 19 previous studies looking into the effectiveness of tilt-in-space. It was found that little evidence existed for function derived from a tilted position, but that tilting posteriorly beyond an angle of 20° reduced tissue interface pressures for those with spinal cord injury.

Tilt-in-Space and Stability

When the tilt mechanism of a chair is adjusted, the weight distribution is altered (Fields, 1992). This is caused by a fixed point of rotation between seat frame and chassis. Where two pivots, or centers of rotation, are used, the center of mass moves far less, keeping the stability characteristics more constant. This mechanism may also be referred to as a

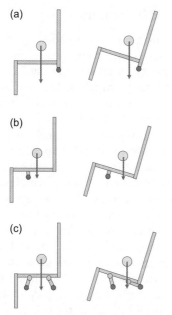

FIGURE 22.2 Larger circles indicate center of mass of person; arrows indicate where this falls relative to pivot point(s) on wheelchair chassis, shown as small circles. (a) fixed pivot (rear): center of mass moves backward (and up) as seat is tilted, reducing rearward stability; (b) fixed pivot (front): center of mass moves backward as seat is tilted, reducing rearward stability; (c) floating pivot: center of mass maintains its position relative to chassis; note that there are variations on the linkage mechanism but having the same effect.

"floating pivot" and allows for a shorter wheelbase (distance between front and rear axles) compared to a fixed point tilt. (See Figure 22.2.)

Tilt-in-Space Base

The tilt-in-space base is a common term to define a chassis onto which seating can be fitted or interfaced, coupled with the ability to tilt to a variable angle. Many of these bases also have the ability to vary the angle of recline (seat to back angle). Some bases are able to be adjusted in this way with ease (which may or may not be preferable) and some require use of a tool.

Buggies

There are a variety of buggies available for small children having additional postural needs. As well as extra supports, many buggies include a tilt-in-space system. The transition from a buggy to a wheelchair can be a difficult issue for parents as it can be perceived as confirmation of disability since transition from a buggy is normally to a walking child.

Straps and Harnesses

There are a large number of straps and harness available for use in wheelchairs. Beyond a basic safety belt fitted across the pelvis, one should take great care in prescribing such equipment as it can act, or be seen to act, as a form of restraint. If the person is able to move and this does not cause any particular issues, why contain that movement? Clearly there are reasons of safety; for example, preventing feet falling into the wheels of the chair, but this should be explored in detail prior to prescription. It is often the case that straps are fitted because straps have always been fitted, and that straps are only used because they are fitted.

The following questions may be helpful: What is the aim of having a strap or harness? Will it, realistically, achieve its aim? Is the person happy with its use? What are the chances of it being used as intended? What are the chances of it being used *not* as intended? Will it be used at all? If satisfactory answers can be provided, a harness may be indicated.

It is worth reviewing the section "Consent to Assessment and Treatment" in Chapter 17. Where the fitting of a strap or harness could be construed as a form of restraint (e.g., a forearm strap to prevent the arm falling off the side of the armrest because there is a risk of entanglement with the spokes of the wheel), a risk assessment should be used to cover the situation and must include a risk/benefit analysis.

Adaptations, Modifications, and Specials

There is a substantial range of components and devices available commercially that negate the need for manufacture of commonly used items. It is optimal to use such equipment where possible, rather than to manufacture, as costs are reduced and the relevant regulations should have been covered.

However, one of the most crucial roles for the clinical engineer in the field of rehabilitation engineering is in adapting and modifying equipment, and in some cases producing specials, to meet particular clinical needs.

Adaptations maintain the original purpose and function of the piece of equipment whereas the term *modification* implies that the piece of equipment has been taken outside its original purpose. Specials are bespoke, one-off items of equipment designed and manufactured to meet the needs of an individual. They may comprise readily available components but will use them in a unique manner.

Powered Seat Raise/Lower

On some powered chairs it is possible to fit a raiser unit which alters the height of the entire seat unit. Some systems focus on raising the seat up to allow improved social integration and/or access to cupboards or bookshelves normally out of reach in sitting. Others focus on lowering the seat to the ground, typically children's chairs designed for use in a play or school environment where being able to communicate/play with peers at an appropriate level is important.

Standing Wheelchairs

A small number of wheelchairs are designed around a standing mechanism. This offers the user the ability to move from sitting to standing, according to function and social circumstance. This can work very well for some but the complexity of the necessary mechanisms adds weight, which is critical in a manual wheelchair in terms of self-propulsion and in a powered chair in terms of battery capacity. The chairs tend to be more bulky which can create problems with access. Adding even minimal postural support can be very difficult, or even contra-indicated, as the body changes shape between sitting and standing, and the mechanism of levers does not articulate in the same way as the hips, knees, and ankles, resulting in postural supports being aligned correctly in one orientation but not in the other.

Such chairs can be anything up to 10 times more expensive than a chair without a standing mechanism which, for most people, removes the possibility of even considering the acquisition of a standing chair.

Sports Chairs and Adaptations for Sport

There are many manual wheelchairs made for the sole purpose of a particular sport (e.g., racing, basketball, rugby, etc.). Being highly tuned to a specific activity will preclude routine everyday usage and so it is unreasonable to expect one wheelchair to meet all requirements. Powered wheelchairs can be fitted with additional components (bumper guards) to allow participation in some ball games. It can also be necessary to alter seating components to allow participation in sport, particularly boccia (Ibrahim, 2012).

Transportation

Wherever possible, wheelchair users should be transferred to a standard vehicle seat for the purposes of transportation. There are instances where this is either unsafe due to lack of postural support or not possible because the person cannot transfer without a hoist. It is then a requirement that the wheelchair be suitable for use in transportation. In 2010 a group of experts developed by consensus a best practice guideline in this respect. It is recommended that reference be made to this work for further information (Tiernan et al., 2010).

Over the last 10 to 15 years there has been a marked increase in the number and quality of vehicles adapted to transport individuals seated in their wheelchairs. These are known as wheelchair accessible vehicles, or "WAVs," tend to carry one person, and are usually based on a small van being modified by specialist firms. They can be relatively inexpensive to purchase. Access can be via a ramp, which may also come with a winch, or via a powered lift either at the rear or side of the vehicle. Difficulties can arise where the person is tall, more in passing through the door aperture than in relation to the height available inside. Width and length can also cause problems in some cases.

Maintenance, Repairs, and Insurance

Well-used wheelchairs require regular attention to their mechanical and electrical systems. Costs may not be insignificant in the case of more complex or technologically advanced equipment. In some countries this is paid for by the state but where purchase of equipment is undertaken privately, the likely costs of ongoing maintenance and repairs should not be overlooked.

As well as considering insurance protection for oneself in the event of an accident, it may also be appropriate to consider third-party insurance cover. A powered wheelchair is capable of substantial, if unintentional, damage to both property and people.

CUSTOM-CONTOURED SEATING

Background

Custom-contoured seating is molded specifically for an individual, interfaced to a wheelchair, shower chair base, or static seat, and is not reusable by another person. It is indicated where modular or "off the shelf" systems cannot give the required level of support or shaping, and is typically used for those having established asymmetry in the pelvis, hips, and/or spine, coupled with low postural ability, movement disorders, and/or high levels of discomfort.

Materials

There are a variety of materials available, as described in Table 22.1. All capture body shape using either a process of direct fitting to the patient (used only for "instant" systems and by some matrix fitters) or, far more commonly, vacuum consolidation (bead bags with a valve through which air is either introduced to soften the bag or extracted to produce a firm shape). In the case of the latter technique, the shape is captured either through the taking of a plaster cast "positive," or a digitized image (Figure 22.3). It is not possible simply to scan body shape directly as tissues are unloaded and corrective forces have not been applied.

The digital image may be used with a CNC machine to mill either a carved foam seat or a high density foam "positive" over which a thermoplastic or interlocking material sheet can be draped.

Lynx is able to be expanded and contracted (within limits) without the need to add or remove cells, whereas with Matrix this is not possible. Second Generation Matrix addresses this by having the capability to fit expanding cells. It is also available with flexible components which can be used to introduce trunk

TABLE 22.1 Materials Used in Custom-Contoured Seating

	Molded foam	Interlocking (Matrix, Second Generation Matrix, Lynx)	Thermoplastics (ABS, HDPP, HDPE)[a]	Instant systems (foam-in-place system (FIPS); bead seats)
Material	Combustion modified foams of varying densities, stiffnesses, and performance characteristics	Flat sheets of interlocking components having a distance between cell centers of between 35 and 55 mm	Sheet material, usually 10 mm thickness for ABS or 6 mm for HDPP/HDPE	FIPS: Constituent chemicals of visco-elastic foam Bead: Polystyrene beads/bonding agent
Construction method	(1) Foam carved out to body shape from body casts using either (a) a pantograph arrangement comprising a tracer on bean bag and router bit on foam, or (b) digitized image and CNC router (2) Foam molded over an impression of body shape (1) and (2) Waterproof membrane; supported by thermoplastic or aluminum shell; removable covers in a flame retardant fabric	Either (a) direct fit with sheet hung from frame on which patient sits, or (b) draping of sheet over plaster cast taken from body casts; supportive aluminum tubing structure; removable covers in a flame retardant fabric	Heated sheet material is draped over a (reinforced) plaster cast of body shape with a vacuum-forming machine, then trimmed and covered in a flame retardant fabric	Both are rapid setting FIPS: Chemicals mixed in clinic and poured into polythene bag behind patient to form around body shape Bead: Bead bag is shaped around person; bonding agent poured and molded while setting Covers must be flame retardant
Speed of construction (1 being fastest)	3	4	2	1
Interfacing to chassis	Variety, attached to thermoplastic shell	Variety, attached to tubular aluminum structure	Variety	Variety
Heat dissipation	Poor	Good	Poor; improved by boring series of ventilation holes through material	Poor
Ease of casting	Very good	Very good	Very good	Difficult as time for molding limited to a few minutes
Potential for adjustment	Poor, due to shape being irregular and three-dimensional	Extensive, but being ultimately limited by labor costs and skill of fitter	Least adjustable of all; some materials have very narrow period of transition between solid and liquid, making them particularly difficult to work with, ABS having the wider transition period	Very poor

(Continued)

TABLE 22.1 (Continued)

	Molded foam	Interlocking (Matrix, Second Generation Matrix, Lynx)	Thermoplastics (ABS, HDPP, HDPE)[a]	Instant systems (foam-in-place system (FIPS); bead seats)
Weight	Comparable with interlocking systems once shell and interface added to foam	Comparable with a molded foam system	ABS molds are comparable with foam; others are significantly lighter due to thinner material and reduced interface	Dependent on interface method
Absorption of odor	Yes, over time, and despite waterproofing and use of covers	Minimal (except in covers but these can be laundered or replaced)	Yes, over time, and despite waterproofing and use of covers	Yes, over time, and despite waterproofing and use of covers
Structural integrity	Foam degrades over time, particularly where applied loads are high, leading to reduced performance; thermoplastic shells can fracture (rare); waterproofing can fail leading to rapid degradation of foam	Edges of sheet can loosen if lack support from framework; tubing can fracture (rare)	ABS can crack, although this is less likely with thicker walls; all are dependent on material not becoming too thin during draping	Foam degrades over time, particularly where applied loads are high
Areas of use	Wheelchairs, static seating	Wheelchairs, static seating	Wheelchairs, toilet seats, shower chairs	Wheelchairs, static seating
Costs (1 being lowest)	3	4	2	1

[a]Thermoplastics materials: ABS (acrylonitrile butadiene styrene), HDPP (high density polypropylene), HDPE (high density polyethylene)

support anteriorly without impeding transfers. A skilled fitter will be able to manipulate any of the interlocking materials rapidly to fit body shape.

There is a perception (a natural reaction) that foam seats are more comfortable than either interlocking or thermoplastic molds. However, one should remember that with custom-contoured seating, the applied load is distributed over a large area, thus reducing point loads and improving comfort.

Thermoplastic molds can be lighter than interlocking and foam systems but this is dependent on material, material thickness, and method of interfacing. Lighter weight is an advantage in the slightly less common scenario of someone being able to self-propel but having the need of a custom-contoured seat. There is very little scope for adjustment post-production and heat retention can be a problem. Some molds have aeration holes drilled at intervals.

Increasingly, hybrid seats are being prescribed, i.e. those where the seat base is made from one material and the seat back another. Perhaps the most common hybrid is the molded foam base and interlocking back. This provides protection, to a degree, against mispositioning of the pelvis, together with a

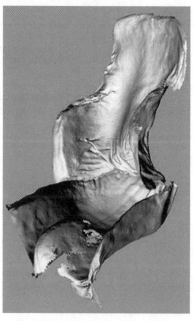

FIGURE 22.3 Scanned digital image (before processing).

perception of comfort, coupled with the potential for adjustment in the back where fine tuning is often required due to the multijointed and often unstable spine/trunk/head unit. It would, in theory, be possible to create any combination of materials to meet a particular need.

Instant systems are quick to use but offer very little opportunity for adjustment of shape before the material cures. The person must be able to remain still and, in the case of foam-in-place, tolerate the heat given off by the chemical reaction of the foam constituents. Instant systems are generally suited to less complex shapes and while initial costs are low, these can increase where a significant amount of finishing is required post-cast.

Material choice should be guided by the particular needs and circumstances of the patient and their carers, coupled with what is available locally in terms of materials and experienced engineers (Figure 22.4). A further consideration is whether to use a one- or two-piece system. A one-piece system offers greater structural integrity, which may be critical for very strong individuals, whereas two-piece systems offer greater scope for adjustment.

Taking a Shape

The single most important determinant of a successful outcome is the configuration of the molded material, hence one must be equipped with a full set of assessment data, including relevant background information, functional considerations, and physical limitations. These limitations are used to determine an optimized postural position, further information on which may be found in Chapter 19 "Posture Management."

The skill lies in negotiating the best compromise between function and posture, within the limits set by patient and carer. This takes time to determine and may require an element of iteration, such as slowly adjusting the material until an optimal shape is achieved.

Guidelines for pelvis and leg positioning:
- Position casting chair upright to allow gravity to effect a shape in the casting bag
- Check and recheck pelvic position against assessment recommendations
- Ensure that the pelvis is supported posteriorly at the sacrum
- Position legs according to any joint range limitations
- Extend medial and/or lateral thigh support for control at point furthest away from hip (fulcrum)
- Ensure that the front edge of the seat is shaped to allow for tight hamstrings

Guidelines for trunk positioning:
- Position the casting chair in posterior tilt to allow gravity to effect a shape in the back casting bag

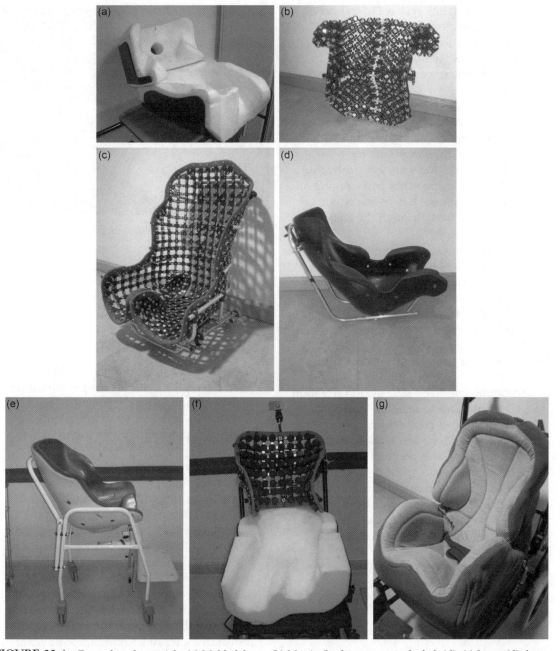

FIGURE 22.4 Examples of materials: (a) Molded foam, (b) Matrix (backrest as part of a hybrid), (c) Lynx, (d) thermoplastic wheelchair seat, (e) thermoplastic shower chair/commode, (f) hybrid (molded foam seat and Lynx backrest), (g) custom-contoured seat with covers.

- Derotate the trunk relative to the pelvis, according to assessment recommendations
- Correct trunk lateral bending as far as possible without compromising pelvic position
- Completely accommodate fixed component of kyphosis: ensure there is enough depth of casting bag and soften the bag to let the shape sink in
- Ensure the shoulders are not protracted and that the shape is not encouraging an enhanced kyphosis where one is not present: pull back through the chest and shoulders
- Check person can be removed; it is all too easy to cast a seat that provides ideal support but from which it is not possible to extract the torso

Guidelines for head positioning:

- Check alignment for breathing, swallowing, and vision
- Allow for any lateral offset caused by scoliosis
- Accommodate fixed side flexion/rotation
- Use tilt-in-space to facilitate position

Guidelines for feet positioning:

- Support where the feet want to rest relative to the knee and hip (consider tight hamstrings)
- Consider how this might conflict with the wheelchair frame and wheels
- Trial a removal of support where excessive whole body extension is an issue

Guidelines for arms positioning:

- Support the weight of the arms which may otherwise pull the trunk into an enhanced kyphosis
- Consider support from a tray or bean bag

The casting process is physically demanding of the person and the assessors, and requires a great deal of time and patience to achieve satisfactory outcomes (Figure 22.5). It is necessary

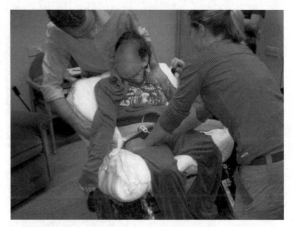

FIGURE 22.5	Example of casting process.

to be calm, resourceful, and, at times, authoritative to achieve the desired aims.

Case Study Examples

1. A child with a deteriorating condition, having a highly complex shape, tending to get hot, and having frequent issues with management of continence: may be advised to proceed with an interlocking system.
2. A child with cerebral palsy, having a moderately complex body shape, being prone to developing pressure ulcers under the ischial tuberosities, and having parents who place a high priority on comfort: may be advised to use a molded foam seat and back.
3. An adult with cerebral palsy, having a moderately complex shape, self-propelling a manual chair (indoors), frequently needing to remove the seat from the chassis to allow transit in a car, and having no significant postural changes or weight gain/loss in past few years: may wish to pursue the option of one of the lighter thermoplastic molds, mostly because of the need for self-

propulsion. In this circumstance it would also be appropriate to explore whether powered mobility would be viable.

4. An adult with cerebral palsy, displaying powerful, full-body extension when communicating, tending to get hot and to bruise easily, presents more of a challenge as a one-piece system would provide increased structural integrity but less in the way of protection against bruising. There are a number of possible options:

   a. Thermoplastic mold with holes bored for ventilation and extra padding to protect skin

   b. One-piece interlocking system with extra padding

   c. Two-piece interlocking system mounted to a seat frame having a dynamic backrest, i.e., one that has a spring loaded and damped joint between seat and back, with the aim of accommodating powerful movements

   d. Another combination according to precise circumstances of the person and what services and skills are available locally.

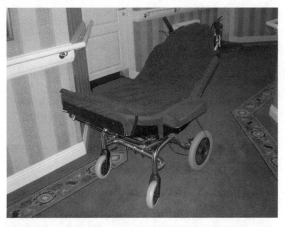

FIGURE 22.6 Platform seat to facilitate mobility for someone unable to attain a seated posture.

Where there are *very* significant issues with pain, the process of casting and fitting may be intolerable. Extreme lower limb contractures may render the person unseatable in the traditional sense. In these instances it may be appropriate to provide a small, padded, and upholstered platform on a wheelchair base to facilitate mobility (Figure 22.6).

Contraindications

There are times where custom-contoured seating may appear appropriate but where there are additional confounding factors that negate its use. Where someone is in a rehabilitation setting, they may change shape, increase joint range, or gain increased postural ability while they are waiting for seating to be manufactured, rendering it unsuitable at the point of issue. Rapidly growing children are a particular challenge and there is a point at which one must accept a certain amount of wastage. People having rapidly deteriorating conditions, such as motor neuron disease or multisystems atrophy, may deteriorate faster than the manufacturing processes for seating.

WHEELCHAIR STABILITY

Overview

This is a key concept for wheelchair provision. There is a danger, in our risk-averse society, that we pay too much attention to making things safe and not enough to making them functional. If the primary purpose of a wheelchair is to enable someone to travel from A to B, then surely ease of propulsion must be a significant consideration. A failure to do so may lead to dysfunctional mobility, reduced quality of life, and reduced participation in society.

A very stable wheelchair is unlikely to fall over, but it is also likely to be difficult to push, either for the occupant, attendant, or both, in the following respects:[1]

- Propulsion
  - With the center of mass being far forward of the rear-drive wheels, the front castor wheels bear a higher proportion of the weight than if the center of mass were farther back (see Figure 22.7). This causes two problems:
    - Being small, castor wheels offer greater resistance to motion and so a greater pushing force is required.
    - Castor wheels are less free to spin in their stems when they are loaded with increased weight causing the wheelchair to resist turning.
  - An added complication for the occupant in pushing a very stable wheelchair is that he or she must reach back for the wheels, reducing biomechanical efficiency and increasing the risk of shoulder injury.
- Reduced straight line stability, caused by:
  - An increased tendency for the wheelchair to roll down a slope that is being traversed due to the moment created by the center of mass being ahead of the rear axle line, coupled with the natural tendency of the castor wheels to swivel (Figure 22.8); with powered wheelchairs, loss of traction to the drive wheel at the top of a slope (that is, again, being traversed) will have the same effect, resulting in the wheelchair turning and slipping down the pavement toward the road.
  - Imperfections in the ground surface being more likely to throw the chair off line because the overall mass, being centered

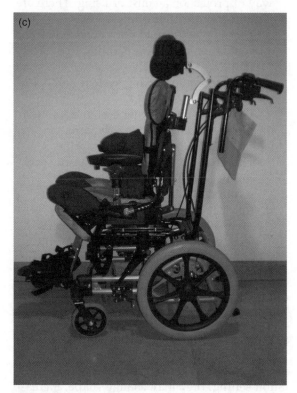

FIGURE 22.7 An altered center of mass within the frame increases the weight taken by the castor wheels (thin rimmed circle indicates combined center of mass of wheelchair and occupant). (a) Center of mass forward in frame, (b) center of mass further back in frame, (c) clinical example of seating unit bringing center of mass forward.

[1]This discussion relates mostly to the traditional large rear wheel, small front castor arrangement. The same considerations will be relevant to alternative configurations, but will need transposing.

forward of the rear axle, has more effect due to a larger moment than if the center of mass were placed closer to the rear axle.

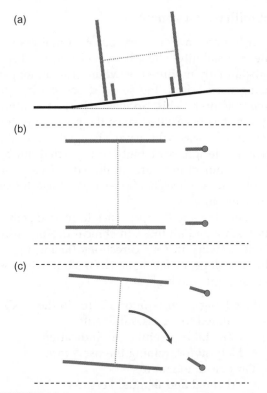

FIGURE 22.8 Wheelchair tends to roll down a slope being traversed. (a) Front view showing large rear wheels with axle and front castors; (b) plan view static on slope; small circles indicate castor stems; (c) plan view rolling across slope, rotation of castor stems turning the chair down the slope.

- Turning, due to the increased leverage required caused by the center of mass being at a distance from the fulcrum (rear axle)
- Negotiating tight turns due to a long wheelbase (distance between front and rear axles)
- Negotiating curbs or obstructions, caused by the action of tipping requiring more leverage

It is necessary, then, to optimize the balance between stability and instability, a process that should be carried out jointly with the patient and their family or carers. Where there is a specific problem of stability with a particular wheelchair, it is worth considering if an alternative model might be more suitable.

Further information is available in the MHRA publication Guidance on the Stability of Wheelchairs (2004), which can be downloaded freely within the United Kingdom.

Risk Assessment

A risk assessment should be used to document considerations of stability. As mentioned previously, it is all too easy to be over-cautious, and while it is important to optimize levels of safety, it is also important to facilitate mobility, that is, to balance risk with direct benefit or potential for benefit. To help justify decisions taken, risk assessment can be used in a positive and facilitatory manner. This is covered in more detail in Chapter 17 "Introduction: Medical Engineering Design, Regulations, and Risk Management."

"Active" Wheelchair Users

People with very good upper body control and strength, typically paraplegics, are able to propel a manual wheelchair extremely effectively, but this is only possible because the wheelchair has been set up to be distinctly unstable. This instability is controlled with balance of the trunk, arms, and head to adjust the center of mass during controlled maneuvers. The occupant has also learnt the skills through practicing techniques and experimenting, sometimes making misjudgements and falling out of the chair. This is a necessary part of the process. We do not suddenly become good at something. It is necessary, then, to allow wheelchair users the chance to develop their skills and to support them as their equipment needs adjusting, or even replacing.

Effect of Specific Body Configurations and Movement Patterns and Disorders

It will be necessary in certain circumstances to have different approaches to stability; for example, double amputees may need to have the rear wheels set back to offset the loss of mass at the front of the chair; people with very extended hip positions may be more suited to mid-drive configurations; people with uncontrolled movements, such as Huntington's chorea, may need to prioritize stability over maneuverability.

Anti-Tipping Levers/Wheels

These are devices that are attached to the wheelchair frame, almost always at the rear, to "catch" the wheelchair, should it start to tip. They are very useful aids but are not without their problems:

- When pushing, the attendant's feet can catch the levers, particularly on narrow chairs.
- Access up and down curbs can be blocked: while most anti-tippers can be raised, there is then a risk that they are left in the "off" position.
- They can be adjusted to be completely ineffectual, leading to false confidence.
- Many can be removed altogether, without the need for tools.
- Their performance is reduced where the wheelchair is on a slope facing up the incline as the line of the center of mass is drawn backward: if the wheelchair starts to tip, the anti-tipping levers will be easier to overcome.

Used appropriately, and in controlled circumstances, they can be an extremely effective means to facilitate improved maneuverability while maintaining safety.

Stability Assessment

It is necessary to carry out a formal assessment of stability when the wheelchair has been substantially modified or where it is no longer being used within the scope described by the manufacturer. This applies where a seating system has been fitted since the position of the center of mass will have changed from the original design, so stability tests carried out by the manufacturer are nullified. This also applies to a wheelchair prescribed for a lower limb amputee.

There are a few approaches to formal stability testing and all those in clinical use measure static stability in four directions (facing up or down a slope, or facing across a slope with the left or right side higher):

- Fixed angle ramp (often 12/16° in the U.K.)
  - Limited to a pass/fail result
  - Can still be useful as an indication
  - Helpful for training the user/carer
- Variable angle ramps
  - Powered and manual versions
  - Allow the determination of a precise angle of instability
  - Helpful for training the user/carer
- Force plate systems
  - Take readings of force under each wheel at two angles, typically at 0° and 5° to the horizontal, which together with measurement of certain dimensions of the wheelchair allow calculation of angles of tip in all four directions
  - Reduces manual handling for operative
  - Reduces user anxiety in relation to being placed on a steep ramp
  - No opportunity for training the user/carer

It is arguable that all these methods have flaws. First and foremost, none measure dynamic stability and at the time of writing there are no known, validated methods for assessing dynamic stability, although research continues in this extremely complex subject area: assessment in motion gives rise to a startling array of variables. Secondly, there are any number of intrinsic and extrinsic variables that will affect stability: inflation pressure of tires, angle of tilt or recline (where these are adjustable), fluctuation in weight of the occupant, additional equipment carried on the wheelchair (typically bags hung on the back of the chair).

In any assessment, then, one must consider the errors that are likely and make provision for these in any recommendations for use. A further method of testing stability is to tip the wheelchair on its front and rear wheels by hand. This is more likely to produce a false positive result (i.e., that the wheelchair is sufficiently stable), because the wheels are on a flat surface rather than a slope, causing the center of mass to be more likely to fall within the wheelbase. No measurement is taken and so no hard data can be recorded. On the other hand, in the absence of any form of measuring equipment, this method does give a very rough indication and can be used with care if its limitations are realized.

To summarize: Any decision on wheelchair stability is a clinical judgment on risk versus benefit and should be recorded appropriately. All methods of stability measurement have flaws and these must be recognized. Training should be provided to occupants and attendants as appropriate.

References

Engström, B., 2002. Ergonomic Seating: A True Challenge When Using Wheelchairs. Posturalis Books.

Fields, C.D., 1992. Living with tilt-in-space. TeamRehab Rep. 25–26.

Ibrahim, D., 2012. Seating Requirements to Maximise Performance in Boccia. Posture Mobility. 29:1 7–12.

Medicines and Healthcare Products Regulatory Agency (MHRA), 2004. Guidance on the stability of wheelchairs. Med. Health. Products Regulatory Agency. DB2004(02).

Michael, S.M., Porter, D., Pountney, T.E., 2007. Tilted seat position for non-ambulant individuals with neurological and neuromuscular impairment: a systematic review. Clin. Rehabil. 21, 1063–1074.

Tiernan, J., Appleyard, R., Arva, J., Bingham, R., Manary, M., Simms, C., et al., 2010. International Best Practice Guidelines: Transportation of People Seated in Wheelchairs. Fourth International Conference on Posture and Wheeled Mobility. (available from www.pmguk.co.uk/bpg-transportation-comment.html)

Powered Wheelchairs

Ladan Najafi* and David Long[†]

*East Kent Adult Communication and Assistive Technology (ACAT) Service, [†]Oxford
University Hospitals NHS Trust

INTRODUCTION

Where an individual is unable to self-propel a manual wheelchair due to limitations in muscle strength, muscle control, joint range, or fatigue, the use of a powered wheelchair may need to be considered. Although usually more costly than manual wheelchairs, they are expected to increase an individual's independence and to improve quality of life.

There are three basic categories:

1. Indoor: Designed to be compact and to turn in tight spaces, usually being unsuitable for any form of outdoor driving due to reduced stability, less powerful motors, and batteries having less capacity
2. Indoor/outdoor: Dual use, but more biased toward the indoor environment and limited to smoother surfaces outdoors
3. Outdoor/indoor: Dual use, but more biased toward outdoor use, being more capable over rougher surfaces, having greater range, and having greater top speed; their potential for use indoors is governed by the amount of space available

In the United Kingdom, the state provides electrically powered indoor (wheel)chairs (EPIC) and electrically powered indoor/outdoor (wheel)chairs (EPIOC). There is occasional reference to electrically powered outdoor/indoor chairs (EPOC) but these are not provided by the state.

Research indicates that provision of powered mobility in children and adults has resulted in significant improvements in several social components such as expressive behavior, cooperation, interacting with family, in the quantity of motor activities, and in the quality of interactive and symbolic play (Rosen et al., 2009; Nilsson et al., 2011).

Appropriate postural management is a prerequisite in successful assessment and provision of a powered chair (see Chapter 19 "Posture Management"). It is then important to establish an access site that is consistent and reliable and that will enable the user to safely control the powered wheelchair. This process is similar to assessment of accessing all other assistive technologies (see Chapter 17 "Introduction: Medical Engineering Design,

Regulations, and Risk Management"). An additional factor when assessing for powered mobility is to determine that the user is able to initiate or cease a movement as required because safety may be compromised if they are unable to stop the chair in a timely manner. This may be caused by impaired cognitive function, poor muscle control, including involuntary movements, or a medical complication, such as uncontrolled epilepsy.

WHEEL LAYOUT

The type of powered wheelchair will have direct impact on maneuverability, which is an important factor to consider during prescription for those users where space is an issue (Figure 23.1). Maneuverability depends on the position of the drive wheels. In rear-wheel drive (RWD) the drive wheels are behind the center of mass and the castor wheels are in front. In mid-wheel drive (MWD), the drive wheels are directly below, or very close to, the center of mass, the front and rear castor wheels being designed to be in contact with the ground at all times; note that at least one of the axles must be sprung so that the chair can negotiate uneven ground without "beaching." In front-wheel drive (FWD) the drive wheels are in front of the center of gravity and the rear castor wheels behind. Koontz et al. (2010) constructed an environment in which 90°,

180°, and 360° turns were carried out both with and without barriers. The purpose was to determine the minimum space required to perform specified maneuverability tasks. They concluded that MWD powered wheelchairs required the least space for the 360° turn and that FWD and MWD performed better than RWD in all other tasks.

In many services, though, RWD continues to dominate, partly due to cost, partly to custom of practice, and partly because it is more intuitive to learn, particularly for people having been used to driving a car. It is important to note that there are differences in maneuverability between commercially available electric wheelchairs belonging to the same category (Pellegrini et al., 2010) and therefore a trial period following an assessment is advised.

The choice of wheel layout may also be driven by the person's posture: tight hamstrings may preclude some front- and mid-wheel drive chairs as the battery box tends to be placed forward in the chassis, preventing the feet from being placed behind the normal support position. The provision of a RWD chair to obese and bariatric patients, having a center of mass in a more forward position, may give rise to loss of traction at the drive wheels, a high rate of wear at the castor tire/bearing/fork/stem, and forward instability. A front-wheel drive chair would usually be the more preferable option. The same issues can arise with a bespoke seating system or the carriage of additional medical equipment (e.g., a ventilator), both of which can alter the position of the center of mass.

It is also necessary to consider changes in center of mass that occur under dynamic conditions. The effects of an adult moving within the seat when driving off a low curb, for instance, will be greater than with a small child. A tall adult has a longer trunk and so the moments caused by movement of the shoulder girdle, arms, and head are amplified compared to someone who is shorter. The

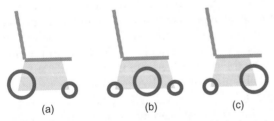

FIGURE 23.1 Powered wheelchair wheel layouts. (a) Rear-wheel drive, (b) mid-wheel drive, (c) front-wheel drive.

center of mass of an amputee will be further back and will reduce rearward stability. This may be particularly important when the chair is accelerated suddenly from rest when pointing up a slope. Many other examples exist and it is the job of the clinical engineer to ensure that they are considered.

POWERED ADJUSTMENT TO POSITION

Over the last 50 years, powered wheelchairs have become extremely diversified, allowing users with physical disabilities the option of different postural positions (Edlich et al., 2004) through the following components:

- Tilt: Orientation of the whole seating unit at an angle posterior to the vertical, allowing gravity to assist in maintaining a sitting position, or, less commonly, anterior to the vertical enabling a forward lean position to be adopted, or to facilitate a standing transfer
- Recline: Rearward movement of the backrest, pivoting at the base, to allow an alternative position but which will usually induce sliding in the seat and the potential to develop pressure ulceration; recline is usually combined with tilt
- Elevation of the feet: To assist with management of edema and for comfort, *not* for placing the hamstrings on passive stretch; often referred to as elevating leg rests, or ELRs
- Elevation of the whole seating unit: To allow access to objects on higher shelves or to bring the user's head in line with those standing nearby
- Lowering of the whole seating unit: To allow access to objects dropped on the floor or, for younger children, to allow them to be

at the same height as their peers sitting/playing on the floor
- Passive standing: There is a wealth of literature on the benefits of standing for people with physical disabilities, but it should be noted that standing wheelchairs are not without their problems which center around loss of postural support during the process of moving from sitting to standing, and the increased engineering complexity which can result in higher production and maintenance costs, as well as increased weight

ASPECTS OF CLINICAL ASSESSMENT SPECIFIC TO POWERED WHEELCHAIRS

To ensure that as many of the user's needs as possible are met, all factors must be considered (see Chapter 21 "Introduction to Mobility and Wheelchair Assessment") which, specific to this context, include:

- Environment: Steps, ramps, door thresholds, width of corridors, tightness of turns, heights of work surfaces, access to the toilet
- Transport: Most powered wheelchairs cannot be readily folded or dismantled to fit into a small car boot
- Control options: The method by which the user will control the chair must be determined
- Vision: The user must have sufficient visual ability that they can see hazards such as other people, furniture, steps; in some conditions, such as stroke, people can suffer with "neglect" on the affected side which means they cannot see hazards in certain fields of vision; it is possible to drive a powered wheelchair with visual impairment if compensatory strategies are in place, e.g., turning the head to check for objects out of

the field of vision, or learning and committing to memory a route from A to B

- Epilepsy: The onset of an epileptic seizure can cause loss of control of a powered wheelchair—this might result in coming to a sudden stop when crossing a road, or pushing and holding the joystick in the full forward speed position; the condition is sometimes nocturnal or may be suitably controlled with medication

CONTROL INTERFACES FOR POWERED WHEELCHAIRS

There are two types of control interface available: proportional (Figure 23.2) and switched. Proportional control enables the user to drive the chair in whichever direction the joystick is displaced, and the greater the displacement the faster the chair drives. This is not possible when switched controllers are used where it can be extremely slow to change direction and where the chair moves only at a preselected speed, regardless of displacement.

Correct configuration of a system enables use of a powered chair to its full potential while maintaining safety and control. If parameters such as speed, acceleration, and deceleration are not programmed appropriately, the user may

feel insecure. As an example, the chair may turn too quickly, giving rise to a feeling of insecurity, and which may make the person reluctant to drive the chair. Each powered wheelchair must be set up and programmed to meet the individual's needs and preferences, according to the environment in which they drive the chair (i.e., indoor or outdoor). Most powered chairs can be programmed using a computer or a handheld programmer that plugs into the system directly.

When driving with a joystick and traveling in a straight line it is possible to make fine adjustments to direction to take account of veer caused by uneven surfaces. This highlights one of the main disadvantages of switched controllers, which cannot compensate for unwanted veer or turn as they do not have the required level of control. Take the example of an individual having been set up to drive their chair with a clock face scanner and a single switch. If the chair veers to right, they must release the switch to stop driving forward, wait for the scanner to offer the left direction (it scrolls through the directional options), select the left direction to correct the veer, wait for the chair to move sufficiently to the left, release the switch to stop driving left, wait until forward is offered on the scanner, select forward, and finally carry on driving forward (until a further change of direction is required when the whole procedure must be repeated).

Veer is caused by gravity affecting the castors that are swiveling in the direction of the slope on which the chair is being driven (Langner and Sanders, 2008) and is most evident in RWD and FWD configurations. Even the direction of the pile of carpeted floor will cause a powered wheelchair to veer. Veer can be corrected to a very limited extent through programming. Some manufacturers have addressed this problem through the use of additional electronic components, for example, using a gyro system that detects and corrects position.

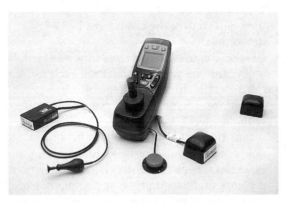

FIGURE 23.2   A selection of control options.

As described, proportional control is mostly the preferred option but it is sometimes the case that the only way to enable independent mobility is through switches.

SPECIALIZED CONTROLS

The most common method for control of a powered wheelchair, the proportional joystick, is typically positioned on either side of the chair or in midline, for driving with the hand. Joysticks may also be positioned for access by chin, tongue, or foot (see section on Access to Electronic Assistive Technology in Chapter 24).

Where an individual is not able to use a standard setup an alternative option, commonly known as specialized controls, will need to be considered. This is necessary where, for example, the user is unable to exert sufficient force to operate a standard joystick, has limited range of active movement, or has only one reliable access site but requires access to more than one assistive device. Specialized controls may provide more efficient access to powered mobility and other assistive devices (see also Chapter 24 "Electronic Assistive Technology").

It is critical that a thorough assessment is conducted to identify precisely what is required from the controls. Of particular importance is to understand any physical limitations. It is also necessary to modify the approach according to the age and cognitive abilities of the user. The input of carers and family members can be invaluable.

The following list includes commonly used examples of specialized controls:

- Switch arrays: The setup uses several switches with each one allocated to a specific direction. These can be positioned separately as required for ease of the individual's access. Switches may be mechanical or electronic (see Chapter 24 "Electronic Assistive Technology"). The powered chair control system can be set up in either momentary or latched mode. In momentary mode, the chair is only driven while the user is in contact with the switch; for those users who cannot maintain the contact but are able to press and release the switch quickly, latched mode can be used. The latched option requires a detailed assessment because it is important that the user, or attendant, can stop the chair immediately, as and when required. When latched option is considered, the chair is usually programmed so that the switch has to be pressed at regular intervals (e.g., every 10 seconds) to reactivate the latch. It is necessary to provide an emergency stop switch for reasons of safety. This would be activated either by the user and/or an attendant.

- Scanner using a single switch: A sequence of directions, often in the form of a clock face, is offered to the user at a set interval of time. The user is required to press the switch when the desired direction is offered. The interval between offered directions can be adjusted to the individual. Single-switch scanning is time consuming, cognitively demanding, and requires timing skills. However, for some users with limited motor skills and one reliable access site, single switch can be faster and less effortful than using two-switch scan.

- Scanner using two switches: One switch is used for scanning through directions and the other to select the desired direction and drive the chair, thus it can be faster or more controlled than timed scanning.

- Sip and puff: The technology is used for those with good oral motor skills where no other means of access can be established, e.g., for people with very high level spinal cord injuries. The setup can be customized for each user: for example, hard puff = forward, soft puff = right, soft sip = left, and hard sip = reverse. This can also be set up with both latched and momentary modes. As this is a form of switched system, the same limitations exist. One must also have sufficient controlled head movement to be able to negotiate the end of the tube.

- Mini/light joysticks: These are designed for those individuals with restricted movement and poor strength where fine motor control is intact, e.g., people with muscular dystrophy. This can be set up and positioned to be used by hand, chin, or tongue. The smallest type requires force exertion of less than 10 grams with the joystick itself being approximately the same cross-sectional size as a match stick and about 10 mm in length, thus requiring minimal movement to reach full displacement. As a result, it is necessary to have very fine motor skills.
- Tablet control: This is a flat touch proportional control that can be operated by only touching and remaining in contact with the surface of the tablet.
- Heavy-duty joystick: For those who, due to their physical impairment, may damage standard joysticks due to the exertion of very large amounts of force and extreme gross motor movement caused by a lack of fine motor skills.
- Foot control: For those who are unable to use their hand or head but have reasonable, reliable, and consistent movement to activate a foot control system that may be a switch, joystick, or a combination.
- Finger pot steering control: This is a proportional control that does not require any physical contact to drive a wheelchair. The position of the user's finger is detected, as if it were a joystick shaft. If the user's finger is in the center of the control, this is recognized as joystick in neutral and the wheelchair will be stationary. It will then drive in the direction that the finger moves.

- Assisted driving options: In normal development, children experience mobility at an early age (i.e., crawling at around 7–10 months of age). Assisted driving options are intended for those children and young adults with complex physical and/or cognitive disabilities who may not be able to drive a powered wheelchair independently and otherwise are not able to experience mobility. This may be in form of using a track system through infrared technology such as the Smart Wheelchair (Nisbet et al., 1988) or Sensing Collision Avoidance Device (SCAD) through ultrasonic technology (Langner, 1996). SCAD creates a safe environment for the user by detecting obstacles and helping avoid collisions, which could otherwise increase anxiety about driving. Sometimes, use of such technology is a first step for introduction of driving and as the child develops the appropriate skills, it may be appropriate to change to a standard or specialized controls option.

Controls should be tailored to the individual, a range of possibilities being considered for each situation. There may also be the requirement to take into account other

FIGURE 23.3 Access with a chin joystick and integrated access including an environmental control.

equipment which may include integrated systems (see the section "Integrated Systems" in Chapter 24 "Electronic Assistive Technology"), as shown in Figure 23.3.

POWERED ASSISTANCE TO MANUAL WHEELCHAIRS

Push-Rim Activated Power-Assisted Wheelchair

The push-rim activated power-assisted wheelchair (PAPAW) augments the power applied to the push rims by the occupant through the use of motors and batteries. They can either be retrofitted to an existing wheelchair or purchased as part of a complete system. When the push rims are moved, the effect is amplified, resulting in less effort being required to achieve motion. Some systems also amplify the braking effect from resistance applied by the occupant to the rims. A microprocessor is used to determine the amount of acceleration or braking effect based on the input received from its sensors. It can also be tuned to provide differential input to each wheel or an increased coasting effect between pushes (Cooper et al., 2004).

Powered wheelchairs can be difficult and clumsy to control in an indoor environment. The PAPAW allows the control of a manual chair indoors but provides assistance to propulsion outdoors, meaning that in some instances two wheelchairs are not required. Since the PAPAW augments rather than replaces manual propulsion, the motors and batteries need not be so large as those contained in a fully powered wheelchair. This has the effect of reducing weight, which allows the use of lighter weight wheelchair frames and easier maneuvering of the wheelchair into a car, either by the occupant or an attendant. Degeneration of the shoulder joint is a common problem among people having

propelled manual wheelchairs for many years (Cooper et al., 2004). The PAPAW may help to delay the onset of such injuries or may help those who already have an injury.

Add-on Power Packs

Manual wheelchairs can be heavy to push for the attendant, particularly where the occupant is an adult. A powered wheelchair is one solution to this problem but it is (1) expensive, (2) difficult to drive as an attendant due to not being seated in the moving chair, and (3) difficult to drive in an indoor environment compared to a manual chair, as already mentioned. Instead, it is possible to fit a device known as an "add-on power pack" to a manual chair, which comprises a battery, motor/gearbox, and drive wheel which are fitted as one unit between the rear wheels of the chair using a special bracket. A simple lever fitted to one of the push handles controls speed. Some systems offer a reverse function.

The system is particularly effective at enabling transit over greater distances but must be detached or disengaged to allow turning in very small spaces, such as in the home or small shops.

References

Cooper, R.A., Cooper, R., Schmeler, M., Boninger, M., 2004. Push for Power. Rehab Manage. 32–36.

Edlich, R.F., Nelson, K.P., Foley, M.L., Buschbacher, R.M., Long, W.B., Ma, E.K., 2004. Technological advances in powered wheelchairs. J. Long. Term. Eff. Med. Implants. 14 (2), 107–130.

Koontz, A.M., Brindle, E.D., Kankipati, P., Feathers, D., Cooper, R.A., 2010. Design features that affect the manoeuvrability of wheelchairs and scooters. Arch. Phys. Med. Rehabil. 91 (5), 759–764.

Langner, M., 1996. The development of special mobility systems at Chailey Heritage. First Symposium on powered vehicles for disabled persons.

Langner, M., Sanders, D., 2008. Controlling wheelchair direction on slopes. J. Assist. Technol. 2 (2), 32–41.

Nilsson, L., Eklund, M., Nyberg, P., Thulesius, H., 2011. Driving to learn in a powered wheelchair: the process of learning joystick use in people with profound cognitive disabilities. Am. J. Occup. Ther. 65 (6), 652–660.

Nisbet, P.D., Loudon, I.R., Odor, J.P., 1988. The CALL Centre Smart Wheelchair. Proceedings of the First International Workshop on Robotic Applications to Medical and Health Care, Ottawa.

Pellegrini, N., Bouche, S., Barbot, F., Figère, M., Guillon, B., Lofaso, F., 2010. Comparative evaluation of electric wheelchair manoeuvrability. J. Rehabil. Med. 42 (6), 605–607.

Rosen, L., Arva, J., Furumasu, J., Harris, M., Lange, M.L., McCarthy, E., et al., 2009. RESNA position on the application of power wheelchairs for paediatric users. Assist. Technol. 21 (4), 218–225.

Electronic Assistive Technology

Donna Cowan, Jodie Rogers†, Ladan Najafi**, Fiona Panthi††,
Will Wade***, Robert Lievesley†††, Tim Adlam****, and
David Long††††*

*Chailey Heritage Clinical Services, †East Kent Adult Communication and Assistive Technology
(ACAT) Service, **East Kent Adult Communication and Assistive Technology (ACAT) Service,
††East Kent Adult Communication and Assistive Technology (ACAT) Service, ***ACE Centre North,
†††Kent Communication and Assistive Technology Service (Kent CAT), ****Bath Institute of
Medical Engineering, ††††Oxford University Hospitals NHS Trust

INTRODUCTION AND ASSESSMENT

Electronic assistive technology (EAT) includes equipment such as computers, environmental control systems, communication aids, and powered mobility. All of these devices provide different functions. Some of them can be integrated into a single device; for example, a communication aid will often have facility to control an environmental control system, and computers can be used as communication aids. The range of commercially available solutions suitable for consideration both as equipment and as access methods is constantly evolving as many mainstream devices have applications directly intended to meet the needs of someone with a disability, or have accessibility options built into the standard software. In assessing for the optimal access method, a team should always look for the most efficient and reliable method available for the client for that given activity. It follows, then, that different solutions may be required for different activities.

EAT and Posture Management

A prerequisite to achieving successful access and control of technology is appropriate postural management. The individual needs to be in a position that is supported, comfortable, and promotes function. A functional seating position is one that enhances function and postural control while simultaneously decreasing effort, spasticity, or involuntary movements

(Pope, 2007; Pountney et al., 2004; Trefler and Taylor, 1991; Steen et al., 1991). When sitting in a supported position, an individual can attend to a task without the distraction of trying to control their posture. If the client is comfortable, they will be more likely to maintain their position for an appropriate period of time, ensuring that the access method can be placed in the same position repeatedly which means they will always be able to reach and activate it.

As a result, it is often recommended that, where possible, an individual's postural management is addressed prior to assessment for access to EAT (see Chapter 19 "Posture Management").

Assessment

Once posture management needs have been addressed, motor, sensory-perceptual, cognitive, linguistic, and psycho-social skills need to be assessed (Cook and Polgar, 2008) along with a discussion around what the client wishes to achieve using the equipment. This information can be gathered in a number of ways including observation, assessment, and reports from family and other professionals and/or teams associated with the client. Examples of assessment charts and templates are available from a variety of sources, including on the Internet, and forms are covered in Cook and Polgar (2008) to assess for this range of technology.

Observation of the client will reveal their range of movements and over which part of the body the client has the greatest control. This will form the starting point for determining how the interface will be activated. Consideration of movement patterns such as asymmetric tonic neck reflex (ATNR), extensor patterns, tremor, and so on, need also to be known so that accidental activation is avoided, or if this cannot be avoided that the client's

safety is not compromised (e.g., if considering powered mobility).

Any visual or hearing impairment needs to be known as this can affect the shape, size, and position of an interface and the number of options that a client can access at any one time. Cognitive issues must be considered so that the client remains safe and that the activity is presented at an appropriate level. While EAT can help fill the gap between a client's physical skills and their cognitive skills, providing them with a device that they don't understand adequately will not enhance their independence. In some instances clients find it difficult to learn new skills, remember instructions, or have an awareness of danger.

When looking for activities for an assessment, consideration of the client's interests and identifying a subject that motivates them is necessary to promote best efforts. This is particularly important when assessing children and, in particular, those who cannot communicate verbally.

The assessment is a complex process that requires a cycle of initial assessment, implementation, and evaluation. Furthermore, as technology develops, more suitable devices may become available or, as an individual's abilities change, the device may no longer meet their needs. A process of regular review should ideally take place to ensure that the user's needs continue to be met in the long term.

The end goal of any assessment is to provide a solution that has value, therefore assessment is not the end point for determining a solution. To promote confidence in the use of the device, trialing of options should be offered wherever possible along with appropriate instruction on how to use the device. This should include advice on how to troubleshoot minor issues and also what to do in the event of a breakdown that cannot be solved locally. In this way, the use of the device can be assimilated into everyday lifestyle with confidence and ease.

Where clients have complex or multiple disabilities, it is highly recommended that a multidisciplinary team approach is adopted when assessing for both equipment and access as this offers a mix of skills that allows holistic consideration of a client's needs. It may be necessary to consider whether the task should be simplified or broken down into small steps to facilitate best use of equipment, indicating that therapeutic input is required alongside technical input.

Selection of an Appropriate EAT Device and Access Method

Having carried out an assessment, the information gathered can then be matched against a range of devices. Other aspects such as where the device is to be used, what support is available, and the user's own preference must also be taken into account.

Successful use of assistive technology is dependent on identifying an appropriate means of accessing and controlling that device. Technology is continually developing and a growing multitude of access devices are available, making the selection of an appropriate device for an individual an inherently complex process. Assistive technology (AT) models, such as HAAT (Cook and Polgar, 2008), SETT (Zabala, 2005), and Matching Person to Technology (Scherer et al., 2007; Scherer, 2008), propose a predictive feature matching based approach. This systematically matches the individual's needs, skills, and abilities with the characteristics of AT devices. It requires an in-depth knowledge of both the individual and the AT, and a multidisciplinary approach involving the AT clinician, the user, and their families to ensure all aspects are considered (Herman and Hussey, 1998; Hoppestad, 2006, 2007).

It should always be kept in mind that clients with complex disabilities are likely to have in place EAT from other services or a private purchase. Any piece of equipment provided should fit into the collection already provided and should not adversely affect usage of another device. If a user with a single reliable voluntary movement has been issued a switch to operate an environmental control, consideration should be given to providing an integrated access system if providing them with a second piece of equipment (see the section "Integrated Systems").

ENVIRONMENTAL CONTROL SYSTEMS

Environmental control systems enable a person with disabilities to have control over a range of appliances installed around the house. In general these systems consist of a central controller operated by the user, using their most reliable method of access (e.g., direct touch, switches, voice activation, eye gaze). The central controller generally outputs either infrared or radio frequency codes to a range of peripheral devices.

Using these systems a person with the most complex physical disabilities, with perhaps only one reliable movement (i.e., able to operate a single switch), can have independent access to a wide range of activities. As well as providing a degree of independence for the user, these systems can also provide support for the family or carers as they are no longer solely responsible for ensuring the comfort of their family member. In some instances it can give additional confidence to enable a carer to leave the house for short periods of time without having to worry, as if anything happens the person at home can seek help immediately.

In general these systems can offer access to the following areas of function:

- Leisure: Devices such as a television, music appliances, a computer, or a page turner can be operated.
- Communication: Components can include intercom systems and hands-free telephones; some systems are able to output a number of phrases for those with communication difficulties.
- Comfort: Altering the temperature and light levels of a room are possible, e.g., remote control of heating and overhead lighting, motorized windows, control of the position of a chair or bed, curtain and blind control, and mains operation for devices such as lamps, fans, or heaters.
- Security: Support from others can be sought using pagers or call alarms. Allowing entry to the home can be screened using door openers/release mechanisms with intercom via television or telephone for interrogating whoever is at the door.

The central controller can take a variety of shapes and sizes and the interface and format of information presented to the user can be altered to meet the individual's sensory and cognitive needs (Figures 24.1 and 24.2).

A typical example of a menu is where the user is presented with a display of options perhaps relating to a single room. Within that room they might have control of intercom, alarm, lighting, a mains socket, curtains, television, and telephone. If the user can only operate a single switch this can be mounted appropriately such that access is maintained. When activated, the system scrolls through a menu until the option of choice is reached (e.g., television operations). The user can then press the switch to select that option. If the user has a visual impairment the options can be read out loud as the system scrolls through the menu. If the user cannot recognize text, symbols can be used. The speed of scrolling can be adjusted to meet the reaction times required from the user to make a selection. Having made a selection such as television operations, a second set of options may open up. This set will include the operations required to control the television, such as channel up, channel down, volume up, volume down, and so on. Again, the user is taken through this list and they select the option required when offered.

Systems can also be programmed to deal with certain known "scenarios"; for example, when the phone rings the system can be set to automatically mute the television or the music appliance (if it is on) and then switch automatically to the phone option in the menu and output a given message if the user cannot

FIGURE 24.1 Primo infrared controller offering different interface options for users. *Source: By kind permission of Possum Ltd.*

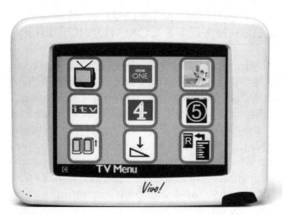

FIGURE 24.2 Vivo infrared controller offering different interface options for users. *Source: By kind permission of Possum Ltd.*

speak. Similarly, if someone knocks at the door it can be programmed to switch on the intercom, and switch on the television to the channel where the door CCTV camera can be viewed, so that the person inside the home can be assured of who is calling before allowing entry.

In the United Kingdom, environmental control systems are provided through NHS services, however, some additional items such as internal door openers for users of powered mobility have to be purchased privately. Eligibility criteria also vary.

These systems need to be regularly maintained as they often have safety critical systems incorporated, such as calling for help or operating entry and exit to the house.

Assessment

The assessment process for such a system should be undertaken as a multidisciplinary process. Like all areas of assistive technology, the user and carer need to be at the center of the assessment. There needs to a clear understanding of what motivates the user and what they wish to have control over in their environment. This then needs to be aligned with their physical and sensory needs to enable the team to identify the correct hardware and software. The user's cognitive skills need also to be taken into account as there may be a mismatch between what the user/carer expects of the system versus what they can actually achieve: with some acquired conditions or injuries, cognitive deficit can occur. Language skills also play an important part in determining the most suitable piece of equipment.

The assessment of the method of operation of the controller and where it is best placed or mounted requires careful consideration. Often for adults this is one of the few opportunities to have a switch assessment for alternative access of their equipment. Clients with highly compromised physical abilities frequently require postural or specialist seating to enable them to be in a position whereby they can access a reliable voluntary movement. When out of their seat, their access option may need to change (i.e., they may be unable to control the device in the same way when lying compared to sitting).

There needs to be consideration of all the equipment in use and how this can be best used to meet the needs of the client without affecting any other area of function. An example would be if the user has a single switch operating a voice output communication aid (VOCA). If they have another device requiring single-switch access, the user will need support switching between the VOCA and the environmental control system. However, by considering a device that allows integrated access they can retain their independent use of both devices and, as a result, the goal of improving independence is achieved. Sometimes this requires different agencies or health services to work together. In some instances devices that the user already owns have infrared output capability (e.g., VOCAs). In these cases consideration should be given to using the VOCA as the central controller

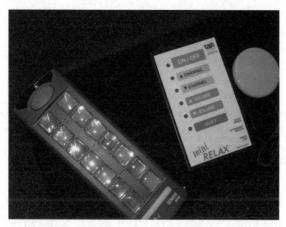

FIGURE 24.3 Simple scanning (right) and large buttoned programmable infrared operated remote control (left).

instead of supplying a separate device. This helps to rationalize the equipment in use, creating the most economic and acceptable form of system.

NHS services provide systems for meeting complex needs but some users' needs might be relatively simple. Younger children, for example, do not require control of such a broad range of household functions, although it is supportive of their development to be able to control certain aspects of the home environment, just as typically developing children do as they grow. This might include operating a food mixer in the kitchen by pressing a switch while another person holds and moves the mixer. Alternatively they might control devices such as sensory toys or some functions of the television. Simple devices are available to facilitate this and provide a useful introduction to environmental control. Adults may have problems seeing the small buttons on a standard remote control or find them too complex. Again, there are a range of simple remote controls, with or without large buttons, to meet this level of need (Figure 24.3).

AUGMENTATIVE AND ALTERNATIVE COMMUNICATION

Communication is a vital part of everyday life. Living with a communication impairment can reduce social participation and the overall quality of life. The United Nations Convention on the Rights of Persons with Disabilities[1] highlights the right of disabled people to freedom of expression and opinion through communication of their choice, including augmentative and alternative communication (AAC). Communication impairments in children can affect speech and language development, cognitive, literacy development, and access to education. As a result, timely AAC intervention is critical (Drager et al., 2010; Fried-Oken et al., 2011; McNaughton and Bryen, 2007). In adults with acquired communication disorders following a stroke or traumatic brain injury, reduced or loss of participation in work and leisure activities can lead to social isolation.

In the World Report on Disability (WHO, 2011), Professor Stephen Hawking states that "we have a moral duty to remove the barriers to participation." AAC intervention endeavors to help break down some of these barriers to reduce the detrimental effects of the communication difficulty in a person's life. Enabling people to develop social networks gives the potential for greater independence. In some cases this enables people to follow their aspirations: Lee Ridley, who has cerebral palsy and is known as "The Lost Voice Guy," uses his communication aid on stage. During an interview he said "I've always got a buzz out of making people laugh and I guess stand up comedy was always my dream job" (The Independent Northern Comedy Guide, 2012).

AAC refers to a wide range of techniques that supplement or replace speech and

[1]*www.un.org/disabilities/convention/conventionfull.shtml*

handwriting. There are many strategies, techniques, and tools that can enable people with complex communication needs (CCN) to communicate more effectively. These strategies may be unaided, whereby the individual uses their own body movement to communicate, such as gestures, facial expression, eye pointing, and signing. Aided communication systems require tools and equipment, such as pen and paper, picture or symbol based communication books, and voice output communication aids. Aided communication systems are generally categorized into low, light/medium, and high tech AAC.

Low Tech AAC Systems

These are nonelectronic and generally inexpensive. They are often paper-based and designed specifically to provide symbol/picture/photo communication books or charts for individuals with no or impaired literacy skills. Alphabet charts, E-tran frames (Figure 24.4), and other text-based systems can be provided for literate individuals.

FIGURE 24.4 An E-tran frame (low tech): The communicator's eye points to a block, then a color, to indicate the desired letter; the communication partner confirms the letters and words as the message is created.

Access Methods for Low Tech AAC

Methods are categorized into direct and indirect:

- Direct access: The communicator is able to point directly to the desired letter or symbol using a body part such as finger, hand, elbow, foot, knee, or head. Sometimes a pointing aid, such as a head pointer or a mouth stick, can be used for ease of access.
- Indirect access: Those with CCN may also have other physical disabilities that could prevent them from using a direct method of access. In these circumstances, partner-assisted scanning can be used. The conversation partner points to or reads out the letters, words, or symbols on the communication page. The communicator indicates the desired item by an eye blink, gesture, or any other previously agreed method, until the message is conveyed.

The responsibility for achieving effective communication is shared, with the communication partner taking a significant part in working out the final message. As a result, the conversation partner needs to be aware of the potential for misunderstandings or misinterpretations.

A low tech system can be a very powerful method of communication and may, in some circumstances, be the only suitable method. Even when an individual uses an electronic communication aid it is good practice to use low tech AAC as an adjunct since the electronic technology can fail (Figure 24.5). On some occasions, use of low tech AAC might be more practical.

Light/Medium Tech AAC Systems

Battery operated VOCAs, such as recorded message devices, are defined as medium or light tech. These communication aids range from providing a single spoken message to a number of spoken messages. At a glance, these

FIGURE 24.5 Symbol-based communication book (low tech): The communicator navigates their way around the book and points at the desired words and phrases. Where the communicator is unable to point directly, the communication partner reads out the words and phrases. The communicator then selects the required word or phrase by nodding, eye blinking, or other methods of confirmation that has been established between the two.

FIGURE 24.6 The MegaBee™ is a medium tech, battery-operated device that works similarly to the E-tran frame, except that the communication partner presses keys that correspond to the color of the blocks and colors of the letters. The message is then displayed on the screen in view of the communicator. Quicker communication can be achieved by using abbreviations to display messages.

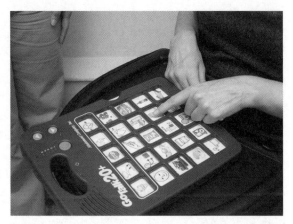

FIGURE 24.7 GoTalk20+ is a battery-operated VOCA that has the capacity to record 100 messages using communication overlays over five levels. The communicator selects the cells to play the message.

devices are limited in the number of voice output messages they can offer, but for some users they are the most suitable aids and offer an effective method for communication. Light tech AAC can be accessed via direct or indirect methods (Figures 24.6 and 24.7).

High Tech AAC

High tech VOCAs can be comprised of vocabulary packages installed on standard computers or dedicated communication aid systems sometimes referred to as speech-generating devices. Dynamic communication display is an important feature of high tech AAC systems. The communicator is able to make a selection of a particular topic or category (e.g., sport), which automatically displays another page of the relevant items within that category, such as football or tennis, together with other related vocabulary. Dynamic pages reduce the time spent on finding words or phrases, and composing messages. It enables links to a multitude of relevant vocabulary items and pages. Another feature is word prediction which can enable quicker retrieval of words and phrases and help individuals with impaired spelling skills. High tech AAC has the advantage of being able to provide a wealth of vocabulary, literally thousands of words, which could not be provided by low or light tech AAC.

There are a number of important considerations in the implementation of high tech AAC which must be addressed by the multidisciplinary assessment team during the assessment process. These factors include language and voice options, ease of use of the device, reliability, and the availability of technical support in addition to the time required to generate a message, personal preferences, family perceptions, and support (Baxter et al., 2012; Lund and Light, 2007; Rackensperger, 2012). Access methods and integration of technologies are areas that need to be addressed by the practitioner when assessing a person for an appropriate VOCA.

Vocabulary Packages

High tech AAC devices contain symbol or text-based systems. Deciding on the most appropriate vocabulary package for an individual is based on language and literacy skills, cognitive abilities, and sensory impairments. The specialist AAC team will have knowledge of the range of communication devices with preloaded language software such as systems that use Semantic Compaction/Minspeak[2] whereby language is organized by a relatively small set of multimeaning pictures and icons offering access to a large vocabulary. Selection of language software will also be guided by evidence-based practice.

It is important to note that every AAC system will need to be set up or modified to meet the individual's needs and considerations given to language development where appropriate. Regular updating and

[2]www.minspeak.com/students/AboutBruceBaker.php

development of the vocabulary is essential to ensure the communicator has access to language enabling them to talk about recent events and new activities. However, AAC does not have to be comprehensive: some individuals with progressive neurological conditions, such as motor neurone disease, may only wish to include language regarding their palliative care at a later stage of the disease.

Generally speaking, all low, light, and high tech AAC systems require planning and interventions for today *and* tomorrow, an integral component of the participation model (Beukelman and Mirenda, 2005), a systematic process for conducting AAC assessments and intervention.

Physical Features

These considerations include portability of the device, ruggedness, speech, volume, aesthetics, battery life, and mounting. There may be other considerations that need to be taken into account that are usually led by the communicator, such as lighting conditions and the effects on the screen in the environment that the individual is intending to use the device.

Access Method

Prior to assessment of VOCAs, a person's postural management needs must be addressed. Accessing VOCAs requires the development of physical skills, whether direct or indirect. It takes practice to develop these skills, which are generally taught separately to those required to access the VOCA. In children with complex disabilities, physical skills are usually taught through fun activities such as playing games. This is to ensure that lack of motor skills are not misinterpreted as lack of communication skills, and vice versa. As part of the process, positioning the device to enable ease of access and optimum function must be also assessed (see Chapters 19 "Posture Management" and this chapter's section "Access to Electronic Assistive Technology" for more details).

Integration Features

VOCAs can be integrated with computers, environmental control systems, and mobile phones (e.g., for texting). Increasingly, most dedicated VOCAs are on computers running Windows and are used as computers as well as communication aids, and have built-in environmental control modules.

There are a number of high tech AAC systems on the market. They may be keyboard based for those who are literate and have the ability to type, handheld, or less portable where usually they can be positioned on a desk or mounted onto a wheelchair.

The last few years have seen mainstream technology being used increasingly in place of dedicated communication aids (Figure 24.8). However, dedicated equipment is still preferred and more appropriate for the following groups of people:

- Those unfamiliar with Windows/iOS-based devices who may prefer a simple VOCA with a standard keyboard, such as a Lightwriter
- Those who are familiar with a device and are already able to communicate with a specific vocabulary package
- Those who require an alternative method of access: Not all methods of access can be used with all applications on mainstream devices
- Those who require simple setups: Some dedicated communication aids have built-in interfaces for access such as a switch module which allows the direct connection of switches to the VOCA; mainstream devices such as tablet PCs may require additional components or software to enable switch access

Conclusions

Through the assessment process the skills of the individual are matched to the features of the AAC system, regardless of whether it is low/light or high tech AAC, and more than one device might be considered in the early stages of intervention. Not all people with CCN have the same skills and difficulties. Cognition, sensory and motor skills, language, literacy, and the ability to learn can differ considerably from person to person. In addition, psycho-social considerations, such as the communicator's attitude to technology, can play an important role in the decision-making process. Some communicators may prefer to use light or low tech AAC. It is also worth noting that high tech AAC may place considerable learning demands on the individual and their family in supporting them.

A trial period should be part of an assessment process whereby the communicator can receive training and plenty of opportunities to practice with their AAC device, along with

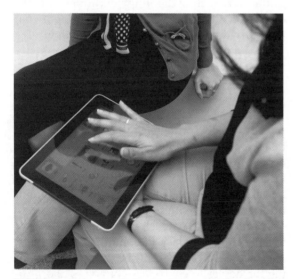

FIGURE 24.8 An iPad with Sensory Software's Grid Player app with grid set Symbol Talker A, a symbol-based (Widget and SymbolStix) communication system that enables the communicator to express a range of messages, opinions, chat, and so on.

regular support provided by their family/ carers and AAC practitioners. Ongoing evaluation and reviews can establish whether the communication device is appropriate and whether it continues to meet the individual's needs.

ACCESS TO ELECTRONIC ASSISTIVE TECHNOLOGY

The term *access* refers to the means by which an individual interfaces with the assistive technology; for example, an individual using a computer is typically using a standard keyboard, mouse, and display. However, a person who has suffered an illness or accident that has resulted in a disability may not have full use of their upper limbs, and is therefore unable to use a computer in this way. Alternative access technology allows those individuals with disabilities who cannot use standard controls to achieve independent control of their chosen device. The access method needs to be considered in the provision of all EAT devices, and whether it is for computer access, high tech augmentative, and alternative communication (AAC), environmental controls, or powered wheelchair driving, the underlying principles are the same.

This section gives a basic overview of alternative access, the process involved in selecting an appropriate EAT access method, and an overview of some of the more commonly used access devices.

Principles of Alternative Access

The aim in the provision of alternative access is to enable the individual to independently control their chosen device. This control needs to be both reliable and effective for the individual to be able to use the device functionally. Within the literature there are no agreed measures of efficacy for EAT access. Quist and Lloyd (1997) state that the underlying principles in the successful use of alternative access should include best fit to the user's abilities, minimum effort, minimum learning, and maximum output. For example, a user of AAC will wish to be able to communicate their message quickly in a conversation, without experiencing undue fatigue from the effort of doing so.

When considering alternative access methods, the AT clinician also needs to ensure that the method being considered does not have any contraindications for the user, such as inducement of pain, harmful patterns of movement, or increased tone, all of which can have long-term implications.

Components of Alternative Access

Access can be divided into several component parts, as follows:

- Control interface, or input device: The hardware that an individual uses to operate a device
- Selection set: The set of items from which the user is making a choice; these can be letters, pictures, or symbols, and can be presented in visual, auditory, or tactile formats
- Selection method: The method by which the individual makes a selection using the control interface; the selection method can be divided into two categories, direct and indirect
  - Direct selection: User is able to directly select their choice using direct access; this could include using their hand to touch a screen, using a mouse, or eye-pointing
  - Indirect selection: User has to take intermediary steps to make their selection; the most common method is scanning, whereby items from the selection set are each presented in turn

and the user waits to select their chosen item

- Input devices that enable direct selection should be considered first as there are no intermediary steps involved, enabling quicker and less cognitively demanding access
- Output: This is how the information is fed back to the individual; it could be in the form of text, pictures, or sounds and speech

Overview of Access Methods

There is a wide range of access devices and systems available and these are continually developing. The following subsections give an overview of some of the devices currently used.

Pointer Control

- Ergonomically designed mice: These are an adaptation of the "standard" mouse design. They may be mini, ergonomically shaped, or upright but will operate in the same way as that of a standard mouse.
- Trackballs: The user rolls the ball in the desired direction to correspondingly move the onscreen cursor. These are available in many different shapes and sizes and can be operated with just a thumb, single finger, or whole hand.
- Joysticks: The user pushes the joystick in the desired direction to correspondingly move the cursor. These are available in a range of shapes and sizes and require varying degrees of force. There are joysticks available to be controlled by movements of the hand, chin, tongue, or foot.
- Touchpads: A pad with a tactile sensor, where movement of the finger on the pad controls the cursor.
- Mouse emulators: Usually consist of software that allows the user to mimic the

functions of a mouse, such as directing the cursor and mouse clicks, when using an alternative access method. An example of this is Windows® MouseKeys, which allows the user to control the cursor using a number pad.

- Dibbers, mouth sticks, and head pointers: These are physical pointers that the user holds or wears and allows them to press the keys or touchscreen of their device.
- Head trackers: The user's head movement, tracked via a camera, controls the movement of the onscreen cursor (Figure 24.9). These systems are typically infrared (IR) based, where the user wears a dot that reflects the IR emitters, or can be via video tracking. To perform mouse "clicks" the user can use either a switch or dwell selection where the cursor is held still (for a specified length of time) over the chosen icon to select it.

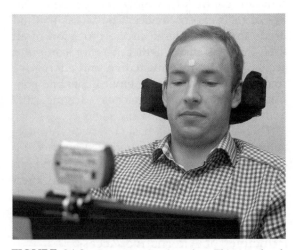

FIGURE 24.9   SmartNav head tracker. This is a head-operated mouse that consists of an infrared camera placed above the monitor. The user wears a reflective dot on their forehead or glasses. The camera follows the user's head movement which controls the position of the onscreen cursor. This can be used with an onscreen keyboard or mouse control software to enable the user to fully operate their computer hands-free.

- Eye gaze technology: The most common commercial eye tracking systems are video based using pupil/corneal reflection techniques. The system calculates the direction of the user's gaze without the need for the user to wear any additional component such as a lens or glasses (Figure 24.10).
- At least two reference points are required for the gaze point calculation. By measuring the corneal reflections relative to the center of the pupil, the system can compensate for a certain degree of head movement (Duchowski, 2003; Donegan et al., 2005). The degree of compensation varies in different systems, a factor that is very important to take into account when assessing an individual with complex disabilities. For example, users with cerebral palsy may require a system that can accommodate their involuntary movements, but this is not usually a consideration when assessing those with spinal cord injuries. The corneal reflections are typically from an infrared or near infrared light source. The advantage of this source is its invisibility to the eye, and therefore not distracting to the user.

- Touchscreens: The user is able to move the cursor or activate an icon on the screen by touching it directly. There are two types of touchscreens, resistive and capacitive. Resistive touchscreens work on the basis of pressure applied to the screen. They can be used with a finger, fingernail, or stylus. Capacitive touchscreens work by using the conductive properties of an object, typically the skin on the finger or other parts of the hand.

Keyboards

There is a wide range of keyboard options available such as large keys keyboards, mini keyboards, one-handed keyboards, and key guards to aid key selection (Figure 24.11). The keys can be presented according to an individual's needs and preferences. Options include an ABC or frequency-based layouts, instead of QWERTY, or can be in contrasting colors for those with visual impairment.

Switches

Switches come in a large range of shapes and sizes and are typically activated by limb or head movement (Figure 24.12). Switches can only perform one action and so the user may

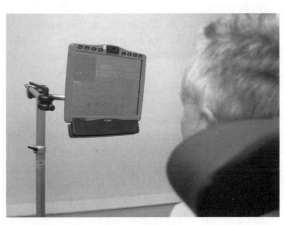

FIGURE 24.10 Eye gaze system.

FIGURE 24.11 A range of alternative keyboards and mice.

FIGURE 24.12   Example switches that vary in shape, size, activation force, activation travel (i.e., the distance by which they must be displaced for activation), and auditory feedback.

need to use a range of switches or use scanning to perform a task.

- Mechanical switches: These need to be physically moved, depressed, touched, or released to initiate a command. They vary in activation force, shape, size, and travel (the physical displacement required to activate the switch) and some provide auditory feedback.
- Electronic switches: These do not require physical contact from the user. Examples include proximity and fiber-optic switches. Proximity switches are activated when the user is close but not necessarily touching the switch. Fiber optic switches emit an invisible beam that initiates a command when interrupted; for example, a user activates the switch by blinking.

Software and Control Enhancers

- Speech recognition software: This is software that allows the user to control their device using their speech and is most commonly used for computer control where

the user can dictate into a document or perform computer tasks using a range of voice commands.
- Accessibility options: These are options that exist in the operating software of a device that can be adjusted to suit an individual user's needs. Options can include adjusting how the display is set up and how the mouse or keyboard operate.
- Word prediction software: This is used to increase the text output when typing and is already widely used in mobile phone technology. It is available for use with computers, and some AT devices, such as communication aids, have word prediction software built in.
- Screen readers: These interpret and read out the content of the display, using a synthesized voice. Screen readers can be used by those with visual impairments to aid computer access.
- Screen magnifiers: This can be presented as whole screen magnification or the magnification of sections of the screen. The sections are displayed in a second box and change in correspondence to the onscreen cursor, i.e., as you move the cursor over a section of the screen, that section is magnified.
- Onscreen keyboards: A graphical representation of a keyboard, or other type of selection set, is displayed on the screen. The keys or cells can be activated by switch or pointer control (see previous options) and will then appear as text in the user's chosen application.
- Interface positioning: The position of an individual in relation to their technology interface can have a significant impact on the ease of use of that device. For example, if an individual is unable to place their legs under their desk and therefore has to lean forward to use their mouse and keyboard, they are likely to find their computer difficult to use. Correct positioning of the device, whether it

is a keyboard, mouse, or switch, is essential in enabling the user to achieve optimum function with minimum effort.

Other Considerations

Once an appropriate control interface has been identified, it may be necessary to consider mounting this device so that it is kept in the appropriate position that enables optimum function. Mounting options can include angled rests, desk mounts, floor stands, wheelchair trays, and mounting poles that fit the device to the user's wheelchair (if applicable).

Future of Access

The range of access technology options for individuals with disabilities is rapidly growing. The following are some examples of new technologies that are emerging:

- Brain–computer interface (BCI): This technology is explained in more detail in the section "Brain–Computer Interface." Presently, most human BCI systems used for alternative access use noninvasive electroencephalographic (EEG) based technologies; however the development of functional BCI to serve as alternative access is still an area that requires further research (Fager et al., 2012).
- Eye gaze technology: This technology is widely used for computer and AAC access. Its use in powered mobility is a topic that is consistently brought up by the users of such technology and discussed among professionals. There are, at present, safety issues that require further research to evaluate its suitability for controlling powered wheelchairs.
- Gesture recognition: These can range from interfaces that recognize a few symbolic gestures to recognition of sign language. Similarly, interfaces may recognize static

hand/head poses, or dynamic hand/head motion, or a combination of both. In all cases, each gesture has an unambiguous semantic meaning associated with it that can be used in the interface. Currently there is a range of commercially available software on the market such as Swype, or those already integrated into mainstream devices such as AssistiveTouch on iPhone. Fager et al. (2012) suggest that predictive gesturing promises to substantially reduce the cognitive and physical workload for people with severe disabilities when writing. Further research is required to investigate accuracy and efficiency of these new technologies for those client groups with limited physical abilities.

- Tongue drive system (TDS): This is a wireless assistive technology that can enable those with severe physical disabilities, such as spinal cord injuries, to access computers and environmental control systems, and drive powered wheelchairs using their tongue movements. The system allows the user to send commands by pointing their tongue in different directions. This is still an area of research and currently those systems available are prototypes and for research purposes alone.

INTEGRATED SYSTEMS

Integrated systems use the same control interface to operate more than one device. As an example, an individual may use their joystick to drive their powered wheelchair and, using the same joystick, access other assistive devices such as a communication aid, environmental control, and computer.

The concept of integration of assistive devices is relatively new and the literature goes back only to the early 1990s. With the advances made in electronics and software in recent years, the integration of EAT devices

FIGURE 24.13 iPortal™ is one example of integration where the same access method, the joystick, can be used both for driving and to access an Apple iPod, iPhone, iPad, or, with the mouse mover option, a computer. (Technology has been developed to allow devices running Google's Android operating system to be accessed in the same way; see the section "Using and Adapting Mainstream Technology for Assistive Technology.")

has become more widely available to those individuals with disabilities.

Reasons for Integration

The advantages of integrated control are that persons with limited motor control can access several devices with one access site and without assistance, and the user does not need to learn a different operating mechanism for each device (Ding et al., 2003; Figure 24.13).

The findings of Guerette and Summi (1994) indicate that integrated access may be useful for the following reasons:

- The user has one single reliable control site
- The optimal control interface for each assistive device is the same
- Speed, accuracy, ease of use, or endurance increases with the use of a single interface
- The user or the family prefers integrated controls for aesthetic, performance, or other subjective reasons

When making a decision to either use integrated access or separate control interfaces, commonly known as distributed controls, a number of factors need to be taken into account.

Factors to Consider when Recommending Integrated Access

Although integrated control is beneficial for the reasons already stated, there is always a danger that if one aspect fails, all aspects fail. Breakdown in an integrated system can cause significant disruption for an individual who may be reliant on EAT for mobility, communication, and/or environmental controls. When recommending such systems, a backup system should also be considered; for example, if the single method of access is a joystick, a stand-alone joystick that can be used with the VOCA in isolation from the integrated access may be needed so that the user is not left without a voice for a period of time.

It is important to ensure that accessing one technology is not compromising the performance and efficacy of the other; for example, the skills required to drive a powered wheelchair are different than those required to access a computer.

A comprehensive assessment will take this into account (see "Introduction and Assessment" section at the start of this chapter).

Guerette and Summi (1994) also conclude that integrated access may not be appropriate when:

- Performance on one or more assistive device is severely compromised by integrating control
- The individual wishes to operate an assistive device from a position other than from a powered wheelchair
- Physical, cognitive, or visual/perceptual limitations preclude integration
- It is the individual's personal preference to use distributed controls

• External factors such as cost or technical limitations prohibit the use of integrated controls

Most of these factors are still valid today but with advances in technology some external factors are no longer a deciding factor; for example, more and more mainstream technology now can be integrated with minimum technical support and the prices of specialized controllers are more or less the same as standard controllers.

Another important factor to consider is the ownership of assistive devices and the consequent network of support and inter-service cooperation the user requires to maintain the system. Responsibility for maintenance and troubleshooting problems with equipment can be complex when a single device is used to control a number of pieces of equipment provided through different agencies (Nisbet, 1996). Assistive devices may be owned by various organizations such as wheelchair services, environmental control services, AAC services, and/or privately funded equipment. In such circumstances, the organization recommending and setting up the integrated system must, in agreement with all parties, deliver a maintenance plan to the user.

USING AND ADAPTING MAINSTREAM TECHNOLOGY FOR ASSISTIVE TECHNOLOGY

The Need for Mainstream

Mainstream technology, that is, technology not specifically designed for people with disabilities, is ubiquitous. All individuals use mainstream technology every day, whether they are disabled or not. Many items from microwaves, televisions, and fridges to cars and mobile phones come as standard unspecialized. A service supporting individuals with disabilities, or the users themselves,

have two options when looking at equipment: to purchase mainstream or to adapt. The choice may not become clear without some thought.

Specialist assistive technology products are often developed to meet as many needs as possible by having a large array of configurability. Although this can increase a product's breadth within a population, costs tend to be increased and a relatively high level of knowledge is required for configuration and maintenance.

A small number of services have workshop facilities, staff expertise, and the time available to develop bespoke equipment, which is obviously desirable in some circumstances. A custom-made device for an individual that has been designed to meet a specific set of needs will hopefully last a long period of time with minimal configuration being required. This can, however, lead to problems with repair and maintenance where the client moves area or where the service is no longer available, the client being left helpless with a broken device and a reduced level of independence. The lifetime of any assistive technology often used repetitively every day in compromising situations (e.g., at the front of a wheelchair) is unknown. Where budgets are tight, this may have an impact on what is supplied.

With this in mind, services are increasingly tending to look first at what is available on the consumer market to see if a mainstream product will meet the needs of the client. This applies equally to services having bespoke development opportunities. As an example, small children operating a computer may require a small mouse. It makes more sense to look at hardware that is available on the mainstream market than elsewhere. If what is available is unsuitable, the next question is whether adaptation of a mainstream device is possible. This is not without its difficulties, notably that adaptions may invalidate warranties or

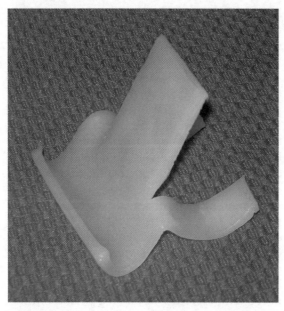

FIGURE 24.14   Example of ShapeLock used to create a prop for an iPod. *Source: Courtesy of Luke Duncan/Flickr: lucianvenutian.*

compromise device safety. Hardware adaptions may include the use of technology such as modeling silicone (e.g., Sugru[3]) and polycaprolactone-based products (e.g., ShapeLock[4]), to help the handling of objects (Figure 24.14).

Other items held by the user can be used to help facilitate access to mainstream technology; for example, *dibbers* or *touch-enabling devices* are terms used to describe technology that when strapped onto or held in the hand of a user can increase functional usage to, say, a touchscreen device or keyboard. This can include everything from small T-bar shaped pipes from local DIY stores held in the hand, to toy fingernail extensions to provide better "point" for a client's finger.

[3]*http://sugru.com/about/*

[4]*http://shapelock.com*

Adapting "High Tech" Technology

Unlike mechanical equipment, it is not always so obvious how to adapt and modify technology based on computer systems for a client's needs. Due in part to legislation and government incentives, all popular operating systems now have a number of common accessibility features built in as standard (Field and Jette, 2007; see Figure 24.15). A range of features include:

- Keyboard access
  - Key repeat/filter keys
  - Key delay until repeat
  - Acceptance delay
  - Sticky keys
- Mouse access
  - Mouse keys
  - Mouse speed/delay
  - Mouse cursor size
  - Click and scrolling speed
- Vision assistance
  - In-built screen reader
  - Magnifier functionality
  - Contrast variability

Many of these options provide a huge array of configurability to allow a large number of people with physical and cognitive difficulties to access a computer. However, access to the computer alone is sometimes insufficient: a client may wish to access a particular program that is not entirely accessible, needing modification to work successfully. One advantage of a modern computer is that developing unique software or scripts within the operating system is usually not as difficult as might be thought. Many tasks can be achieved with relatively little experience in software development. To highlight this we will use two case studies.

1. George is a 6-year-old boy with cerebral palsy who has difficulty accessing a computer. His occupational therapist went through a number of activities, for example, drawing on the computer and typing.

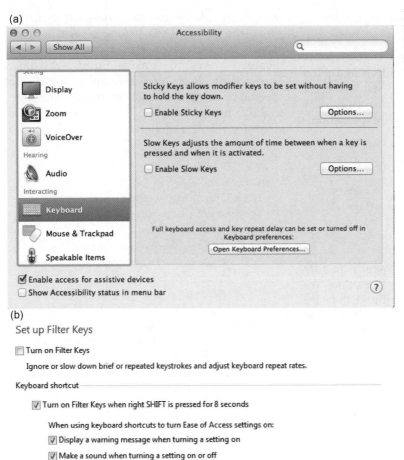

FIGURE 24.15 Examples of keyboard accessibility settings in (a) Mac OS X (10.8.2 Mountain Lion) and (b) Windows 7.

After trying a number of keyboards, both specialist and mainstream, it was decided that typing on a conventional keyboard with a specialist key guard was suitable. Keyboard accessibility settings were assessed and configured using Keyboard Wizard.[5] George could operate the mouse well for drawing, accurately using the left-click button. However, he kept hitting the right-click button accidentally. His therapist wanted a solution that would mean George could learn to use the mouse but be introduced to the right-click

functionality at a later date, that is, he should not have to spend time correcting errors. A number of options were considered, including free software. With limited access to the Internet, his therapist wrote a script in AutoHotkey[6], a free program that allows remapping of any key to another and simple scripting of the operating system at a low level. This script was turned into a self-running executable that the therapist created on her computer and then placed on the user's Startup folder so it would run when the computer was started up.

[5]*www.kpronline.com/kbwiz.php*

[6]*www.autohotkey.com*

It was also configured so that a helper could pause the script.

The "script" in this example is remarkably straightforward. Note this is written in a plain text file. Lines beginning with a semicolon are comment lines, that is, lines that are not read by the computer but there to provide explanation.

```
; Map right button to left button. All presses
of any mouse button act as a left button click
RButton::LButton
; Press cntrl + s keys together to suspend the
programme. Operated by a helper or applied to
a switch
^s::Suspend
```

Note: AutoHotkey scripts generally require AutoHotkey to be installed. However, it is possible to do this on a machine without the software installed by converting to a self-running executable using the "Convert .ahk to .exe" program installed by AutoHotkey.

Here is another example:

2. Susan is a 19-year-old student with an acquired brain injury (ABI) who has been working with her therapist on switching skills. As part of this, it was identified that Susan wishes to control music independently. Knowing that operating a music player fully can be difficult, her therapist develops an intervention program where, for the first few days, Susan presses the switch to hear 45 seconds of music. To hear the next 45 seconds of music, Susan has to hit the switch again. This will help Susan build up her experience of using a switch for repeated action and also help make sure the team has located the switch in the best place. Next, her therapist wishes Susan to start investigating a second switch for other functions in her music player.

There are many ways for Susan to operate a music player with a single switch. Most media programs simply operate on the space key/bar to play music and pressing again to pause the music. Thus linking a switch up to "space" should be enough. However, Susan's therapist wants to achieve timed switching. This is difficult without developing something specific or using specialist hardware. Although slightly more complex, the essence of the script for this is as follows:

```
; Define a variable for length of delay
DelayTimeMS := 45000; 45 seconds in
milliseconds
; Our shortcut will be shift control t. We
could use anything in truth.
^+t::
    ; DetectHiddenWindows means look for
      windows not at the front.
    ; This means Susan could work on some-
      thing else and iTunes is not at the
      front
    DetectHiddenWindows,On
    ; This line, although hard to remember
      (copy & paste!)
    ; means send a space command to iTunes.
      Space to iTunes means Pause/Play toggle
    ControlSend, ahk_parent, {space}, iTunes
    ahk_class iTunes
    ; Our script is now quietly not going to do
      anything for the period of time (45 sec)
    Sleep, DelayTimeMS
    ; Ok now wake up after 45 seconds and press
      space to iTunes
    ControlSend, ahk_parent, {space}, iTunes
    ahk_class iTunes
    ; Hopefully iTunes has now paused
return
```

This script is not without fault; for example, iTunes needs to be running, a playlist highlighted, and not playing to start with.[7]

[7]Improvements can be found here: *http://wllw.de/K14nkE* also *https://github.com/willwade/Scripting-Recipes-for-AT/tree/master/Autohotkey*

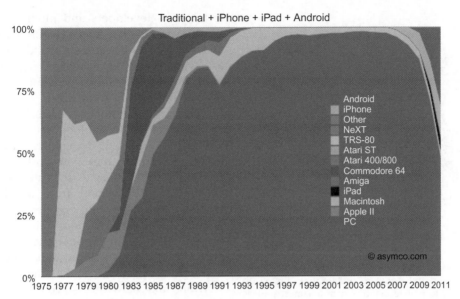

FIGURE 24.16 The rise and fall of personal computing. *Source: Deidu, 2012b.*

Other, more advanced alternatives for scripting a computer are available. Regedit, for example, allows you to edit the registry of windows (Windows Registry).[8] Useful if a user needs to regularly change their accessibility settings, a registry file can change back and forth settings by simply importing registry files each time. Scripting languages such as Visual Basic are also popular on Windows-based operating systems for doing even more complex tasks with applications. Alternatively, languages such as Python[9] allow more cross-compatibility between different operating systems.

The Future of "High Tech" Technology: New Interface Methods and New Challenges

At the time of writing, traditional desktop PCs are still a mainstay of schools and workplaces. However, outside of these settings there is a considerable change in the way consumers control and access online content. Currently, 29% of adults in the United States own a modern tablet/e-reader (Kleiner et al., 2012). Unlike the traditional computing platform of the past, where the focus of interaction was turning an indirect method of control (a keyboard or mouse) into a graphical user interface element (mouse pointers, cursors, and screen elements to access these), interaction with these newer devices is based on natural user interface methods to control the system (i.e., speech, touch, and gesture). Some are suggesting that the pace of sales in this new arena will lead to traditional desktop-based PCs being outsold by tablet devices (Grabham, 2012; Deidu, 2012a). This is depicted in Figure 24.16.

Along with the redefined user interfaces, there is also a redefining of a number of services and systems that were traditionally whole markets in themselves (Kleiner et al., 2012). Car navigation systems, software distribution,

[8]*http://en.wikipedia.org/wiki/Windows_Registry#.REG_files*
[9]*http://python.org*

FIGURE 24.17 Tecla[10], a system built on an open-source hardware and software architecture aiming to facilitate access to mobile devices, such as smart phones and tablets.

media delivery, telephony, photography, and banking are just a few sectors that have had to change direction to cope with this new interest. In some areas, assistive technology has redefined itself to make use of these new technological opportunities (Figure 24.17).

Developing tablet computing that is responsive, bright, loud, and having sufficient battery life, even compared with high-end laptops, is not without its downsides. To achieve these features, manufacturers have used hardware that is power-efficient and software to match. This means that programs are often compartmentalized or modular, not easily interfacing with each other, creating problems for assistive technology software which needs to provide a single point of control for a user. Projects such as Tecla aim to work on devices having either Google's Android or Apple's iOS system to provide a single point of access for assistive

[10]Tecla KomodoOpenLab. Available at: *http:// komodoopenlab.com/tecla/*

technology developers and modifiers. Also of significance is that methods of access users and services had become accustomed to, such as keyboard debounce, switch boxes, and common interface methods such as USB, have been abandoned in some areas.

The marginal number of sales for companies in the assistive technology sector, combined with an ever-present demand for lower costs, leads to difficulties innovating rapidly (Harwin, 1998). Tecla was created as a social enterprise with support from academia and with the aim of creating an open-source hardware and software solution. This means that a company can sell a product and provide support to end users in a traditional sense, but if required, services can make alterations to the software and hardware since the structure is freely available. Companies in the assistive technology sector have traditionally had a highly conservative approach to development (Harwin, 1998), making sure any intellectual property is carefully guarded. This new model allows for faster development and reaction to user needs. As the pace of change in technology ever increases, this model allows professionals who have the skills to provide innovative and well-supported solutions.

BRAIN—COMPUTER INTERFACES

A brain—computer interface (BCI) is a device that enables messages and commands to be conveyed directly from the brain to the external world via a computer. Ordinarily, intentions at the brain are converted into physical actions via the nervous and musculoskeletal systems. However, disease or old age may cause these systems to become impaired, leaving a person unable to interact with the environment as they would wish. BCIs offer the possibility of bypassing the body's motor systems and returning functionality to those who may otherwise be severely disabled by their impairments.

The technology is, at the time of writing, in its infancy, and for the vast majority of people in need of assistive technologies the more established methods of access are more suitable. However, conditions such as motor neurone disease (MND) or a brainstem stroke may lead to a person developing locked-in syndrome. In these cases, a fully functioning healthy brain may be trapped within a body that is unable to move, and BCIs offer the only possibility of interaction with the external environment.

This section of the chapter gives a basic overview of how BCIs work and what has been achieved to date. Those interested are referred to the suggested readings at the end of the chapter.

Detecting Intentions at the Brain

A BCI must first be able to detect intentions at the brain, and second convert these intentions into actions on the external environment. The first task is extremely difficult. The brain is a complex organ made up of over 100 billion neurones, and the mechanisms by which it works are far from being fully understood. While techniques such as positron emission tomography (PET) and functional magnetic resonance imaging (fMRI) are able to provide some useful information about brain activity, they currently have the disadvantage of being large and expensive. For this reason, most BCI research has instead centered on using surface electroencephalography (EEG) to detect intentions at the brain.

Surface EEG

Messages are transmitted within the brain by electrical activity in the neurones. Each time a neurone is activated, its electrical

potential alters by approximately 100 mV. To successfully detect and understand intentions at the brain, a BCI needs to acquire this signal, process it, and then recognize patterns within it.

The signal is acquired by placing electrodes on the scalp. The electrodes must be extremely sensitive, as by the time the signal reaches the scalp it is attenuated to only a few microvolts (Figure 24.18).

The signal must then be processed to remove artefacts caused by the electrical activity associated with muscle (electromyography/EMG) or eye (electrooculography/EOG) movements, which would interfere with the EEG signal.

Once processed, the signal must be interpreted. However, at such a distance from the

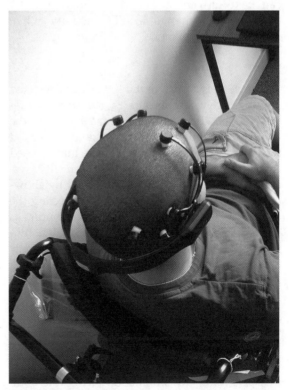

FIGURE 24.18 Emotiv EPOC neuroheadset in use.

brain it is impossible to distinguish the activity of an individual neurone. Each electrode will be picking up activity from billions of neurones and only an overall idea of activity within a certain area of the brain can be ascertained. For this reason pattern recognition is needed, and two different methods have proved to be successful with BCIs: sensorimotor rhythms and the P300 signal.

Sensorimotor Rhythms

Sensorimotor rhythms are particular frequencies of EEG signal detected near the motor and sensory areas of the brain. It has been shown that when a movement is planned (but before it is carried out), there is a decrease in sensorimotor activity. Conversely, after a movement has been carried out there is an increase in sensorimotor activity. BCI systems can take advantage of this knowledge by detecting the change in EEG activity when movements are planned.

P300

The P300 is a peak in the EEG signal detected approximately 300 ms after a stimulus is presented. The effect is strengthened by using the "oddball" paradigm: a strong signal is evoked from a low-probability stimulus interspersed among many high-probability stimuli. BCI systems can take advantage of this effect by showing users a number of options and detecting the peak in EEG signal when the desired option is viewed. Unlike sensorimotor rhythms, where a user must learn how to imagine movements that create detectable changes to the EEG signal, the P300 is evoked automatically.

Device Control

Once intention has been detected at the brain, the next task of a BCI is to convert this intention to interaction with the real world. Two examples of devices that can be successfully controlled by BCIs are wheelchairs and communication aids.

Powered wheelchairs can be controlled using sensorimotor rhythm pattern recognition. The BCI is set up to distinguish when a user thinks about moving their left arm and when they think about moving their right. These actions are then programmed into the left and right controls of a powered wheelchair, allowing a user to steer just by using their thoughts. It should be stressed that at the time of writing clinical application has been limited, mainly because this application is so new and also because there are clear safety implications.

Communication aids can be controlled by BCIs that use the P300 signal to spell out words. A grid of letters is displayed on a computer screen. The user concentrates on the letter they want, while the computer randomly highlights different rows and columns in turn. Each time the desired letter is highlighted, a P300 signal is evoked automatically, allowing the correct letter to be selected.

The Future

BCIs are currently only able to offer limited benefits, even to people with severe impairments, due to the relatively slow speed of interpreting the EEG signal. However, if the technology continues to develop, there are great possibilities for what can be achieved in the future.

One area of research is the surgical implantation of electrodes directly onto the brain cortex to detect the activity of individual neurones. It has been shown that such systems can detect intention to the level of the individual words a person would like to speak, or the specific hand movements they would like to make. Combined with sophisticated engineering, it is possible to imagine future BCIs giving precise control over robotic arms or full speech, simply by thought.

DISABILITY AND SMART HOUSE TECHNOLOGY

People with disabilities, physical and cognitive, use technology to offset their impairments and enhance their ability to function. Common examples of assistive technology include wheelchairs, walking sticks, and reminders. These devices are useful to the disabled person and enable them to do things they would not otherwise be able to do. However, these are simple devices that perform their specific role without coordination with other systems. They are technology that is additional to the technological context in which the person lives.

An alternative approach to assistive technology is to adapt the *environment* in which the disabled person lives, building the ability to offset disability into building infrastructure. When such infrastructural technologies are integrated with control and coordination systems in a domestic setting, then a "smart house" (or "smart home") is created.

A smart house is a domestic dwelling fitted with sensors, controllers, and assistive devices that can offset functional losses caused by physical or cognitive impairment, and enhance quality of life.

A smart home requires several key technologies to function:

- Sensors to detect the activity or status of the occupant
- Control system to assess the sensor data and decide on appropriate interventions
- Actuators to effect the desired interventions
- A network to link the sensors, control system, and actuators
- Communications technology to enable information from the smart home to be sent to external stakeholders

Sensors

Sensors are used to detect and measure the activities of the occupant of a smart home so that the controller can provide appropriate support when it is needed. Common examples include passive infrared motion sensors as used in security systems, light level sensors, floor pressure pads, and bed occupancy sensors. These sensors can be used to determine what activity the occupant is engaged in or whether a problem has occurred that needs to be addressed. The information from the sensors can be used in the short term to direct an immediate response to a situation, or in the longer term to adjust the system's response to better suit the individual, or to detect and track trends that may indicate a slowly developing problem that can be flagged and responded to by an external stakeholder such as a doctor or social worker.

Controllers

The control system in a smart home is what differentiates it from an environmental control system or a telecare system (see the next section "Telecommunications in the Provision of Healthcare"). The smart home controller should be able to integrate information from a variety of sources, including the installed sensors, and make decisions about interventions based on that information. This computing platform may be centralized or distributed around the building. Simple controllers use rules-based systems to make decisions such as:

IF "The bath water is 150 mm deep" THEN "the occupant should be prompted to turn off the taps"
IF "the bath water is 300 mm deep" THEN "the bath taps should be turned off"
        or

IF "there has been no activity for six hours" THEN "the occupant's daughter should be informed"

More complex systems will integrate multiple sensor inputs and make probabilistic judgments about the safety of the occupant or a task that he is trying to complete. The COACH system controller (Mihailidis, 2008), for example, provides context-sensitive voice and video prompts to assist people with dementia with completing a hand-washing task, based on inputs from a video camera.

An area requiring further research is the programming and configuration of smart house controllers by nontechnical stakeholders. At the moment, configuring a smart house is the job of a technician at best, but an expert in information technology may also be required. There is a great need to develop means of configuring smart house controllers that can be understood and carried out by people without technical training such as the occupants themselves, social workers, and sheltered housing managers.

Actuators

So what can a smart home do? This depends on who the home is designed to support and the nature of their needs. To support a person with a spinal injury who uses a wheelchair for most of the day, a smart home will enable that occupant to reach things that might ordinarily be out of reach, perform tasks that are difficult from a wheelchair such as opening and closing doors and windows, and control lighting. Where there is no cognitive disability, the difference between a smart home and an environmental control system may be smaller. Where the occupant has a cognitive disability, such as dementia or a brain injury, then the type of support provided is likely to be more complex. The system will need to offset the cognitive disability and its symptoms which may include memory loss or

difficulty with orientation in time and space. Examples of interventions in this context include voice prompts to tell an occupant to turn off a cooker where smoke has been detected and reminders to go back to bed at night time if the occupant is confused about the time of day. Other examples include lighting the way to the toilet at night to assist with route finding.

Networks

Many different network systems have been used to link sensors, controllers, and actuators together. Some have been based on existing automation systems within buildings and others on more generalized networking systems. Buildings' automation systems applied to smart homes include KNX, EnOcean, and LonWorks. These networking systems are robust and designed to be reliable for many years in the infrastructure of a building. They are typically available in wired and wireless versions. Ethernet, Wi-Fi, ZigBee, and Z-Wave are also being used increasingly for smart home installations. ZigBee is of particular interest as it is based on the IEEE 802.15.4 standard and is designed for the low-power networking of many small devices such as would be found in a smart home. Z-Wave differs from the other wireless systems in that it is proprietary and not an open standard.

Ethics and Autonomy

The design and implementation (Adlam et al., 2009) of smart homes raises some ethical questions regarding the autonomy and privacy of the occupant (Bjoerneby et al., 1999; Mahoney et al., 2007). It is important that systems with such potential to be intrusive and controlling are not imposed on people. They should be sensitively designed and configured, taking into account the needs, desires, and autonomy of the individual (Adlam et al., 2004). When designing a smart home system, what do the people who will be living with it think about the design? It is important to take the time to find out what they think at the beginning of the process, rather than assumptions being made (Orpwood et al, 2005).

TELECOMMUNICATIONS IN THE PROVISION OF HEALTHCARE

Telehealth and Telecare

The idea behind telehealth and telecare is to use the technological advances of recent years to provide improved and more efficient delivery of healthcare to people in their own homes, reducing the need for attendance at a clinic or costly residential care. This is particularly relevant to older people and those with long-term conditions (Davis, 2012). One of the main drivers for this is that people are living longer and with increasingly complex health needs, putting increased strain on already stretched budgets.

The concept behind telehealth is remote patient monitoring. Point-of-care technologies are used to take various physiological measurements, such as blood pressure or heart rate. If a measurement falls outside what is expected, action can be taken and hopefully this would occur prior to the situation becoming critical, thereby reducing the rate of hospital admission (Schwartz, 2012).

Telecare is also remote patient monitoring, but focusing on the personal safety of the individual. It uses elements of a "smart" home (see the previous section "Disability and Smart House Technology"), such as a sensor to detect if the patient has fallen, to transmit important information to the relevant person. It can also include a personal alarm which the patient can trigger if they get into difficulty (Schwartz, 2012).

Videoconferencing

The use of videoconferencing is starting to emerge as a means of providing healthcare. In some instances it is possible to carry out an assessment and/or to make changes to someone's care using this facility, which would not be possible with a telephone/teleconference. It allows multiple people to meet and to share a problem and, with the help of a visual display, to find a way to solve that problem. Dedicated videoconferencing suites are available in some organizations but low-cost and widely available technologies are now available to most people in their own homes.

There are limitations, of course. Camera placement is crucial and may not reveal the full picture; for example, views from different angles or environmental context. Getting the technology operational for everyone involved in the appointment can take time and is dependent on sufficient broadband speed and technological ability. Some organizations do not permit the use of the widely available technologies for fear of a breach of data security.

Despite these limitations, it seems likely that we will see an increase in the use of this technology as it becomes adopted more commonly across different organizations.

Remote Access

Many information technology support services already solve problems by taking control of a computer remotely, even within the same organization. This reduces significantly the need for travel to individual computers. This same technique can be applied to electronic assistive technology, allowing a clinician, engineer, or technician to work on a patient's computer from their office, negating the need to spend time travelling, thereby reducing costs. At the time of writing, use of this technique is not widespread but it is anticipated that it will increase in the coming years.

References

Baxter, S., Enderby, P., Evans, P., Judge, S., 2012. Barriers and facilitators to the use of high-technology augmentative and alternative communication devices: a systematic review and qualitative synthesis. Int. J. Lang. Commun. Disord. 47 (2), 115–129, March–April 2012.

Beukelman, D.R., Mirenda, P., 2005. Augmentative and Alternative Communication, Supporting Children and Adults with Complex Communication Needs. third edition. Paul H. Brookes Publishing Co. Inc.

Cook, A., Polgar, J., 2008. Cook and Hussey's Assistive Technologies Principles and Practice. Mosby, Inc.

Davis, D., 2012. 3millionlives programme: turning the spotlight on telehealth and telecare, Department of Health (U.K.), Allied Health Professions Bulletin no. 91 gateway 18271.

Deidu, H., 2012a. When will tablets outsell traditional PCs? (Asymco) Available at: <www.asymco.com/2012/03/02/when-will-the-tablet-market-be-larger-than-the-pc-market/>.

Deidu, H., 2012b. The rise and fall of personal computing (Asymco) Available at: <www.asymco.com/2012/01/17/the-rise-and-fall-of-personal-computing/>

Ding, D., Cooper, R.A., Kaminski, B.A., Kanaly, J.R., Allegretti, A., Chaves, E., et al., 2003. Integrated control and related technology of assistive devices. Assist Technol. 15 (2), 89–97.

Donegan, M., Oosthuizen, L., Bates, R., Daunys, G., Hansen, J.P., Joos, M., et al., 2005. D3.1 User requirements report with observations of difficulties users are experiencing. Communication by Gaze Interaction (COGAIN), IST-2003-511598: Deliverable 3.1. Available in PDF format (last accessed June 2012) at <www.cogain.org/results/reports/COGAIN-D3.1.pdf>, <www.cogain.org/wiki/File:Pdficon_small.gif>.

Drager, K., Light, J., McNaughton, D., 2010. Effects of AAC interventions on communication and language for young children with complex communication needs. J. Pediatr. Rehabil. Med. 3 (4), 303–310.

Duchowski, A.T., 2003. Eye Tracking Methodology: Theory and Practice. Springer-Verlag, London.

Fager, S., Beukelman, D.R., Fried-Oken, M., Jacobs, T., Baker, J., 2012. Access interface strategies. Assist. Technol. 24 (1), 25–33.

Field, M.J., Jette, A.M., 2007. Institute of Medicine (U.S.) Committee on Disability in America. In: Field, M.J., Jette, A.M. (Eds.), The Future of Disability in America. National Academies Press, Washington (DC) (U.S.); 2007. 7, Assistive and Mainstream Technologies for People with Disabilities. Available from: <www.ncbi.nlm.nih.gov/books/NBK11418/>.

Fried-Oken, M., Beukelman, D.R., Hux, K., 2011. Current and future AAC research considerations for adults with acquired cognitive and communication impairments, <www.ncbi.nlm.nih.gov/pubmed/22590800> - Assist Technol 24(1):56–66.

Grabham, D., 2012. Microsoft: tablets will outsell desktops next year: Windows 8 is "both an old and new bet" (techradar.computing), Available at: <www.techradar.com/news/software/operating-systems/microsoft-tablets-will-outsell-desktops-next-year-1087076>.

Guerette, P., Summi, E., 1994. Integrating control of multiple assistive devices: a retrospective review. Assist. Technol. 6 (1), 67–76.

Harwin, W.S., 1998. Niche product design, a new model for assistive technology. In: Placencia Porrero, I., Ballabio, E. (Eds.), Improving the Quality of Life for the European Citizen: Technology for Inclusive Design and Equality. IOS Press, pp. 449–452.

Herman, J.H., Hussey, S.M., 1998. Module III Assistive Technology Provision Module Part I: Assessment. In: Herman, J.H., Hussey, S.M. (Eds.), Fundamentals in Assistive Technology, third ed. RESNA, Arlington, VA, pp. 1–38.

Hoppestad, B., 2006. Essential elements for assessment of persons with severe neurological impairments for computer access utilizing assistive technology devices: A Delphi study. Disabil. Rehabil. Assist. Technol. 1 (1-2), 3–16.

Hoppestad, B., 2007. Inadequacies in computer access using assistive technology devices in profoundly disabled individuals: an overview of the current literature. Disabil. Rehabil. Assist. Technol. 2 (4), 189–199.

Kleiner, Perkins, Caufield, Byers, 2012. 2012 KPCB Internet Trends Year-End Update (Slideshare.net) Available at: <www.slideshare.net/kleinerperkins/2012-kpcb-internet-trends-yearend-update>.

Lund, S.K., Light, J., 2007. Long-term outcomes for individuals who use augmentative and alternative communication: Part III--contributing factors. Augment. Altern. Commun. 23 (4), 323–335.

McNaughton, D., Bryen, D.N., 2007. AAC technologies to enhance participation and access to meaningful societal roles for adolescents and adults with developmental disabilities who require AAC, 2007 Sep. Augment. Altern. Commun. 23 (3), 217–229.

Nisbet, P., 1996. Integrating assistive technologies: current practices and future possibilities. Med. Eng. Phys. 18 (3), 193–202.

Pope, P.M., 2007. Severe and Complex Neurological Disability. Elsevier.

Pountney, T.E., Mulcahy, C.M., Clarke, S.M., Green, E.M., 2004. The Chailey approach to postural management: an explanation of the theoretical aspects of posture management and their practical application through treatment and equipment Chailey Heritage Clinical Services.

Quist, R.W., Lloyd, L.L., 1997. "Principles and uses of Technology. In: Lloyd, L.L., Fuller, D.R., Arvidson, H.H. (Eds.), Augmentative and Alternative Communication: A Handbook of Principles and Practices. Allyn and Bacon Inc, Boston, pp. 107–126.

Rackensperger, T., 2012. Family Influences and Academic Success: The Perceptions of Individuals Using AAC. Augment. Altern. Commun. 28 (2), 106–116.

Scherer, M., 2008. Institute for Matching Person and Technology, Inc. <www.matchingpersonandtechnology.com/> (last accessed on 13/10/2012).

Scherer, M., Jutai, J., Fuhrer, M., Demers, L., DeRuyter, F., 2007. A framework for modelling the selection of assistive technology devices (ATDs). Disabil. Rehabil. Assist. Technol. 2 (1), 1–8.

Schwartz, J., 2012. What impact does telehealth have on long term conditions management? The King's Fund. www.kingsfund.org.uk/topics/telecare-and-telehealth/what-impact-does-telehealth-have-long-term-conditions-management

Steen, Raddell, Lanshammer, Fristedt, 1991. as referred to in Glennon, S. and Decoste, D. (1997) AAC Assessment Strategies: Seating and Positioning. In the Handbook of Augmentative and Alternative Communication, pp. 193. Singulair Publishing Group.

The Independent Northern Comedy Guide, 2012. Giggle Beat. Feb 4, 2012 (last accessed June 2012), <www.gigglebeats.co.uk/2012/02/lost-voice-guy-to-make-stand-up-debut-tonight/>

Trefler, E., Taylor, S., 1991. Prescription and Positioning: evaluating the physically disabled individual for wheelchair seating. Prosthet. Orthot. Int. 15, 217–224.

World Health Organization (WHO), 2011. World Report on Disability, produced jointly by WHO and the World Bank, <whqlibdoc.who.int/publications/2011/9789240685215_eng.pdf> (last accessed June 2011).

Zabala, J., 2005. Using the SETT Framework to Level the Learning Field for Students with Disabilities. <www.joyzabala.com/uploads/Zabala_SETT_Leveling_the_Learning_Field.pdf> (last accessed on 16/06/2012).

Further Readings

Introduction and Assessment

Greenwood, R.J., 2003. Handbook of Neurological Rehabilitation. Psychology Press E, Sussex.

Nisbet, P., 1998. Special Access Technology. Call Centre, Scotland.

Environmental Control Systems

Greenwood, R.J., 2003. Handbook of Neurological Rehabilitation. Psychology Press, East Sussex.

Cook, A.M., 2008. Assistive Technology: Principles and Practice. Mosby Elsevier.

Augmentative and Alternative Communication

Ace Centre, Speech Bubble (last accessed November 2012) <www.speechbubble.org.uk/>.

Binger, C., Kent-Walsh, J., 2010. What Every Speech and Language Pathologist/Audiologist Should Know Mentative and Alternative Communication. Allyn and Bacon, Boston.

Communication Matters. <www.communicationmatters.org.uk/>.

Access to Electronic Assistive Technology

Church, G., Glennen, S., 1992. The Handbook of Assistive Technology. Singulair Publishing Group, Inc.

Cook, A., Polgar, J., 2008. Cook and Hussey's Assistive Technologies Principles and Practice. Mosby, Inc.

Communication by Gaze Interaction (COGAIN) website (last accessed June 2012) <www.cogain.org>.

Majaranta, P., Aoki, H., Donegan, M., Witzner Hansen, D., Hansen, J.P., Hyrskykari, A., et al., 2012. Gaze Interaction and Applications of Eye Tracking: Advances in Assistive Technologies. First edition IGI Global.

Communication by gaze interaction (COGAIN) website, Last accessed June 2012 <www.cogain.org/wiki/COGAIN_Reports>.

Abilitynet, 2012. My computer my Way <www.abilitynet.org.uk/myway/> (last accessed 16/06/2012).

Emptech: <www.emptech.info/> (last accessed 16/06/2012).

Integrated Systems

Abilitynet, 2012. My Computer My Way <www.abilitynet.org.uk/myway/> (last accessed November 2012).

Church, G., Glennen, S., 1992. The Handbook of Assistive Technology. Singulair Publishing Group, Inc.

Communication by gaze interaction (COGAIN) website, Last accessed November 2012 <www.cogain.org/wiki/COGAIN_Reports>.

Cook, A., Polgar, J., 2008. Cook and Hussey's Assistive Technologies Principles and Practice. Mosby, Inc.

Emptech: <www.emptech.info/> (last accessed November 2012).

Majaranta, P., Aoki, H., Donegan, M., Witzner Hansen, D., Hansen, J.P., Hyrskykari, A., et al., 2012. Gaze Interaction and Applications of Eye Tracking: Advances in Assistive Technologies. First edition IGI Global.

Brain Computer Interfaces

Berger, T., Chapin, J., Gerhardt, G., McFarland, D., Principe, J., Soussou, W., et al., 2008. Brain-Computer Interfaces: An International Assessment of Research and Development Trends. Springer Science and Business Media.

Graimann, B., Allison, B., Pfurtscheller, G., 2010. Brain-Computer Interfaces: Revolutionizing Human-Computer Interaction. Springer, Heidelberg.

Tan, D., Nijholt, A., 2010. Brain-Computer Interfaces: Applying our Minds to Human-Computer Interaction. Springer-Verlag, London.

Disability and Smart House Technology

Abowd, G.A., Bobick, I., Essa, E., Mynatt and Rogers W "The Aware Home: Developing Technologies for Successful Aging", Workshop held in conjunction with American Association of Artificial Intelligence (AAAI) Conference 2002. Alberta, Canada, July 2002, (2002). workshop publication Accepted of Collection:, Proceedings of AAAI Workshop and Automation as a Care Giver.

Boger, J., Hoey, J., Poupart, P., Boutilier, C., Fernie, G., Mihailidis, A., 2006. A planning system based on Markov decision processes to guide people with dementia through activities of daily living. Inf. Technol. Biomed. IEEE Trans. 10 (2), 323–333, April 2006.

Dewsbury, G.A., Taylor, B.J., Edge, H.M., 2002. Designing dependable assistive technology systems for vulnerable people. Health Informatics J.104–110. 10.1177/146045820200800208, June 2002 8.

Diane, F., Mahoney, R.B., Purtilo, F.M., Webbe, M.A., Ashok, J., Bharucha, T.D., et al. Working Group on Technology of the Alzheimer's Association, 2007. In-home monitoring of persons with dementia: Ethical guidelines for technology research and development, Alzheimer's and Dementia 3, July 2007, pp. 217–226, ISSN 1552-5260, DOI: 10.1016/j.jalz.2007.04.388.

Martin, S., Kelly, G., Kernohan, W.G., McCreight, B., Nugent, C., 2008. Smart home technologies for health and social care support. Cochrane Database Syst. Rev. 2008 (4), CD006412.

Mihailidis, A., Bardram, J., Wan, D., 2006. Pervasive Computing in Healthcare. pub CRC Press, Nov 2006, ISBN-13: 978-0849336218.

Mihailidis, A., Cockburn, A., Longley, C., Boger, J., 2008. The acceptability of home monitoring technology among community-dwelling older adults and baby boomers. Assist. Technol. 20 (1), 1–12, 2008 Spring.

Orpwood, R., (no date). "Smart Homes", International Encyclopedia of Rehabilitation, <http://cirrie.buffalo.edu/encyclopedia/en/article/155/>.

Orpwood, R., Gibbs, C., Adlam, T., Faulkner, R., Meegahawatte, D., 2005. The design of smart homes for people with dementia - user-interface aspects. Universal Access Inf. Soc. 4 (2), 156–164, pub. Springer-Verlag, December 2005, ISSN: 1615-5289.

Park, K.H., Bien, Z., Lee, J.J., Kook Kim, B., Lim, J.T., Kim, J.O., et al., 2007. Robotic smart house to assist people with movement disabilities. Auton. Robots 22, 2 (February 2007), 183-198. DOI = 10.1007/s10514-006-9012-9 <http://dx.doi.org/10.1007/s10514-006-9012-9>.

Web Resources

Brain Computer Interfaces
Links to interesting BCI sites which have information about current projects:

www.tobi-project.org/

www.decoderproject.eu/

www.bci2000.org/BCI2000/Home.html

YouTube videos
www.youtube.com/watch?
    v = G71mTc1hiP0&feature = fvsr
www.youtube.com/watch?v = gnWSah4RD2E

Clinical Gait Analysis

David Ewins* and Tom Collins[†]

*Queen Mary's Hospital and University of Surrey, [†]Queen Mary's Hospital

INTRODUCTION

Gait analysis or assessment is the systematic study of human locomotion. Often formally based on the facilities offered by a gait laboratory, clinical gait analysis is typically provided by multidisciplinary teams including medical consultants, physiotherapists, clinical scientists, and other scientific and technical staff for the benefit of patients whose mobility has been affected by a wide range of neuromuscular and skeletal conditions including amputation, cerebral palsy, spina bifida, spinal cord injury, stroke, or talipes (IPEM, 2012).

Clinical applications of gait analysis include comparison with an age appropriate normal database to identify and quantify abnormalities in movement; investigation of the causes of abnormalities (i.e. the specific skeletal or neuromuscular impairments) to guide patient management and rehabilitation programs; and as a record that allows comparison before and after some intervention. Gait analysis is also used extensively in research and to aid development of improved clinical practice.

An assessment of a patient's gait may require the compilation of data from multiple sources, demanding several hours for processing and interpretation, or may make use of one or two tools for rapid, real-time feedback to patients during a relatively short appointment (IPEM, 2012).

The "tools" or methods used in gait analysis include visual observations made by a skilled clinician; the use of video, stop watch, and measured distances; systematic review through the use of observational gait analysis techniques; and instrumented kinematic, kinetic, electromyography, and energy expenditure measures. These data are often collected in parallel with other relevant measures; for example, joint range of movement, muscle strength, and patient self-assessment questionnaires.

Selecting the best methods requires an understanding of the limitations. The clearest example of a limitation is that gait data are often collected in a laboratory that is not a representative walking environment.

Although there are differences in how gait laboratories operate, there are similarities in the approach taken from referral to final report. A typical approach is illustrated in Table 25.1. Key issues are the timing of the

TABLE 25.1   Typical Gait Assessment Procedure

BEFORE APPOINTMENT	
Referral received and discussed at team meeting	Referral form requests information such as: Patient details Relevant medical history Reason for referral Type(s) of assessment required, e.g., video only or full data set; if this detail is not included or is unclear the referrer is usually contacted
Appointment sent	Appointment allocated and subsequent tasks completed: Acknowledgment sent to referrer Written appointment sent to patient along with an information sheet and appropriate mobility questionnaire Hospital Patient Administration System (PAS) notified
Laboratory setup	Before patient arrives in the laboratory, the environment is checked and equipment set up for the test procedures that are anticipated
AT APPOINTMENT	
History discussed	Patient asked for more details regarding medical history Mobility questionnaire reviewed to confirm main concerns with (and aspirations for) gait and goals of assessment/interventions
Measurements taken	Most suitable methods of assessment are confirmed based on referrer's requests, patient history, and state of mobility Explanation given to patient of what they will be required to do Any concerns with assessment methods are discussed
Measurements checked	Data are checked for completeness and assessed as to whether they depict the "usual" characteristics of the patient Time scale of reporting is explained to patient
AFTER APPOINTMENT	
Tidying	Data are stored (backed up) and assessment forms are completed
Data processed and interpreted, and report written	Analysis carried out on all data to answer referrer's questions/make recommendations Report of summary of findings and conclusions written and sent to referrer (further discussion may occur with referrer to focus the analysis) PAS notified of attendance

analysis, as this may form only part of the patient's clinical pathway, and the need for a well-defined clinical question(s) to be addressed.

In this chapter, following a short summary of normal gait, the main components of gait analysis are reviewed briefly, with comments on their role and limitations.

NORMAL GAIT

The standard pattern of walking is termed *gait*. As with any abnormality of the human body, to understand problems with gait it is first essential to understand "normal" gait.

When walking without impairment, it can be seen that a person will repeat a basic cycle

TABLE 25.2 Phases of the Gait Cycle

Phase	Description	% of cycle
STANCE		**60**
Initial contact	The moment the foot contacts the ground (heel contact), leg reaching in front of the body, extended for stance, ground reaction applies plantarflexion moment to foot.	
Loading response	Double limb support, ankle plantarflexes, knee flexes for shock absorption, weight transferred onto leg, ends when contralateral foot lifts off.	10
Mid-stance	Single limb support, knee extends for contralateral swing clearance, hip extends as body progresses until weight over forefoot, and heel starts to rise.	20
Terminal stance	Single limb support, heel continues to rise, hip hyperextends as body moves ahead of leg until contralateral foot contacts.	20
Preswing	Double limb support, weight transferred off leg, knee flexes, hip progresses toward neutral, foot applies propulsive plantarflexion, ends when foot lifts off.	10
SWING		**40**
Initial swing	Thigh begins to advance with hip flexion, knee flexes further and ankle dorsiflexes to neutral for foot clearance, ends when foot opposite contralateral leg.	13
Mid-swing	Leg advances as hip reaches maximum flexion, knee begins to extend to continue foot progression and neutral ankle maintained for foot clearance, ends when tibia vertical.	14
Terminal swing	Hip flexion is maintained and knee continues to full extension, with ankle held in neutral, as leg prepares for contact.	13

Source: Adapted from Pathokinesiology Service and the Physical Therapy Department, 2001; Perry, 1992.

of limb movements. The gait cycle can be defined in a variety of ways, emphasizing different aspects. Table 25.2 gives a common breakdown of the gait cycle based on splitting the cycle into two phases: the stance phase and the swing phase. These phases are defined for one leg, with stance being the phase when the foot is in contact with the ground and swing being the phase when the foot is not in contact. The percentages stated are approximate proportions of the entire cycle and are typical of a mature gait. The actual values vary with individuals and with the speed of walking.

Temporal-spatial parameters define various aspects of gait in terms of distance and/or time. They are useful because they can indicate abnormalities and are relatively simple to measure with just a stop watch and tape measure (though in practice they are usually derived from other measurement techniques such as kinematic analysis). Key distance parameters are shown in Figure 25.1 and, combined with the amount of time spent in each phase of the gait cycle, can give useful indications of lack of symmetry or reduced stability. To allow comparison between people (particularly comparing with the "normal") it is useful to express the times as percentages of the entire cycle. Typical percentages are given in Table 25.2.

Speed of walking can be expressed and calculated in various ways. The instantaneous speed is simply given by stride length per

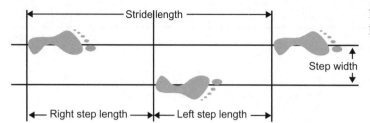

FIGURE 25.1   Definition of key spatial parameters.

cycle time (typically m/s), but the average speed is generally more useful (found from time to cover a certain distance or from averaging several strides). While each person does have a natural speed, in reality speed is constantly adjusted to match the conditions. Similar to speed, cadence is an expression of steps per minute.

Normative values of temporal-spatial and other gait parameters are usually established by each laboratory as there is not a complete consensus due to the variability of facilities and techniques.

Gait and associated parameters do change to some extent with age (Sutherland and Olshen, 1998). A child will typically begin walking around the age of 11 months but at this early stage there is no clear gait pattern as the body learns and responds to experience. Gradually, as the muscles and nervous system develop, the child's steps will become more regular, the width of steps will narrow (reflecting increased stability), and the learned control will lead to an emergence of the mature gait pattern. Temporal-spatial parameters gradually change until mid/late teens, then, without pathology, the individual's temporal-spatial characteristics will remain relatively constant until the age of about 70 years. In old age the gait trends are to slow down, with relative increase in stance phase duration and shortened stride length for stability (Pathokinesiology Service and the Physical Therapy Department, 2001; Kirtley, 2006). To facilitate the interpretation of changes in gait with age, particularly in children, some workers advocate the use of normalization techniques; for example, forces with body weight, and step length with leg length or height (Stansfield et al., 2003).

PATIENT FUNCTIONAL SELF-ASSESSMENT QUESTIONNAIRES

A gait laboratory environment is usually quite atypical so it is essential to have other information on the difficulties a patient encounters with mobility in "real life" situations. There are also many aspects that cannot really be tested but should still have an impact on decisions about treatment. Taking into account a patient's aspirations, concerns, and assessment of their own condition, together with their views on the success (or not) of past interventions, can be invaluable if treatment is to be effective. Elements of these can be met from a well-constructed referral form and when taking a clinical history from the patient or their carers. However, questionnaires can add important information, particularly if completed by the patient/carer at home without feeling under the pressure that might be perceived in the clinical environment. There is not one definitive questionnaire in use, but examples for pediatric assessment include the Gillette Functional Assessment Questionnaire (Novacheck et al., 2000) and the Pediatric Outcomes Data Collection Instrument (American Academy of Orthopaedic Surgeons, 2005).

CLINICAL OR PHYSICAL EXAMINATION

Gait assessments often require a physical examination to be made of the patient. The aims of this examination include:

- Assessment of the range of motion (ROM) of joints, generally measured as maximum and minimum angles
  - Passive range is that found by the assessor manipulating the joint and can indicate joint contracture and/or a fixed reduced length (shortening) of associated tendomuscular units
  - Active range is that achieved by the patient using their own selective muscle control and can highlight limited muscle strength when compared to passive range, or problems with selective control
- Evaluation of the level and type of muscle tone (resistance to passive stretch): Comparison of rapid and slow manipulation of a joint can show the difference between fixed or dynamic tendomuscular shortening due to an overactive antagonist muscle(s)
- Evaluation of the extent of deformity of a joint or bone in terms of alignment or abnormal formation
- Assessment of muscle strength
- Assessment of balance and posture

As with most elements of the gait assessment, care has to be taken in the collection of data as even patients who are cooperative may be affected by the actual measurements. In the case of the physical examination this can lead to, for example, increases in tone that may not be evident during gait. Further, one has to consider the relevance of the data in the context of the gait cycle; for example, how much would a reduction in the maximum range of motion seen in a given joint actually impact on walking? Further information on the physical examination can be found in Gage et al. (2009).

OBSERVATIONAL GAIT ANALYSIS AND VIDEO VECTOR

Observational gait analysis (OGA) is the standard approach used in any gait assessment. In some cases it can be enough on its own, allowing determination of the problem, but typically it is used in conjunction with other methods or used to inform the clinician as they decide what other assessment methods are necessary and possible. For example, if the patient walked into the laboratory but struggled to stand up for any length of time it might indicate that kinematic assessment was not suitable as it may take too long to set up while the patient is standing.

In its simplest form OGA just involves the clinician watching the patient walk. Substantial experience is required for this to be useful, as walking is such a complex process with many levels of movement on different planes. In reality it is not possible to assess everything that is wrong with a patient with just the naked eye. It can also be impractical to ask a patient to keep walking until the clinician feels they have seen everything, so OGA generally relies on the support of video-based equipment.

There are several advantages of using such equipment: only a few traverses of the laboratory are necessary (useful if a patient cannot walk long distances); recorded footage can be played back at a slow speed allowing more time to observe the gait events; if the patient has an irregular gait individual cycles can be analyzed repeatedly; the clinician can review the walking when the patient is no longer there (useful if new questions arise when making recommendations); and gait can be compared between sessions.

Scoring/notation systems have been developed to aid the OGA process and to allow some standardization and consistency. These systems generally lead the observer through the gait cycle looking at the major elements at each stage and making notes regarding abnormalities or deviations in movement. They are not intended to allow definitive assessment with immediate identification of problems and solutions; instead they are intended as aids to focus the assessor's attention so that they notice key features and then use clinical experience and judgment to decide on treatment. Examples of OGA systems are the Edinburgh Visual Gait Score (Read et al., 1999) and the Rancho system (Pathokinesiology Service and the Physical Therapy Department, 2001). The book accompanying the Rancho system gives detailed explanations of how to identify the deviations and also attempts to explain the range of possible causes and the significance of each deviation. Likely causes include impaired strength/motor control, ROM deficits, sensory deficits, and pain (Pathokinesiology Service and the Physical Therapy Department, 2001).

In addition to basic video footage, with real-time processing of data from a floor mounted force plate, it is possible to generate a vector representation of the ground reaction force (GRF), which can be overlaid on the sagittal and coronal video views (Figure 25.2). This means that as well as seeing visible characteristics there is also a representation of how much load the subject is applying through each leg and the direction of the GRF. Applications of video vector include providing feedback on gross gait abnormalities to patients; for example, limb weight bearing to amputees (Cole et al., 2008) and in the tuning of ankle foot orthoses to optimize the alignment of the GRF to distal joint centers (Butler and Nene, 1991).

(a)

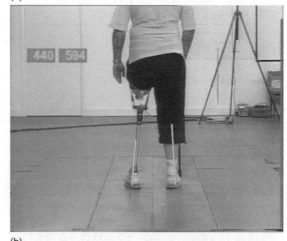

(b)

FIGURE 25.2 Screenshot illustrating video vector lines (white): (a) coronal view during standing and (b) sagittal view during walking, just after right toe-off. *Source: Cole et al. (2008).*

KINEMATICS AND KINETICS

Kinematics is the study of motion of objects in space without reference to forces (Meriam and Kraige, 1998). Many gait laboratories have motion capture systems that can track the movement of passive or active "markers" in space. By appropriate arrangement of the

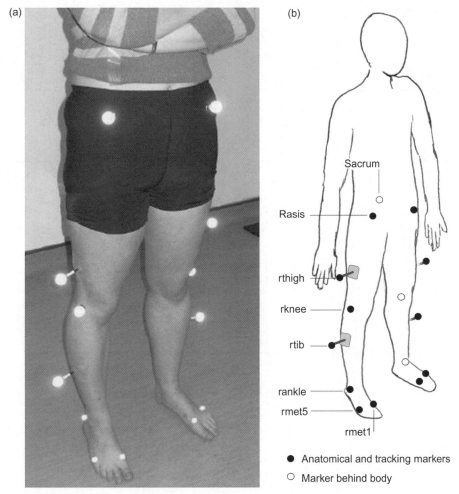

FIGURE 25.3 Example of a modified Helen Hayes marker set using passive (retroreflective) markers. Labels shown for the sacrum and right side only. This model consists of seven segments: the pelvis, and left and right of each of thighs, shanks, and feet. An example of segment definition with determination of joint centers can be found in Davis et al. (1991).

markers on a patient, the motion of body segments can be determined. From these the movement of the segments relative to each other (or to some global coordinate system) can be calculated so providing a measure of inter-segmental joint angles. For illustrative purposes, an example of a "modified Helen Hayes" marker set is shown in Figure 25.3, and typical sagittal plane joint angle plots in

Figure 25.4(a) from a volunteer with a "mature" gait walking at a self-selected normal speed.

Kinetics is the study of forces and the changes in motion that they cause. Kinetic gait analysis is useful as it reveals further aspects of gait that cannot be observed by eye. As indicated, a simple feature shown by force data is symmetry; when combined with video data

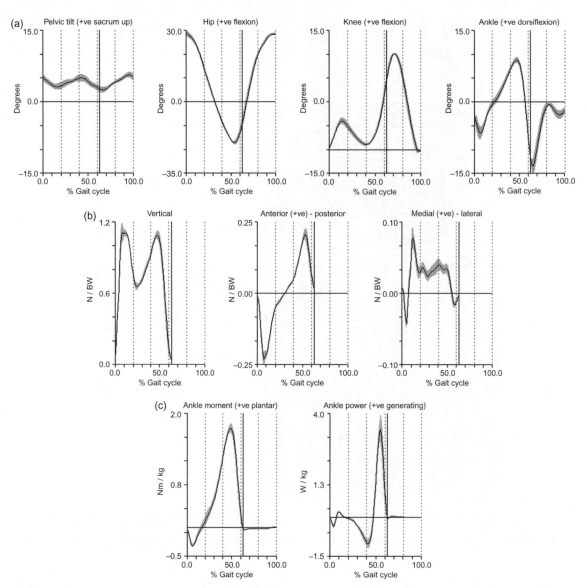

FIGURE 25.4 Example of "mature" kinematic and kinetic data for an unimpaired individual walking at self-selected normal speed (mean ± one standard deviation); solid vertical line at approximately 62% indicates end of stance phase. (a) sagittal plane joint angles, (b) ground reaction forces, and (c) ankle moment and power. (BW: Bodyweight)

this can be a useful tool in gait training. In a more complex application kinetic and kinematic data can be combined with anthropometric data to calculate joint moments and powers. The technique is commonly known as "inverse dynamics." During normal gait the only external forces acting are due to gravity and due to reaction from the ground on the

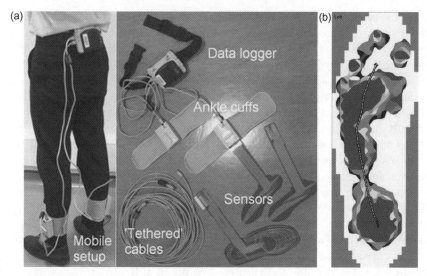

FIGURE 25.5 (a) Example of in-shoe pressure measurement system: Tekscan F-Scan equipment and mobile setup (Tekscan Inc., South Boston, MA, U.S.A.) and (b) example of peak pressure across foot during walking at slow speed: darker shade of gray represents higher pressure and dotted line shows movement of the center of pressure.

feet. The force due to gravity is effectively a constant so there is only need for a method to measure the GRF. The method usually employed and generally regarded as the "gold standard" is the use of force plates. Figure 25.4(b) shows typical GRF data collected from force plates during walking. Figure 25.4(c) shows typical joint moment and power plots for the ankle in the sagittal plane.

The basic principles and limitations of kinematics and kinetics (including inverse dynamics) are outlined in more detail in Subchapter 25.1 (at the end of this chapter) as these are relatively dominant techniques in current practice.

The use of force plates for measurement of GRF (net loading across a foot) has already been introduced, but in some cases actual pressure profiles or maps across the plantar surface of the foot are required; for example, when assessing a diabetic patient for appropriate footwear. In such cases, the use of pressure mats (barefoot) or insoles (shod) would be considered. A typical insole system is illustrated

in Figure 25.5(a) with an example of data presentation in Figure 25.5(b).

In terms of presentation of kinematic and kinetic data, the most common form is to show data by joint and plane across the gait cycle in graphs that contain age appropriate normative data. Some basic commentary on clinical data interpretation is provided in the "Data Interpretation" section, but an extension to the standard presentation includes the use of joint angle–angle plots and joint (angular) velocity–angle plots. These approaches form part of what is sometimes called *dynamical systems theory* through which aspects of (joint) coordination and control can be considered (Glazier et al., 2003).

ELECTROMYOGRAPHY

Electromyography (EMG) is the process of measuring the electrical activity of muscles. Understanding muscle action is key to understanding biomechanical function as every movement of the human body is generated or controlled by muscles. This is a clear example where technology is essential for assessing function as muscle activity cannot be seen by eye and is very complex. Even with technological help it can be difficult to evaluate what is actually happening.

Different types of electrodes are available for EMG investigations. Fine wire electrodes that sit within the muscle are particularly relevant when trying to determine the activity of deep lying muscles. Their invasive insertion does preclude routine use in many gait laboratories and they may be impractical when gross body movements are occurring, such as in gait. For this type of application it is possible to use surface electrodes. EMG signals have the appearance of noise with the energy of the signal being a function of muscle activity. Frequency content is typically in the range of 10 to 350 Hz for surface electrodes (10–1000 Hz for needle electrodes) (Perry, 1992) and amplitude in the order of 2 mV, although these figures vary with distance from the muscle.

As with kinematic marker sets, there is no universally agreed standard for the placement of surface electrodes and for processing of data. However, the SENIAM guidelines (Hermans et al., 1999) are often used for placement, and it is common to report data normalized to the gait cycle and to either a maximum voluntary contraction or maximum seen in the cycle. Some authors report EMG data in a simple on/off form, which can be helpful (Knutson and Soderberg, 1995), but definition of thresholds is not universal.

Surface EMG shows activity from a broad range of motor units and may well show activity from several different muscles. This limitation, combined with the problems of determining threshold magnitudes, means that useful and realistic application of surface EMG is limited. As a negative test, surface EMG can show if there is very little activity, which may

be useful when, for example, a clinician is considering botulinum injections for supposed overactive muscles. Surface EMG is also used to give timing for when muscles or muscle groups are particularly active during the gait cycle. EMG data do not indicate if muscles are contracting eccentrically or concentrically, hence if data are to be used for timing, they must be used in combination with other sources of information to show how levels of muscle activity correspond to body function. A final limitation that must be considered when using EMG clinically is that patients often have deformities leading to muscles not lying in standard locations. Hence, any guidelines of standard electrode locations may actually result in different muscles being assessed. To tackle this issue the assessor may need to palpate the area while the patient performs some specific action, to see which muscles are tensed and where they lie. There are therefore considerable limitations involved in using surface EMG in a clinical context, however, as long as these limitations are understood, EMG can be used as a helpful tool in the process of diagnosis and treatment.

ENERGY EXPENDITURE

In general, the methods presented so far describe how the components of the human body perform structurally and mechanically. Clinical use of this information relies on subjective interpretation of what ideal should be strived for or what is "better." For example, kinematic data would show if one person's leg moves differently compared to people in general, but how does one decide if this is a problem or not? In some instances there are clear mechanical outcome measures of improvement; for example, if someone's leg now moves in a way that means they no longer trip and fall then this is a clear improvement. However, as individuals, we tend to find that an important factor is how much effort an activity takes. Abnormality becomes a problem when it requires too much energy and limits what a person can do.

A combination of inverse dynamics and EMG data does allow some assessment of effort in terms of muscle power, but only for individual components and only the net effect.

The gold standard for energy expenditure is direct calorimetry, but as this normally requires calorimetry chambers this is not practical for gait analysis and so indirect methods are normally used. The most common is measurement in oxygen uptake and/or carbon dioxide production (in ml), measured using a Douglas bag or spirometer. Data are often normalized to body mass (kg) and expressed relative to time (ml/kg/min) as oxygen consumption or to distance (ml/kg/m) as oxygen cost. Indicative data for a range of pathologies can be found in Waters and Mulroy (1999).

Equipment to determine oxygen update can be expensive and encumbering; the face mask typically required to capture air exchange can be uncomfortable and concerning for some patients. As an alternative, it has been found that under certain conditions there tends to be a linear relationship between heart rate and oxygen uptake. Equipment to measure heart rate is readily available, due to use in fitness training, and is very simple and discrete to use. Two examples of methods for using heart rate in assessing energy expenditure are the physiological cost index (PCI) and the total heart beat index (THBI) (Hood et al., 2002). These methods do have limitations; for example, in the original work on PCI a steady-state walking heart rate was a requirement and the effort to walk must be below maximum. For many subjects with gait impairments steady-state walking heart rate may never be reached; walking requires maximum effort so fatigue occurs before steady state.

Extensions to the basic heart rate idea include combining heart rate with measures of activity (e.g., the Actiheart device made by CamNtech Ltd., Cambridge, U.K.), however, evidence for the validity of such devices on an individual basis in typical gait assessment sessions is limited.

DATA INTERPRETATION

For some situations, such as the case of orthotic tuning, only a subset of the data introduced above will be needed, but in others, such as presurgical assessment, a fuller set may be necessary.

A particular problem in data interpretation is that gait abnormalities may be caused by a variety of factors and are often inter-related. For example, a dropped foot following a stroke may have as its primary cause an inability to voluntarily dorsiflex the foot. Over time, this primary issue may lead to a secondary problem of fixed tendomuscular shortening. Both primary and secondary concerns will often lead to compensation mechanisms being applied; for example, in the case of a dropped foot, increased hip and knee flexion may be adopted by the patient to increase swing clearance.

In some respects the approach to interpretation will depend on the question. For example, if it is to (simply!) quantify changes then there are a number of indices that might be considered to reduce the multiple and continuous data sets to a single number indicating overall pathology. Examples are the Gillette Gait Index, the Gait Deviation Index, and the Gait Profile Score/Movement Analysis Profile (Baker et al., 2009). However, if more in-depth advice is required, for example, to inform surgical decisions for a child presenting with cerebral palsy, then the situation is more difficult. In many ways, compared to data collection techniques, interpretation for surgical planning is not particularly advanced. Presently most clinical teams will use comparison to normal databases (bearing in mind limitations of the collected data as outlined earlier), together with their own clinical experience and bio-mechanical expertise to determine the underlying specific impairments the patient presents with, the evidence for these, and the strength of the evidence. This can be a time-consuming process, and although elements of the process can be automated, it is an area that merits further work. An approach to interpretation of gait data can be found in Baker et al. (2010).

ROLE OF THE CLINICAL MOVEMENT ANALYSIS SOCIETY U.K. AND IRELAND (CMAS)

As indicated earlier gait analysis normally requires input from a multidisciplinary team. Bodies such as the Chartered Society of Physiotherapy (CSP), British Orthopaedic Association (BOA), and Institute of Physics in Engineering and Medicine (IPEM) therefore have an interest in the education and training of those who practice in movement analysis. The recent policy statement by IPEM (2012) is indicative of this.

The Clinical Movement Analysis Society U.K. and Ireland (CMAS) was formed to encourage professional interaction, develop and monitor operational standards and training, and to stimulate and advance scientific knowledge in the field of clinical motion analysis.

CMAS has compiled a set of standards that cover many aspects of clinical gait analysis performed by centers across the United Kingdom. These incorporate the clinical environment, calibration of equipment, collection and reporting of clinical data, and administration protocols. The standards will play an important role in ensuring a consistent quality of service is being provided by CMAS accredited movement analysis laboratories.

The work of CMAS is meant to complement more general operating procedures relating to clinical governance that should be in place at each site undertaking gait analysis.

SUMMARY

Gait analysis can play an important role in planning and monitoring clinical interventions that can range from simple (real-time) feedback for patients to multilevel orthopedic surgery. Evidence for the influence of gait analysis in surgical planning is limited, but generally positive (Wren et al., 2011).

The selection of analysis tools requires an understanding of the aims of the analysis and the limitations of each option. For them to be useful they have to be valid, representing something correctly from which you draw justifiable conclusions, and therefore reliable—the measures are repeatable. It is important to understand what a particular method is expected to show and what implications the results of testing will have; incorrect application of clinical data can be a dangerous thing.

Developments continue to be made in equipment for data collection; for example, video and inertial sensor based markerless kinematics. These should make the process more rapid and less intrusive, and better support collection of data in real situations. A key area of concern, however, remains the relative subjectivity of data interpretation, particularly when this relates to decisions on what will be the most appropriate surgical intervention.

References

American Academy of Orthopaedic Surgeons, Pediatric Orthopaedic Society of North America, American Academy of Pediatrics, Shriner's Hospitals, 2005. Pediatric Outcomes Data Collection Instrument (PODCI).

Baker, R., McGinley, J.L., Schwartz, M.H., Beynon, S., Rozumalski, A., Graham, H.K., Tirosh, O., 2009. The gait profile score and movement analysis profile. Gait Posture. 30, 265–269.

Baker, R., McGinley, J., Thomason, P., 2010. Impairment focused Interpretation (A manual prepared for a tutorial delivered to the Second Joint Meeting of the GMCAS and ESMAC in Miami in 2010). Available from: <www.salford.ac.uk/health-sciences/research/research-programmes/gait-biomechanics/gait-analysis-downloads> (accessed 28.08.12).

Butler, P.B., Nene, A.V., 1991. The biomechanics of fixed ankle foot orthoses and their potential in the management of cerebral palsied children. Physiotherapy. 77, 81–88.

Cole, M.J., Durham, S., Ewins, D., 2008. An evaluation of patient perceptions to the value of the gait laboratory as part of the rehabilitation of primary lower limb amputees. Prosthet. Orthot. Int 32 (1), 12–22.

Davis, R.B., Ounpuu, S., Tyburski, D., Gage, J.R., 1991. A gait analysis data collection and reduction technique. Hum. Mov. Sci. 10, 575–587.

Gage, J.R., Schwartz, M., Koop, S.E., Novacheck, T.F. (Eds.), 2009. The Identification and Treatment of Gait Problems in Cerebral Palsy (Clinics in Developmental Medicine). Mac Keith Press.

Glazier, P.S., David, K., Bartlett, R.M., 2003. Dynamical systems theory: a relevant framework for performance-orientated sports biomechanics research. Sportscience. 7.

Hermans, H.J., Freriks, B., Merletti, R., Stegeman, D., Blok, J., Rau, G., et al., 1999. European Recommendations for Surface Electromyography – Results of the SENIAM project, Roessingh Research and Development.

Hood, V., Granat, M., Maxwell, D., Hasler, J., 2002. A new method of using heart rate to represent energy expenditure: the total heart beat index. Arch. Phy. Med. Rehabil. 83, 1266–1273.

Institute of Physics in Engineering and Medicine (IPEM), 2012. Clinical Scientists in Clinical Movement Analysis: standards for practice. IPEM Policy Statement.

Kirtley, C., 2006. Clinical Gait Analysis: Theory and Practice. Churchill Livingstone.

Knutson, L.M., Soderberg, G.L., 1995. EMG: Use and interpretation in gait. In: Craik, R.L., Oatis, C.A. (Eds.), Gait Analysis: Theory and Application. Mosby, St. Louis.

Meriam, J.L., Kraige, L.G., 1998. Engineering Mechanics: Dynamics. fourth ed. John Wiley & Son., New York.

Novacheck, T.F., Stout, J.L., Tervo, R., 2000. Reliability and validity of the gillette functional assessment questionnaire as an outcome measure in children with walking disabilities. J. Pediatr. Orthop. 20 (1), 75–81.

Pathokinesiology Service and Physical Therapy Department, 2001. Observational gait analysis handbook, Ranchos Los Amigos National Rehabilitation Center. fourth ed. Downey, CA.

Perry, J., 1992. Gait analysis. Normal and pathological function. Thorofare. Slack Inc.

Read, J., Hillman, S., Hazlewood, M., Robb, F., 1999. The Edinburgh visual gait analysis interval testing (G.A.I. T.) scale. Gait Posture. 10, 63–64.

Stansfield, B.W., Hillman, S.J., Hazlewood, M.E., Lawson, A.M., Mann, A.M., Loudon, I.R., et al., 2003. Normalisation of gait in children. Gait Posture. 17, 81–87.

Sutherland, D.H., Olshen, R.A., 1998. The Development of Mature Walking. Mac Keith.

Waters, R., Mulroy, S., 1999. Review: the energy expenditure of normal and pathologic gait. Gait Posture. 9, 207–231.

Wren, T.A.L., Otsuka, N.Y., Bowen, R.E., Scaduto, A.A., Chan, L.S., Sheng, M., Hara, R., et al., 2011. Influence of gait analysis on decision-making for lower extremity orthopaedic surgery: baseline data from a randomized controlled trial. Gait Posture. 34 (3), 364–369.

Further Reading

Cooper, J.M., Adrian, M.J., 1995. Biomechanics of Human Movement. Brown & Benchmark, Dubuque.

Cram, J.R., Kasman, G.S., 1998. Introduction to Surface Electromyography. Aspen.

Jones, K., Barker, K., 1995. Human Movement Explained. Butterworth-Heinemann.

Low, J., Reed, A., 1996. Basic Biomechanics Explained. Butterworth-Heinemann.

Merletti, R., Parker, P.A., 2004. Electromyography: Physiology, Engineering and Non-invasive Applications. IEEE Press Series, Wiley.

Roberts, G., Hamill, J., Caldwell, G., Kamen, G., Whittlesey, S., 2004. Research Methods in Biomechanics. Human Kinetics Europe Ltd.

Whittle, M.W., 2003. Gait Analysis: an introduction. third ed. Butterworth-Heinemann.

Winter, D.A., 2009. Biomechanics and Motor Control of Human Movement. fourth ed. Wiley.

Resources

Clinical Movement Analysis Society U.K. and Ireland (CMAS), *www.cmasuki.org*

Gait and Clinical Movement Analysis Society, *www.gcmas.org*

International Society of Biomechanics, *www.isbweb.org*

Surface ElectroMyoGraphy for the Non-Invasive Assessment of Muscles, *www.seniam.org*

Principles of Kinematics and Kinetics

David Ewins* and Tom Collins†

*Queen Mary's Hospital and University of Surrey, †Queen Mary's Hospital

KINEMATIC CONCEPTS

Any point in space can be described by three coordinates if a set of three orthogonal (Cartesian) axes are defined, based on some origin point. This set of axes, called the global coordinate system (GCS), is adequate if one only wants to describe single points in space, but objects are typically made up of many points and it would often be too complicated to describe them all relative to the GCS. If an object can be considered as a rigid body (i.e., all points in the object stay in a fixed relationship to each other), then a very useful simplification can be made. For a particular rigid body a local coordinate system (LCS) can be defined fixed to some reference points on that body. All points within the body can be defined relative to the LCS and only the LCS needs to be defined relative to the GCS. Once such systems are established the movement of the body can be fully defined relative to the GCS in terms of translation of the LCS origin and rotation of the LCS axes. If there is more than one rigid body then a LCS can be defined for each and the relative movements between them can be defined in relation to the GCS.

This is the basis of kinematic gait analysis. The human body is treated as a series of rigid bodies or segments. (This is a major simplification as one can tell from touching the skin.

Body segments are not rigid, however to allow practical mathematical analysis such simplifications are necessary.) Each segment can be assigned an LCS and relative movements between each segment (six degrees of freedom: three translations and three rotations) can be defined in relation to some GCS using the mathematical tools that have been developed for kinematics in general.

SEGMENT AND JOINT COORDINATE SYSTEMS

Marker sets are useful in kinematic gait analysis as they allow definition and tracking of movement of the LCS of each segment relative to some GCS in the laboratory. To fit with the theory described in the previous section, the markers are used to create reference points on each object (segment) and these are used to define the origin and three orthogonal axes of each segment's LCS. (Note that different terminology exists; LCS may also be termed *segment coordinate system* or *SCS*.) It would be possible to attach three reference markers in a nonlinear relationship anywhere on each segment and describe relative motions based on these, however for analysis to be understandable and useful it needs to be related to the human anatomy. Joint angles are often described relative to three anatomical planes: sagittal, coronal, and frontal (or transverse). For this to be possible the LCSs must be defined in a way that represents these planes. Various standard approaches have been established across the field of gait analysis and are presented by such reports as Wu and Cavanagh (1995) and Cappozzo et al., (1995). The review article by Cappozzo et al., (2005) provides a very helpful summary.

In very brief terms it has been discussed how segments can be defined in space using LCSs to establish their position and monitor their movement. However, it would be difficult to understand the movements of body segments if they were described only in relation to the GCS. For analysis that is relevant, the relative movement between different segments needs to be described. Fortunately the general field of kinematics has tools for this in the form of translation and rotation matrices. To define the relative position of one LCS to another, a three-term vector can give translation and a 3×3 matrix can define relative rotation, describing the steps that would be needed to rotate from one LCS orientation to the other. Chau (1980) explains that Euler/Cardan angles are the most convenient way to describe this relative rotation. Cole et al., (1993) further explain that Euler/Cardan angles are a set of three independent angles obtained by an ordered sequence of rotations about the axes of some coordinate system. For describing joint angle rotations in the human body, the LCS of the proximal segment is usually chosen as the reference segment. It can be shown that the sequence of the rotations is important as different sequences result in different size angles, hence the sequence must be defined.

Grood and Suntay (1983) described an equivalent method which has become known as the joint coordinate system (JCS). Here a nonorthogonal coordinate system is defined by taking one axis from the proximal segment's LCS and one from the distal segment's LCS, the third is a "floating" axis that is normal to the other two. This method is very useful in that the three rotations about the axes fit clinical descriptions of joint movement. It has been claimed that the JCS allows rotations to be performed in any order but Cole et al. (1993) point out that this apparent lack of sequence dependency only occurs because the order of rotation is predefined by the choice of axes from each of the LCSs. As this method is clinically relevant it has gained wide acceptance as the best way to represent joint rotations (Schache and Baker, 2007).

Most of the marker sets used in current practice were developed for imaging systems with low resolution so sets had to have as few markers as far apart as possible to avoid merging (Della Croce et al., 2005). One technique used to minimize markers is to employ the same marker for more than one segment. For example, in the (Helen Hayes) system proposed by Kadaba et al., (1990) a marker placed on the lateral joint line of the knee is used to track the thigh segment but also the shank. The resulting model treats the knee as a single point so the two segments can only rotate relative to each other. While this is a reasonable approximation for a basic model it does not allow true expression of the knee movement; in reality the knee also allows relative translation between the two segments. These segments only have three degrees of freedom (DOF).

A 6DOF marker system is one that allows description not only of rotations about three axes but also translations in three directions for all segments. This requires the movement of each segment to be tracked by an independent group of markers, which inevitably means the markers must be closer together. Such sets have only been possible with the improvement of imaging technology. An approach that has been considered by various groups for achieving 6DOF systems is to use clusters of markers mounted on rigid frames that are then attached

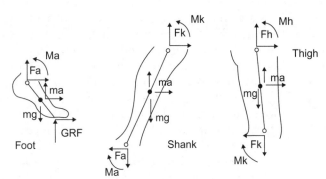

FIGURE 25.6 Free body diagrams for the leg with force and moment approximations applied.

to the body. Cappozzo et al., (1997) suggested design criteria for cluster sets and some other groups have proposed cluster-based marker sets, but each has had various limitations. As concluded by Collins et al., (2009) when comparing a 6DOF model to one based on a modified Helen Hayes marker set, "The 6DOF set showed comparable performance to the conventional set and overcomes a number of theoretical limitations, however further development is needed prior to clinical implementation."

KINETIC CONCEPTS: INVERSE DYNAMICS

An important use of force plate data is in combination with kinematic data to investigate muscle activity by the analysis of joint moments. Obtaining estimations of joint moments requires some involved calculations: the process of inverse dynamics. The principle is to split the body into a series of segments (link-segment model) and treat each as a "free body" with a collection of forces acting on it due to external forces (gravity, GRF) or internal forces (bone/muscle/ligament), Figure 25.6.

The process relies on a series of assumptions:

- Each segment is treated as rigid with a fixed length throughout movement.

- Each segment can be modeled as having a point mass that is fixed relative to the geometry of the rest of the segment (center of mass).
- Mass moment of inertia remains constant (implied from above).
- Each joint is treated as a simple hinge or ball and socket, with a single center of rotation.

It would be an indeterminate problem to calculate the contribution of force from each internal tissue structure (and is impossible to separate) so the unknown forces and moments must be combined into one net force and one net moment, honoring the positive directions of the global coordinate system. The combined effect of the mass of each segment (approximated from anthropometric data), and accelerations of each segment (from kinematic data), can be represented as "inertia forces" (mass × acceleration, ma, or second moment of area × angular acceleration, $I\alpha$).

For each segment, three equilibrium equations can be set up, using Newton's second law, $\Sigma F = ma$. The equations are vertical and horizontal equilibrium and moment equilibrium. To solve for the leg in the sagittal plane the leg can be split into three segments (thigh, shank, and foot; Figure 25.6). Starting with the most distal segment, the foot, as this has the fewest unknowns (Figure 25.7):

Horizontal equilibrium:

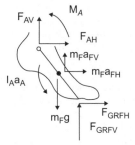

FIGURE 25.7 Free body diagram for the foot.

$$F_{AH} = - F_{GRFH} - m_F a_{FH}$$

Vertical equilibrium:

$$F_{AV} = m_F g - F_{GRFV} - m_F a_{FV}$$

Moment equilibrium about center of mass:

$$M_A = I_A \alpha_A + F_{AV} x_A + F_{AH} y_A - F_{GRFV} x_{GRF} - F_{GRFH} y_{GRF}$$

where notation is defined in the figures and (perpendicular) distances, x_A and y_A, are from the point of application of the force to the center of mass.

The results for the foot can then be used as inputs for the calculations for the shank segment and so on up the body. It is also possible to calculate net joint powers from the product of net joint moment and joint angular velocity. The calculations become considerably more complex for three-dimensional solutions. Figure 25.4(c) shows typical joint moment and power plots for the ankle in the sagittal plane.

This process gives estimates for the net moments in each joint but there are various limitations. The method does not account for joint friction or the shock absorption characteristics of musculoskeletal structures that result in load attenuation. This means that the error in the calculations increases the more proximally they are made. Clearly it is not possible to determine from this the contributions of each muscle. The net force does not account for co-contraction of muscles (often the agonist and antagonist

muscle pairs contract at the same time producing opposing moments to give joint stability, but this has no bearing on the net moment) nor does it indicate the contribution of a muscle that spans two joints, such as the rectus femoris. However, by organizing muscles according to function (e.g., knee flexors) and combining inverse dynamics results with timing information of muscle activity from electromyography (EMG) it is possible to develop a picture of which muscles may be contributing to an action and what the overall affect may be. This can also allow investigation of potential muscle problems.

LIMITATIONS OF KINETIC ANALYSIS

As with all measurement systems it is important to consider the effect the measurement system has on the entity being measured. Force plates are often relatively obvious so when walking over them people may adjust their gait pattern to ensure they stand on the plate (targeting), which means that any recorded information may not be normal. However, in comparison to other measurement methods force plates cause minimal encumbrance to the subject and so have relatively low impact on the quantity being measured. Their limited effect on the subject also makes force plates a relatively safe method of evaluation; as long as the plates are flush with the floor and have a good grip surface they pose no more risk than normal walking conditions.

GRF is very useful and can show clearly when there is abnormality but does not give too much information about the cause of abnormality. In the presentation of inverse dynamics assumptions are made, with various factors neglected, such as attenuation due to muscle properties. This method is more tenuous the more proximally it is taken. For example, slight movement of the predicted hip joint center used in the kinematic modeling can result in large change in moment. Due to these limitations results calculated by inverse dynamics should be used with caution.

References

Cappozzo, A., Catani, F., Croce, U.D., Leardini, A., 1995. Position and orientation in space of bones during movement: anatomical definition and determination. Clin. Biomech. 10, 171–178.

Cappozzo, A., Cappello, A., Della Croce, U., Pensalfini, F., 1997. Surface-marker cluster design criteria for 3-D bone movement reconstruction. IEEE Trans. Biomed. Eng. 44, 1165–1174.

Cappozzo, A., Della Croce, U., Leardini, A., Chiari, L., 2005. Human movement analysis using stereophotogrammetry: Part 1: theoretical background. Gait Posture. 21, 186–196.

Chau, E.Y.S., 1980. Justification of triaxial goniometer for the measurement of joint rotation. J. Biomech. 13, 989–1006.

Cole, G.K., Nigg, B.M., Ronsky, J.L., Yeadon, M.R., 1993. Application of the joint coordinate system to three-dimensional joint attitude and movement representation: A standardization proposal. Trans. ASME: J. Biomec. Eng. 115, 344–349.

Collins, T.D., Ghoussayni, S.G., Ewins, D.J., Kent, J.A., 2009. A six degrees-of-freedom marker set for gait analysis: Repeatability and comparison with a modified Helen Hayes set. Gait Posture. 30, 173–180.

Della Croce, U., Leardini, A., Chiari, L., Cappozzo, A., 2005. Human movement analysis using stereophotogrammetry: Part 4: assessment of anatomical landmark misplacement and its effects on joint kinematics. Gait Posture. 21, 226–237.

Grood, E.S., Suntay, W.J., 1983. A joint coordinate system for the clinical description of three-dimensional motions: application to the knee. Trans. ASME: J. Biomec. Eng. 105, 136–144.

Kadaba, M.P., Ramakrishnan, H.K., Wooten, M.E., 1990. Measurement of Lower Extremity Kinematics During Level Walking. J. Orthop. Res. 8 (3), 383–392.

Schache, A.G., Baker, R., 2007. On the expression of joint moments during gait. Gait Posture. 25, 440–452.

Wu, G., Cavanagh, P.R., 1995. ISB recommendations for standardization in the reporting of kinematic data. J. Biomec. 28, 1257–1261.

Mechanical and Electromechanical Devices

Donna Cowan*, Martin Smith†, Vicky Gardiner**, Paul Horwood†, Chris Morris***, Tim Holsgrove††, Tori Mayhew†, David Long†, and Mike Hillman††

*Chailey Heritage Clinical Services, †Oxford University Hospitals NHS Trust, **Opcare, ***University of Exeter, ††University of Bath

AIDS FOR DAILY LIVING

Aids for daily living are a range of products that support activities generally known as activities of daily living. These are, in the main, self-care activities such as:

- Bathing/toileting
- Dressing and grooming
- Eating and drinking

The ability of people to undertake these activities is routinely taken as a measurement of the functional status of a person.

As with all assistive technology, the key to success is matching the equipment accurately to the client's needs. As a result, it is important to identify the core issues associated with an activity and what the client is likely to accept as assistance, then to match the two accurately.

The range of aids offers support to users with a wide range of conditions, congenital and acquired, incorporating sensory impairment, and physical restriction. It is perhaps most usually associated with older people compensating for changes in physical ability with age, enabling some to remain independent and safer in undertaking these activities.

Occupational therapists will often work with clients to develop strategies that support a different approach to the activity and incorporate the use of an aid. This may be as simple as sitting to put clothes on (or take them off) to prevent falling due to imbalance caused by raising the arms above the head, and using a dressing frame to assist donning of tops and jackets.

Increasingly, this range is becoming mainstream and is finding its way onto the shelves of supermarkets as well as being found in specialist outlets.

Dressing

A wide range of products are available to aid this activity such as dressing sticks and hooks, sock aids, and shoe horns. Clothing itself can also be adapted such as replacing buttons and ties with Velcro, or shoelaces with elastic laces. People require these aids due to reduced dexterity, motor movement, or physical mobility.

Bathing

Aids for this activity range from a bath mat to improve safety (an aid many use now in their homes) to personally molded shower chairs with adaptations to allow for postural impairment and movement patterns. Standard shower seats/chairs provide stability for users and seats can be fitted into a bath to enable safe entry and exit.

Products comprising a long handle attached to a sponge or brush allow users with restricted limb mobility to undertake activities of washing and grooming independently.

In some instances holders are required; for example, for users with limb deficiency, a holder and mount is required to allow a toothbrush to be placed. The user then moves their head relative to the brush rather than having to bring the brush to their mouth.

Toileting

Cleansing aids and wipers are available to promote independent personal care. A range of toilet seats and raisers can be found to ease the difficulty of sitting and then rising from a toilet seat. Steps are also used to promote stability in sitting for young children. The seat surfaces can also be made in a variety of materials to provide a more conforming surface for users and thus increase comfort and surface area contact to promote stability. Personally molded toilet seats/commodes are available for those having particularly pronounced postural impairment. These can also double as shower chairs.

Toilets that offer washing and warm air drying can also be purchased to provide additional support in undertaking this personal activity.

Eating and Drinking

The International Classification of Functioning, Disability and Health (WHO, 2001) defines the task of eating as the ability to carry out the coordinated tasks and actions of eating food that has been served, bringing it to the mouth and consuming it in culturally acceptable ways, cutting or breaking food into pieces, opening bottles and cans, using eating implements, and having meals, feasting, or dining. The aids available to support this include angled cutlery with a range of handles and grips. A single tool that incorporates the use of two of these items (e.g., a fork with a serrated edge) can enable someone with only one functional limb to cut and eat their food independently. Plates shaped to ease manipulation of food are also available. Nonslip mats underneath crockery prevent movement of the plate. Drinking vessels are available that have double grips for those with reduced movement in their hands. Lids on mugs avoid spillage. Mechanical and electromechanical eating aids have also been developed.

FIGURE 26.1 The Neater Eater. *Source: By kind permission of Neater Solutions Ltd.*

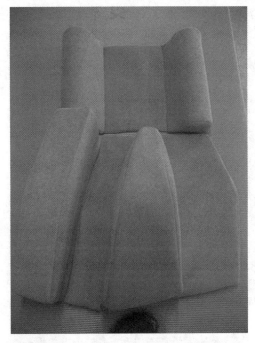

FIGURE 26.2 Two-piece foam lying support.

Writing and Drawing

Grips for pens and pencils enable standard writing implements to continue to be used. Holders for pens and pencils via helmets and mouth sticks are also available. Consideration of the individual is essential in all assistive technologies. In addition, where aids are held in the mouth, consideration of the risk associated with controlling a pen or pencil is required as injury is possible if the user has sudden movements or works in a busy environment (such as a classroom).

There are grips to support access to the computer to press the keys of a keyboard. Standard software packages (e.g., Windows) have settings to accommodate some access needs (e.g., screen magnifier, text narrator, or the ability to offset the effect of tremor). In addition, aids such as key guards and different sized keyboards can be used to provide simple solutions to these problems (further information on this subject is available in Chapter 8, "Electronic Assistive Technology," in the section "Access to Electronic Assistive Technology").

Personalized Devices

Some items, however, are not commercially available. In some cases there is a need to provide personalized solutions for a range of activities. This is usually carried out in specialist centers having multidisciplinary teams able not only to assess the client, but also to provide solutions. The starting point may be a commercial item that may be adapted to meet a particular need. In other cases there is a need to design and manufacture something bespoke. Specialist centers have engineers who are able to synthesize the functional and physical requirements requested by a therapist with the lifestyle requirements of the client who may be a child, young adult, or older person. The engineer's role is to design a solution to bring these sometimes opposing sets of requirements together.

For the design to be successful it is critical that the client and their family/carers are at

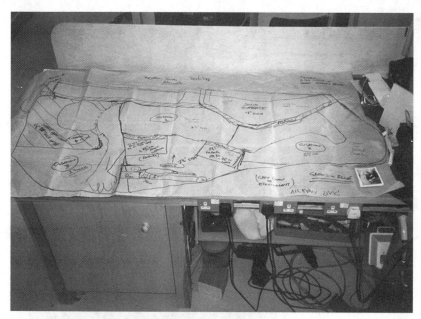

FIGURE 26.3 Personalized lying support design.

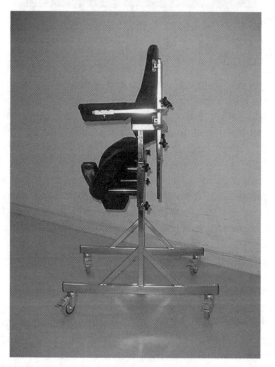

FIGURE 26.4 Custom-made toilet seat for client with arthrogryposis achieving only 5° hip flexion.

FIGURE 26.5 Standard static exercise bike with adaptations to provide additional postural support.

the heart of the design process. They must take an active role in identifying when a solution will work or not. Aesthetics is also important, as to how it will fit into the user's and their carers' daily life.

Choice of materials will be governed by biocompatibility issues, temperature regulation of the user, the environment it will be exposed to, cleaning and/or decontamination requirements, maintenance, and lifespan, coupled with aesthetics.

The engineer is a key element of the team in these circumstances and must not only take into account a number of design and technical considerations, but also risk assessment in terms of how it might be used or misused and what fault conditions may occur (see Chapter 17, "Introduction: Medical Engineering Design, Regulations, and Risk Management," for more on this).

A Potential Pitfall

It is important not to "inflict" assistive technology on people through over-enthusism. One must be sensitive to the personal situation and circumstances of the person and only seek to solve the problems that the person believes exist. One can introduce new ideas but this must be done with care and sensitivity to avoid overloading the person with technology which will often result in wasted resource through lack of use.

PROSTHETICS

Rehabilitation healthcare concerning externally applied prosthetic devices, known as "prosthetics," does not usually require engineering support at a clinical level, the clinical engineer being more likely to be involved in research and development. In certain circumstances, however, a clinical engineer may assist in setting up a prosthesis in an instrumented gait laboratory, but this is limited to those clinics having such facilities.

Demography

The national amputee statistical database was set up in the United Kingdom in 1997.

Figures show that annually, approximately 5000 new referrals are made to the 44 prosthetic centers in the United Kingdom. Of these, 95% are for lower limb amputations, 50% of which are transtibial (below knee) level. There are twice as many males referred as females. The most prevalent cause of amputation is dysvascularity, accounting for 75% of referrals. The second most common cause is trauma, which accounts for 9% of referrals.

Pre-Amputation and Surgery

The amputation level and choice of surgical technique will generally be dictated by the presentation of the affected limb; however, consideration should be given to the length of residual limb remaining below a joint. It is important that sufficient length remains to ensure good muscle power to control a prosthesis and adequate area to distribute socket forces. On the other hand, leaving a residual limb too long may compromise the space required to accommodate prosthetic componentry such as an artificial knee joint.

Amputation through a joint (disarticulation) is likely to heal more quickly than cutting through bone, and this kind of surgery offers greater surface area for force distribution; often it is possible to weight bear through the distal end. However, there are disadvantages to this. Disarticulation can limit space for componentry and compromise aesthetics with sockets appearing bulky around the distal end. One exception to this is disarticulation in pediatric users since the bones do not develop in the same way, becoming generally shorter and less well defined distally.

The human lower limb consists of three main joints—hip, knee, and ankle—and supporting bony structures of the thigh (femur) and shank (tibia and fibula). These provide the stability and articulation required for bipedal motion. In addition, feet form an intricate structure consisting of 26 bones, each of which perform three key functions—shock absorption, compliance over uneven ground, and propulsion—that are achieved effectively at varying speeds. Further reading is suggested on anatomy of the musculoskeletal system and anatomical reference terminology.

Post-Amputation Management

When a person loses a limb it is the job of a professional prosthetist to provide a suitable artificial replacement. With the aid of a prosthesis, the clinician aims to restore function according to the will and needs of the user, enabling the person to safely continue life as normally as possible compared to that prior to surgery.

Prostheses

External prosthetic appliances can be purely cosmetic, but in most cases a more functional solution is required; the complexity of the restoration is essentially dependent on the level and extent of amputation. Energy demand increases with complexity, a problem that is particularly apparent for mobilizing with a prosthesis.

The majority of prosthetic limbs prescribed are modular systems. A socket forms the interface with the patient. Depending on amputation level, prosthetic hip joints, knees, or feet are connected via tubular pylons. Commercially manufactured components are CE marked, but the prosthesis as a whole is considered a custom-made device due to the requirement for a bespoke socket. As a result, compliance with the Medical Devices Directive, according to the guidance for custom made devices, is mandatory.

Sockets

The vast majority of users rely on a socket as a means of interface between the residual limb and distal components of the artificial limb. An alternative method is by osseointegration. This technique involves a component being surgically implanted in the residual bony structures. Time is required for stabilization before an additional procedure to fit an abutment that permanently penetrates the soft tissues. In the case of amputees for whom socket interface is not suitable, the technique offers the possibility to use a prosthesis. However, infection and loosening of the implant means the technique is not widely applicable, with most cases requiring revision surgeries. Research continues to attempt to overcome such limitations in an area of promising advancement. Further reading is suggested on biomaterials and osseointegration.

Socket design is entirely bespoke and could be discussed extensively. It is the most specialized skill of the prosthetist, requiring detailed knowledge of anatomy and a sound understanding of biomechanics as well as a certain "feel factor." There are several well-established styles of design for lower limb sockets: patellar-tendon bearing, total contact, quad, ischial containment, and the Marlo Anatomical Socket. Further reading is suggested if knowledge of socket design is required. In general, the following factors are important when designing a socket and should also be considered if recruiting amputees for research purposes.

The socket must bear load where tolerable and relieve sensitive areas as much as possible. Forces applied in weight bearing, during the gait cycle and while mobilizing with a prosthesis, must be carried by the residual limb and transmitted to proximal joints. Residual tissues are not well suited to this and so surgical quality, the chosen technique, and the residual stump condition have a major bearing on how well load may be tolerated.

Besides the lack of control an amputee will experience from a poor-fitting socket, movement that occurs in relation to the stump is

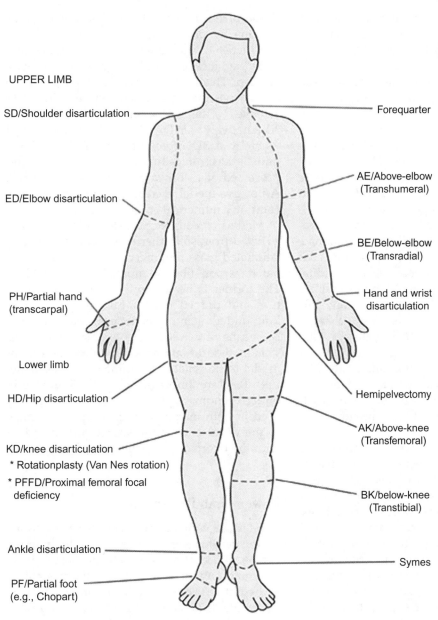

FIGURE 26.6 The main upper and lower limb amputation levels. These may be considered unilateral (affecting a single side) or bilateral (affecting both sides). © 2012 *The War Amps National Amputee Centre.*

UPPER LIMB

SD/Shoulder disarticulation

ED/Elbow disarticulation

PH/Partial hand (transcarpal)

Lower limb

HD/Hip disarticulation

KD/knee disarticulation
* Rotationplasty (Van Nes rotation)
* PFFD/Proximal femoral focal deficiency

Ankle disarticulation

PF/Partial foot (e.g., Chopart)

Forequarter

AE/Above-elbow (Transhumeral)

BE/Below-elbow (Transradial)

Hand and wrist disarticulation

Hemipelvectomy

AK/Above-knee (Transfemoral)

BK/below-knee (Transtibial)

Symes

likely to incur pain. Satisfactory fit of a socket is determined by assessment. Key signs are resistance to rotation about the stump, minimal "pistoning," and an absence of any red areas on the skin that do not disperse readily.

It is not uncommon for amputees to experience blisters, pressure ulcers, fragile or sensitive scar tissue, bone spurs, and phantom limb pain at some stage. If an amputee is experiencing socket issues or discomfort they will often develop compensatory strategies during gait. The actual tactic employed varies between individuals, but a reduction in walking speed is commonly observed.

It is normal for the residual limb to swell and contract to some degree in response to activity, temperature, and hydration. To maintain an optimal fit, amputees wear stump socks in varying thickness and number to accommodate this natural fluctuation. However, the period immediately following surgery is a time when the stump is particularly swollen due to edema. Following surgery, amputees are encouraged back to their feet as soon as possible to optimize their rehabilitation, minimize secondary complications such as flexion contractures, and reduce edema in the residual limb. The time for the residual limb to stabilize varies between individuals and during this period they are likely to require several changes of socket. For this reason they endure a long repetitive cycle of readjustment to new sockets that inevitably load their stump differently. It is difficult to say with any confidence when an amputee will reach steady state but there is satisfactory agreement that 18 to 24 months postamputation is reasonable.

Amputees can manage residual limb volume fluctuations with stump socks. Incorrect use of these can lead to socket problems and create a functional leg length discrepancy if too few or too many socks are worn with the prosthesis. Incorrect length may also be a consequence of the design and/or manufacturing of a prosthesis. Insufficient evidence exists to firmly quantify the amount of leg length inequality required to affect amputee gait, but it is recognized that when present, compensatory strategies emerge that may otherwise be avoided.

Construction

The process of designing and manufacturing a prosthetic socket begins by capturing the shape of the residual limb. Traditionally, this is done by plaster casting, which allows the prosthetist to manipulate the soft tissues and feel the location and degree of pressure applied. This is then back filled with plaster to form a positive cast which can be rectified to achieve the loading/relief required. Alternatively, CAD/CAM may be employed whereby a 3D representation of the residual limb is electronically scanned into a PC and processed via software to rectify the shape. A positive model is created by means of 3-axis computer numerical control (CNC).

Modern prosthetic sockets are largely constructed from solid thermo-softening plastic or laminated glass reinforced plastic (GRP)/laminated carbon fiber reinforced plastic (CFRP). The former is heated to a particular temperature and draped over a cast of the residual limb under vacuum; the latter is manufactured by means of a wet lay-up lamination process. This method of lamination requires the cast or model to be isolated from the structural materials that are impregnated with liquid resin under vacuum. A plastic dummy is commonly used in both methods to form a precise shape at the distal end of the socket corresponding with the interface component being used.

Lower Limb Prostheses

Interface components may also incorporate the suspension method to keep the limb on the person. Silicone or polyurethane based liners are commonly used to improve tissue stability and comfort. These may incorporate a locking pin at the distal end that is released by a push button. Alternative means of suspension may be by vacuum, locking lanyard, straps, or elasticated cuffs. Each method has a bearing on the design of the socket, particularly in the case of self-suspending sockets.

Prescription of lower limb prosthetic componentry is largely based on the weight and

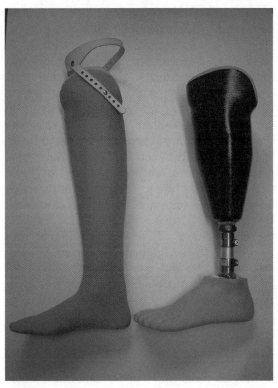

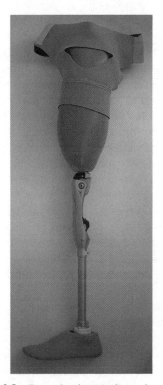

FIGURE 26.8   Example of a transfemoral prosthesis.

FIGURE 26.7   Example of a transtibial prosthesis with cosmesis.

activity level of the user. There is no industry standard for activity levels so manufacturers have developed their own, but they are simple to use and provide a useful reference for the clinician as to what components are suitable for a particular user in terms of their functional demands.

Almost all lower limb prostheses require an artificial foot. Traditionally, prosthetic feet were designed to restore basic ambulation. Evolution from the rigid pylon (historically referred to as the "peg leg") to better reflect natural appearance and function, resulted in the solid-ankle-cushioned-heel (SACH) foot. These remain in wide use today in a variety of forms. However, more active individuals benefit from a foot better suited to their activity levels, particularly if they wish to partake in

sport. The development of feet capable of storing and releasing energy to assist forward propulsion came in response to this requirement. Since the introduction of the "flexfoot" in 1987, which saw a radical departure from earlier versions, there have been a plethora of design variants on the theme. These passive devices are predominantly made from laminated carbon fiber and/or glass-reinforced plastics. Essentially they form a system of elastic leaf springs that deform under the weight of the body, storing potential energy that is released when the load is removed. However, as is the case with most mechanical springs, they are not 100% efficient and suffer some degree of hysteresis. Loss of energy to the system is undesirable because it means there is less available to assist the residual limb through the swing phase of gait. Many prosthetic feet are designed to replace the function of the ankle, which through the action of the plantar-flexors has the ability to generate more energy than it absorbs. The lack of active energy generation by most current prosthetic feet is one of the major barriers to normal gait replication.

To overcome these problems a dynamic solution capable of active power generation is thought to be necessary. Some powered devices have been developed, but these are very expensive, heavy, and bulky.

Prosthetic knees and hips vary considerably in complexity depending on the needs of the user. The most basic form consists of a simple locking mechanism that stabilizes or releases a single axis joint. More functional knee units employ mechanical linkages to improve dynamic stability. These swing freely and may incorporate spring-assisted extension. The requirement for more natural gait replication led to development of swing and stance control mechanisms. There are many different types of hydraulic and pneumatic damping systems incorporated in prosthetic knees that control the rate of swing and adapt to a certain extent to different walking speeds. Progression from simple mechanical systems resulted in "intelligent prostheses" incorporating a microprocessor. These control the damping system in response to feedback from strain gauges, accelerometers, and transducers in real time as well as "learning" the style of a particular user. One such approach applies magnetorheological technology to alter the viscosity of the damping fluid, varying its resistance to angular motion.

Alignment

One of the most critical elements to the success of a prosthesis is how it is aligned. This refers to the orientation and position of the prosthetic joints relative to the socket and proximal anatomy. It has the potential to fundamentally alter a person's gait and ability to use a prosthesis and, as a result, correct alignment must be achieved and maintained for function to be optimized.

In general, lower limb prostheses are set up with the socket flexed to between 5° and 10° in the sagittal plane. A straight vertical line should link the hip, knee, and ankle joint centers, and in the coronal plane the mid-posterior aspect of the socket should fall on a straight vertical line to the center of the heel. Beyond this, alignment must accommodate any excessive varus, valgus, or flexion by shifting the foot laterally, medially, or anteriorly accordingly. Misalignment results in instability, discomfort, and poor gait associated with the generation of unwanted moments that act about the joints.

Amputee Gait

Amputee gait can appear remarkably normal to an inexperienced observer, which is a testament to the skill of the prosthetist and design of components.

Visual gait analysis is routinely used as a means of assessment in prosthetic clinics. For example, when a new foot is fitted, the prosthetist aligns the prosthesis to what they consider optimal. This is a subjective means of assessment but is suitably quick and effective. Various scales have been developed to assist in quantifying the technique for ease of recording and future comparison.

Visual gait analysis is limited when attempting to identify the causes of subtle deviations, for which the use of instrumented gait analysis may be considered appropriate. This powerful means of collecting objective data, measured from several parameters at once, is not free from limitations of its own (see Chapter 25, "Clinical Gait Analysis").

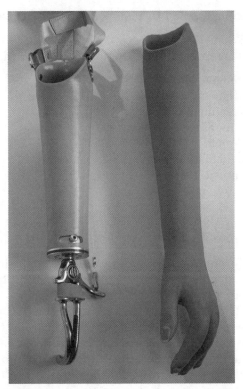

FIGURE 26.9 Example of a transradial hook and cosmesis.

Upper Limb Prostheses

Many upper limb amputees choose not to use prostheses due to their shortcomings. Even the most complex myoelectric limbs are crude in relation to the complexity of the anatomy they are designed to replace. Upper limb prostheses usually fall into three categories: cosmetic, body powered, and myoelectric. Cosmetic prostheses are designed to match the user's sound side as closely as possible and provide very little functional use. Body-powered prostheses are operated by a harness and pulley system employed to flex joints and open and close a terminal device such as a split hook. Myoelectric prostheses detect the user's residual limb muscle movements, which can be employed to control a prosthetic hand.

Advances in materials and innovative design have provided the clinician with components that offer superior function compared with traditional equipment. However, mechanical devices fail to accurately replicate the complex natural structures they replace, primarily because they are mostly passive. This means that amputees compensate by adopting strategies that are energy efficient and functional, but may not necessarily be considered a good outcome from a medical perspective. Further reading is suggested on the World Health Organization medical and social models of disability.

ORTHOTICS

Orthoses, often referred to as splints or braces, are externally applied devices that compensate for impairments or modify the musculoskeletal or neuromuscular systems (ISO, 1989a). Orthotists are the allied health professionals specifically trained to assess, fit, and advise on how an orthosis can be designed and constructed. Orthoses are prescribed to achieve clinical objectives such as to prevent or discourage deformity or to improve function. Orthoses can be "stock" (off the shelf) items taken from a range of sizes, or custom made to measurements or casts of the body. Custom-made orthoses can be molded from plastics, such as polypropylene, and incorporate judiciously placed straps and padding to achieve their goals.

More recently, the use of new materials such as carbon fiber composites, polyolefin elastomers, an expanding range of ethylene vinyl acetate (EVA) foams, and other cushioning materials has helped to replace the traditional use of leather covered metal substructures.

The assessments required to design an orthosis will include passive and active range of motion of joints, muscle control and strength, joint congruency and integrity, level of sensation, and presence of any associated impairments. The range of conditions treated with orthoses is extensive: anything from minor foot problems to complex neurological conditions such as cerebral palsy.

Biomechanical Principles

The means by which orthoses work is biomechanical, through the application of force and lever systems. Typically the forces that orthoses apply are "reaction" forces generated in response to gravity, where muscles are weak, and/or to counter forces of muscle imbalance. Key biomechanical principles underpinning orthotic management are (1) that using longer lever arms means less force is required and (2) that applying forces over larger surface areas means less pressure is applied to the body. To maintain stability, the center of mass must be within the base of support. An appreciation of gait analysis is necessary to understand how orthoses can improve the efficiency of walking.

Functions, Goals, and Design

The functions of an orthosis are as follows:

- Correct/reduce/accommodate deformity
- Protect against further progression of deformity
- Pain relief: Limit motion/weight bearing
- Relocate axial load centrally
- Improve function/activity limitations

All orthoses will have a clinical goal and these can include:

- To relieve pain
- Protect tissues/promote healing
- Correct/accommodate deformity
- Improve function
- Compensate for limb/segment length deficiency

- Compensate for abnormal muscle function

When designing an orthosis the orthotists will consider within the criteria the following:

- Functional objectives
- Biomechanical objectives
- Treatment objectives
- Strength
- Weight
- Reliability
- Patient interface
- Cosmetics

Terminology

Terminology for orthoses can be confusing. The standard system is to use anatomical terminology. This describes an orthosis by the joints that are encompassed in the device (ISO, 1989b). For example, an orthosis enclosing the ankle and foot but finishing below the knee is an ankle-foot orthosis, AFO. Extending proximally to the thigh would then encompass the knee joint, so the term *knee-ankle-foot orthosis*, KAFO, is appropriate. Similarly, a spinal orthosis encompassing the thoracolumbar and sacral spine is referred to as a TLSO. Sometimes the aim or function of the orthosis is included in the name, such as hip abduction orthosis.

Confusion can occur when orthoses are named nominally after people, places, or referred to by trade names. Terms such as *dynamic* are vague and should be avoided.

Foot Orthoses

Insoles, foot orthoses (often known as *functional foot orthoses*), supportive footwear, and shoe modifications can all be used to increase stability during standing and walking. Mild planovalgus or varus deformities can be controlled by foot orthoses that are designed to encourage better skeletal alignment.

Foot orthoses occasionally extend just above the ankle to gain greater control, these designs being termed supramalleolar orthoses (SMOs). Rigid thermoplastic materials are used to provide maximum control and an element of strength and durability. Softer, more accommodating insoles can be molded to redistribute the plantar pressure under the foot utilizing cushioning materials such as EVA or polyurethane foams.

Ankle-Foot Orthoses

To control weakened dorsiflexion muscles (as described in Chapter 18, "Functional Electrical Stimulation"), overactive plantarflexion muscles, an equinus or calcaneus

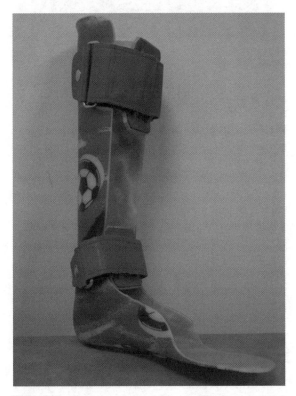

FIGURE 26.10 Example of an ankle-foot orthosis (AFO).

deformity, or moderate to severe valgus or varus deformities, an orthosis needs to encompass the ankle and foot and extend to just below the knee.

Ankle-foot orthoses (AFOs) provide longer leverage and control in the sagittal plane. They may be trimmed to create a flexible, posterior leaf spring AFO, or incorporate hinges to allow dorsiflexion, or made rigid to provide maximum control. AFOs are widely prescribed for flaccid foot drop, hemiplegia subsequent to cerebral vascular accident (CVA, or stroke), cerebral palsy, and other neuromuscular conditions to improve gait efficiency. They are used to immobilize painful joints in musculoskeletal conditions such as osteoarthritis.

AFOs are also used to control foot position and prevent deformity in nonambulant people. They are sometimes used at night for stretching calf muscles and preventing tightening of calf muscles, although if this is found to disturb sleep, softer, resting night splints can be used which maintain a reasonable position through the night with the muscles on less stretch.

Knee-Ankle-Foot Orthoses

A knee-ankle-foot orthosis (KAFO) extends from the thigh and includes the ankle and foot; a hinge is usually included at the knee to enable flexion during sitting but these are often designed to be locked in extension for standing and walking. These devices consist of thermoplastic shells at the thigh and foot/ankle sections with metal joints to hinge at the knee joint.

More recent innovations have developed stance control knee joints that will unlock on the swing phase of gait and automatically lock on the stance phase to maintain the knee securely in extension (see Chapter 18, "Functional Electrical Stimulation" and Chapter 19, "Clinical Gait Analysis" for more information on the gait cycle).

Patients with neuromuscular conditions such as poliomyelitis, muscular dystrophy, or spinal muscular atrophy may find that KAFOs enable them to stand and walk, although gait patterns have to be modified to compensate for a knee locked in extension.

Hip-Knee-Ankle-Foot Orthoses

Hip-knee-ankle-foot orthoses (HKAFOs) can enable standing and upright locomotion in people who are normally unable to maintain standing unaided. HKAFOs extend from the trunk and control the whole lower limbs and pelvis.

People with spina bifida or paraplegia can learn to use custom-fitted modular orthoses such as the Swivel Walker or ParaWalker.

Thoracolumbar Sacral Orthoses

Spinal orthoses (TLSOs) are used to support the trunk by controlling scoliosis or supporting weakened musculature. These orthoses are typically molded from thermoplastics, laminated foam materials, or, less commonly now, fabrics. Spinal orthoses may be provided to improve trunk control, improve sitting posture, or to try to prevent the progression of scoliosis. TLSOs are often prescribed for people with neuromuscular conditions where the aim is to reduce the rate of scoliosis progression and delay the need for surgical stabilization, or when surgery is not possible.

Upper Limb Orthoses

Orthoses can be used to position the arm, wrist, and hand to improve manual ability. Wrist hand orthoses (WHOs) usually hold the wrist in extension, and may extend to control thumb posture to promote a functional position. "Paddle" type designs of orthoses may be used to resist flexion deformities in the wrist

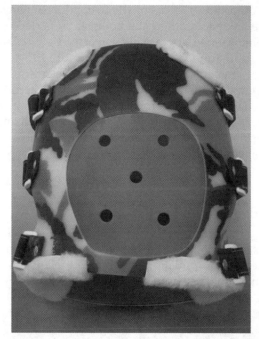

FIGURE 26.11 Example of a thoracolumbar sacral orthosis (TLSO).

and fingers. It should be noted that occupational therapists also mold and provide WHOs, often as part of a rehabilitation program, although those made by an orthotics department will usually be more durable in the longer term due to the more advanced manufacturing techniques available within such a department.

Orthoses for the elbow joint or for the entire upper limb can have the objective of supporting or stabilizing. An example is a brachial plexus injury where the orthosis will encompass the limb from the shoulder to wrist as an attempt to gain function in a flaccid arm.

Head Orthoses

A protective helmet may be indicated when a person is prone to falling, for instance during seizures, to reduce the risk of injury. There are some controversial treatments where helmets are provided as "cranial molding" devices in the treatment of infants with plagiocephaly. Helmets vary in design: some are constructed from dense foams while other designs have harder exteriors, dependent on the objectives to be met.

Summary

Orthoses can be very useful to improve functioning or reduce pain in adults and children with musculoskeletal or neuromuscular conditions and may reduce the rate of some progressive deformities. Appropriate fitting and design of orthoses is required to prevent problems with pressure or rubbing and to ensure that the clinical goals set are maximized. There can be stigma associated with using equipment that identifies people as different to their peers, so the benefits of orthoses for individuals should be carefully assessed and reevaluated over time.

ORTHOPEDIC BIOMECHANICS

Engineering Requirements of Orthopedic Implants

The aim of orthopedic[1] implants is generally to restore normal functional biomechanics to the affected area. This may be in the form of a fracture fixation device or a joint replacement device. The device should work in such a way that once implanted, pain is reduced or eliminated entirely, the body should be able to move through normal ranges of motion, and should be able to cope with the normal loads present in everyday life. An exception to this is fusion procedures,[2] which aim to remove

[1]Orthopedics is the branch of medicine concerned with the musculoskeletal system.

[2]The fusion of a joint in the body through surgical procedure is known as *arthrodesis*.

physiological motion at the operated level to reduce pain and allow the normal load transfer characteristics to be restored.

The body is constantly adapting to the conditions to which it is subjected. If an implanted device shields the skeleton from normal loads, bone may be resorbed and this may lead to injury, pain, or loosening of the implant. Such complications may result in clinical failure and may require further surgery. It is therefore imperative to thoroughly assess the mechanical properties of implants and how they integrate with the natural structures of the musculoskeletal system prior to clinical use.

The design of orthopedic implants requires a certain level of compromise with regard to the geometry and materials that may be used. The implant must fit within the existing framework of the musculoskeletal system, with minimal effect on any other systems present. The materials used must be biocompatible and able to cope with the corrosive environment of the body. This difficult task must be completed bearing in mind that the implant should work, and continue to work, without any adjustment or alteration for many years, under high loading conditions and millions of cycles.

Use of Synthetic Biomaterials[3] in Orthopedics

It is crucial that any foreign material implanted into the body has been thoroughly assessed for biocompatibility (Ratner et al., 1996; Goel et al., 2006). Over time, this has led to a limited number of materials that have been shown to be stable and safe in the body over prolonged periods. Commonly used metals include alloys of titanium, stainless steel, and cobalt-chromium. The compromise

[3]Biomaterial is a material found in the body or synthetically implanted into the body.

in choosing an appropriate alloy is in achieving a suitable level of stiffness, strength, corrosion resistance, and biocompatibility.

Polyethylene is a polymer used as an articulation surface in arthroplasty[4] devices. It can be coupled with either a metal or ceramic component. Increased resistance to polyethylene wear can be achieved using ultra-high molecular weight polyethylene (UHMWPE) and by cross-linking the material (Kurtz et al., 1999; Morrey, 2003; Ries, 2011). Polyaryletheretherketone (PEEK) is a thermoplastic resin that can be injection molded and is commonly used in spinal applications such as posterior stabilization devices and fusion cages. The elastic properties of silicone have been exploited in the one-piece design of the NeuFlex (DePuy Orthopaedics, Inc.) finger joint replacement. Polymethylmethacrylate (PMMA) is a material used as a bone cement to provide fixation for arthroplasty devices. The bone cement is mixed intraoperatively by combining powdered PMMA with a liquid monomer, usually in a vacuum, before injecting it into the required area of the body.

Ceramics are increasingly used as articulation surfaces in arthroplasty devices due to the low wear that can be achieved. Alumina and zirconium are two such materials. Ceramics may be paired in articulation (ceramic on ceramic) or combined with another material, such as a ceramic femoral head articulating on a UHMWPE acetabular component. Hydroxyapatite is a ceramic used for coating implants to provide secondary fixation via bone ongrowth.

It is important to consider the amount of wear in an arthroplasty device, not only in terms of the eventual wearing out of the device, but also because the wear particles produced can be a significant contributor to osteolysis and implant loosening (Morrey, 2003; Wang et al., 2004; Bozic and Ries, 2005).

Mechanical Load Requirements

It must be demonstrated that all devices, whether for fracture fixation, joint arthroplasty, or fusion procedures, be able to cope with the loading environment for which the implant is intended. Preclinical tests are used for this appraisal and generally include both static and dynamic tests. Static tests may be completed to assess the strength, stiffness, and yield point of a device. Dynamic tests may assess wear and fatigue characteristics over the predicted life cycle of the device. There are standard recommendations for such testing published by ASTM International and the International Organization for Standardization (ISO).

In addition to wear, fatigue, and yield tests, it may also be necessary to complete further tests to assess the efficacy of a device, or the effect that an implant may have on the surrounding tissue when used in-vivo.

Approaches to Device Fixation

Orthopedic implants may be screwed, cemented, or press-fitted into place. In addition to these primary methods of fixation, secondary fixation may be achieved through bone ingrowth or ongrowth.

Screws are often used to fix fracture plates and pins to bone. Pedicle screws and rods are used for posterior fusion of the spine. Other orthopedic devices such as dental implants, acetabular cups, and spinal fusion cages may use screws for primary stability.

Bone cement can be used to provide the fixation of an implant. The cement is normally injected into the required space when doughy in consistency, then pressurized to minimize void formation. The implant is subsequently

---

[4]Arthroplasty is the restoration of a joint in the body through surgical procedure.

inserted and held in place until the cement has cured. Cement is less stiff than trabecular and cortical bone (Zysset et al., 1999; Zivic et al., 2012), and may therefore act as a good medium for load transfer between the stiff implant and the host bone.

Cementless fixation can also be achieved by using specific reaming instruments that provide a close fit between the implant and bone. Secondary fixation through bone ingrowth or ongrowth is generally used in cementless devices.

Bone ingrowth can occur when the surface of an implant is porous; for example with titanium beads or mesh. Over time, the bone grows into the pores of the surface and provides stability to the bone and implant interface. Bone ongrowth is achieved through a similar means as ingrowth. An implant may be sprayed with a material such as hydroxyapatite (HA), which promotes bone growth and osseointegration[5] (Solomon, 1992). An HA sprayed surface is generally combined with a textured surface as this is more effective than an HA coating alone (Rodriguez, 2006). It is crucial for both ingrowth and ongrowth that primary stability is achieved. If the displacement of the implant relative to the host bone that occurs during loading (referred to as micromotion) is greater than approximately 150 μm, soft tissue rather than bone will grow into, or onto, the implant surface (Jasty et al., 1997; Kienapfel et al., 1999; Gortchacow et al., 2012), and this can prevent long-term stability through osseointegration.

Common Orthopedic Implants

Fracture plates are commonly used to align bones and allow bone to grow across the fracture (Figure 26.12). Titanium alloys are commonly used for this type of implant due to

[5]Osseointegration is the development of a functional interface between an implant and the host bone.

FIGURE 26.12 Titanium fracture plate.

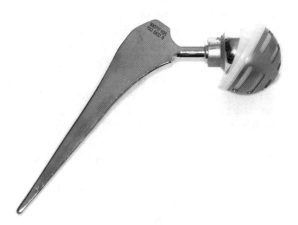

FIGURE 26.13 Double tapered hip replacement device.

their low stiffness compared to other metals used for implants. Small amounts of micromotion induced at the fracture surfaces as loading occurs provide the necessary stimulation for bone repair. Should this micromotion be too large, soft tissue will grow across the fracture first, and the fracture will take longer to fully heal or result in a fibrous union as opposed to bony union (Nordin and Frankel, 2001).

Dental implants are commonly used to reassemble a tooth, or provide a base onto which a prosthesis may be fixed. They are commonly titanium alloy and comprise a screw that is driven into the bone onto which an abutment is fixed. The threads of the screw may be coated to provide secondary fixation via bone ongrowth or ingrowth.

Hip arthroplasty (Figure 26.13) is a common procedure that has good clinical success rates (National Joint Registry, 2011; Swedish Hip Arthroplasty Register, 2011) and a long history of progressive development in both design and materials. Generally, the affected joint is treated by removing the femoral head and neck, reaming into the femoral canal, and inserting a femoral component. This may be cemented into place, or a cementless device may be used. The femoral component may be one piece, or modular in design, which allows an increased number of sizing options. The femoral stem is metal, and the femoral head may be metal or ceramic. The acetabular component often comprises a metal outer shell that is usually cementless, though additional screws may be used for primary stability. Once the shell is in place, the liner is inserted, which may be metal, UHMWPE, or ceramic.

Knee arthroplasty (Figure 26.14) is also a well-established and generally successful joint

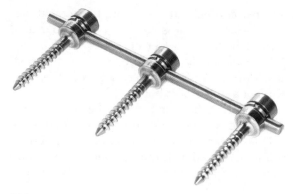

FIGURE 26.15 Pedicle screw and rod assembly.

replacement procedure (National Joint Registry, 2011; Swedish Knee Arthroplasty Register, 2011). A knee arthroplasty device comprises a tibial and femoral component and features metal on UHMWPE articulation. Both components may have a stem for fixation and stability. The tibial plateau is generally comprised of a metal tray with a UHMWPE insert. The femoral component is comprised entirely of metal, with highly polished condyles for articulation on the UHMWPE insert in the tibia tray.

Degenerative disc disease in the spine is common and can be severely debilitating (Cassinelli and Kang, 2000; Adams et al., 2006). An affected level may be fused to restore disc height, correct load transfer, and reduce pain. However, this is carried out at the cost of motion at the operative level. Fusion cages may be inserted into the disc space after a discectomy has been completed. This may comprise one or two cages of metal or PEEK that are combined with allograft bone. Over time, bone grows into the cages and entirely fuses the vertebral level.

Fusion may also be achieved through posterior procedures (Figure 26.15) using screws that are driven into the pedicles of the vertebral levels to be fused. The screws are then linked via a rod. This assembly may be inserted along the pedicles on one side only,

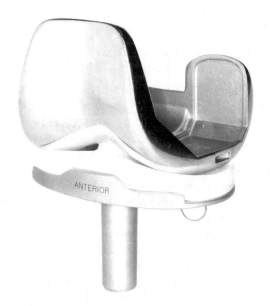

FIGURE 26.14 Total knee arthroplasty device.

or two parallel assemblies may be inserted, which may also be linked together to increase the rigidity of the structure.

Fusion procedures may also combine both anterior and posterior fusion devices to provide more stability than the implantation of one system alone.

MOBILE ARM SUPPORTS

Believed to have originally been developed in California in the 1950s, mobile arm supports (MAS) are devices fitted to a wheelchair, table, desk, or stand that allow people with weak arm muscles to carry out, with less effort, functional tasks such as eating, using a computer or communication aid, and, increasingly, accessing a tablet computer or e-reader. They are also used to aid the turning of pages in a magazine, for drawing or painting, and for grooming (e.g., tooth brushing, shaving, applying makeup). Less commonly, an MAS can facilitate someone in controlling the joystick of a powered wheelchair, but it is necessary to make substantial modification to avoid an excessively wide chair "footprint," and also to carry out a risk assessment in respect of control of the chair.

The device comprises:

- A trough for the forearm fitted to a pivot which tips to allow vertical movement of the hand and rotates in the horizontal plane
- A hinged and balanced linkage comprising a proximal and distal arm (Figure 26.16)
- An adjustable bracket which can be varied in angle and rotation about the vertical plane; this is critical in allowing the device to be adjusted to swing with gravity in specific directions, according to what is required clinically and functionally
- Depending on the structure of what the MAS is to be fitted to, it may be necessary to manufacture a custom mount to hold the angle adjustable bracket

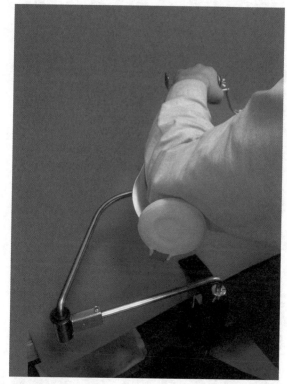

FIGURE 26.16 Operation of the hinged linkage.

- Where hand function/strength is impaired, the distal part of the trough can have attachments fitted to a custom sized and shaped T-bar (Figure 26.17), such as:
  - A pointer for turning pages and accessing keyboards
  - A stylus for keypads/touch screens
  - A self-leveling spoon (food falls off a fixed angle spoon; Figure 26.18)
  - Palm support, where the wrist is weak but fingers have functional movement

Adjustments must be tuned to compensate for muscle weakness and imbalance. Most MASs are passive mechanical devices but powered versions have been developed.

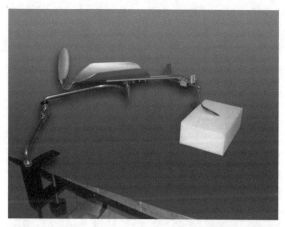

FIGURE 26.17 Detail of the trough and T-bar.

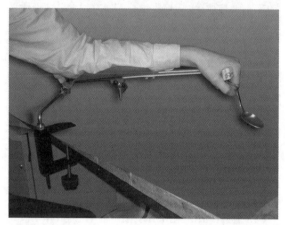

FIGURE 26.18 Occupied MAS with T-bar and self-leveling spoon fitted to a table top.

Applicable Client Groups

MASs are typically used for people having:

- Motor neurone disease (MND), where movement of the shoulder girdle is initially retained while the arms and hands are weakened
- Spinal muscular atrophy (SMA), having, in this context, similar clinical features to those with MND, but being stable in their condition for longer

- Muscular dystrophy (MD), where there is proximal weakness, the hands tending to retain their function for longer, but additional hand support being required as the disease progresses
- Spinal cord injury (SCI), the type of support varying according to the level of injury
- Multiple sclerosis (MS), the type of support varying according to the nature of the impairment and progression of the disease
- Cerebrovascular accident (CVA, or stroke), but mostly in the context of rehabilitation, i.e., improvements in function may develop over time

There is application for many other conditions but these are the most common groups.

Assessment and Trial

It is necessary to carry out a detailed clinical assessment prior to the trial or issue of equipment. Initially, this will be to determine what is required and what is desired. There may be a requirement to arrange additional equipment for the MAS to work successfully, which might typically involve the supply of a stand or an over-chair table with a split top to allow mounting of the MAS on a level surface, the sloped surface being used to angle a magazine toward the client, for example. This will usually involve liaison with a community occupational therapist and typically relates to people with MND whose neck muscles are often weak. Referral to a speech and language therapist may also be required, particularly in respect of facilitating a safe swallow. Critically, there is a requirement for care support for use of the MAS, either from family members or from carers. Without this support, an MAS is very unlikely to succeed in its objectives.

Having gathered the relevant background information, a physical assessment of the arm and shoulder must then be carried out (this assumes that, where necessary, there is satisfactory postural management in place in terms of wheelchair or static seating). The aim is to check for pain-free range of movement in the arm and shoulder such that an MAS can be used. The assessor takes the arm through a standard set of joint ranges:

- Shoulder abduction
- Shoulder internal to external rotation
- Shoulder elevation
- Elbow flexion to extension
- Supination of the forearm
- Wrist flexion and extension

The assessor also checks wrist strength and hand grip, which will guide the prescription of distal components.

Handedness is an important factor as the impairment may have occurred on their dominant side. Additionally, the use of an MAS will be tiring as the person is learning a new skill. This will become more natural as new neural pathways are laid down for the new pattern of movement.

Setting Up

Having made this assessment, an MAS is set up in the relevant context and environment. This usually dictates that the appointment takes place in the client's home as it is almost impossible to mimic this in a clinic setting.

Adjustments are generally made in the following order:

1. Length of trough
2. Orientation of elbow dial/pad
3. Length and shape of T-bar
4. Distal componentry
5. Balance point of trough and T-bar
6. Orientation of angle adjustable bracket (the client is asked to carry out simple tasks, such as to scratch their nose, the assessor observing what adjustments are necessary to make this easier/more efficient)

Stops can be added to limit movement from beyond where, due to weakness, recovery is not possible. Offset swivels suspend the trough, as opposed to it pivoting over a joint, to lower the center of gravity relative to the pivot and to reduce the effort of control.

Additional equipment may be needed to support the head in the form of a collar, of which a wide variety are available. Shoulder cuffs may also be required to stabilize the joint.

Contraindications

There are any number of reasons why an MAS might be contraindicated, but the following are those most common:

- A significant amount of spasticity or tremor, where this is not related to effort, is unhelpful in achieving smooth movement.
- Insufficient range of (pain free) motion in the joint ranges described above.
- While the MAS is designed for people having weak muscles, there must be some muscle power to provide movement.

The task of writing can be difficult as the whole arm moves, making the precise hand movements required for letter formation difficult. Drinking is challenging as the weight of the cup or glass and fluid tend to overbalance the support. Even if the drink could be raised it would be difficult to tip, so a straw would be needed, in which case it would be easier to have the cup positioned on the table with a long straw and nonreturn valve.

Ongoing Care

Once set up correctly for the client, it is unusual for regular adjustment to be necessary. Required more commonly, however, are changes to componentry as a result of disease progression and/or change in task or function.

As any change is likely to unbalance the MAS, a similar assessment and setting up process must be followed.

ROBOTICS

Definitions

A robot is traditionally described (by the Robot Institute of America in 1979) as "a re-programmable, multifunctional manipulator designed to move material, parts, tools or specialized devices through variable programmed motions for the performance of a variety of tasks." This definition is still relevant for a traditional industrial application. However, the application of robotics in a nonindustrial application, such as healthcare or rehabilitation, requires a more advanced definition. "The integration of enabling technologies and attributes embracing manipulators, mobility, sensors, computing (IKBS, AI) and hierarchical control to result ultimately in a robot capable of autonomously complementing man's endeavors in unstructured and hostile environments" (described by the U.K. Department of Trade and Industry's Advanced Robotics Initiative in 1987; Hillman, 2004). From these definitions we see that a robot is not limited to either the humanoid representation of science fiction or the traditional industrial definition. Within rehabilitation, robotics have been applied in the following areas.

Assistive Robotics

Assistive robotics aims to help those with often severe levels of disability to live as independently as possible in a relatively unmodified environment. Examples of applications are:

- Fixed site: The disabled user is able to use the robot (often a variation on a small

industrial robot arm) in a workstation environment set up to optimize specific tasks. These may be of an "office" type or more personal areas such as eating or personal hygiene (Van der Loos, 1995; Hammel et al., 1989). Although many research groups have developed such systems, which have been used effectively, and a small number of commercial products have been available, the limitations are obvious. Though no commercial products are currently available, one of the most cost-effective systems was the Handy 1 (Topping, 2001).

- Mobile: For some, the ideal has been to provide a robotic carer. The most notable implementation of this was the Movar system developed at Stanford University (Van der Loos et al., 1986). More recently, this concept has been extended to the care of the elderly but with the emphasis on communications rather than manipulation.

- Wheelchair mounted: For wheelchair users having compromised upper-limb mobility and dexterity, the ideal might be a "third arm" mounted to the wheelchair. The MANUS (Kwee et al., 1989) iArm by Exact Dynamics[6] is the most successful product in this area and still continues in production. Though popular in its home country of the Netherlands, costs can be prohibitive. The two main obstacles to a wider use of wheelchair robotics (besides price) are of an efficient human machine interface and the problems of integrating the system to a wheelchair without compromising overall width, stability, and battery life of the wheelchair. Two examples of wheelchair mounted robot arms are shown in Figures 26.19 and 26.20.

[6]*www.exactdynamics.nl*

FIGURE 26.19 Robot arm in use to provide increased reach. *Source*: *Assistive Innovations*.

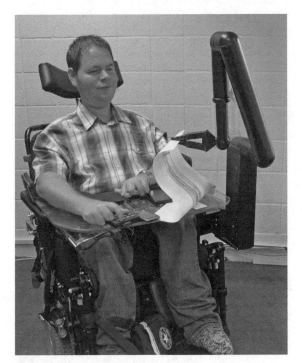

FIGURE 26.20 Enabling access to the work environment. *Source*: *Assistive Innovations*.

Therapy

At the time of writing, the greatest interest in robotics in rehabilitation both in terms of research and commercial products is in providing therapy to the hand, upper limb, and lower limb. Much of this has concentrated on the rehabilitation of stroke survivors. Effective stroke rehabilitation therapy makes use of the ability of the human brain to rewire itself (neuroplasticity) following damage to a part of the brain. It has been found that intensive repetition of movement promotes motor recovery following a stroke (Dipietro et al., 2012). Therapy can be given in the following ways (Lum et al., 2002):

- Passive: Movement is externally imposed by the robot while the patient remains relaxed.
- Active: Patient initiates the movement, but the robot assists along a predefined path.
- Active resisted: Patient must move against a resistance generated by the robot.

As the potential of robot-mediated therapy is explored, the research questions that are being addressed are: Is it an effective therapy? Is it more effective than current exercise therapies by physiotherapists? Is it cost effective?

Apart from physical rehabilitation it should be noted that robotics are also being used for cognitive therapy of children with autism (Robins et al., 2004).

Exoskeletons

At a completely different level of technical complexity are exoskeletons which are now starting to appear on the market. These are robotic devices requiring actuators, sensors, and power packs. Much of the initial research was pushed by potential for military application to assist soldiers in carrying heavy loads over rough ground. However, the application to people with disabilities was recognized early, the challenges being to decrease the weight needing to be carried and finding a way to provide integration with the patient (Zoss et al., 2006; Suzuki et al., 2007).

Other Application Areas

Robotics have also been applied in the areas of smart wheelchairs with long range and short range navigational functions, and smart prosthetics, notably of the knee and the hand. Other areas of potential application are in education and communication.

References

Adams, M., Bogduk, N., Burton, K., Dolan, P., 2006. The Biomechanics of Back Pain. Elsevier Ltd., Philadelphia.

Bozic, K.J., Ries, M.D., 2005. Wear and osteolysis in total hip arthroplasty. Semin. Arthroplasty. 16 (2), 142–152.

Cassinelli, E.H., Kang, J.D., 2000. Current understanding of lumbar disc degeneration. Oper. Tech. Orthop. 10 (4), 254–262.

Dipietro, L., Krebs, H.I., Volpe, B.T., Stein, J., Bever, C., Mernoff, S.T., et al., 2012. Learning, not adaptation, characterizes stroke motor recovery: evidence from kinematic changes induced by robot -assisted therapy in trained and untrained task in the same workspace. IEEE Trans. Neural Syst. Rehabil. Eng. 20 (1), 48–57.

Goel, V.K., Panjabi, M.M., Patwardhan, A.G., Dooris, A.P., Serhan, H., 2006. Test protocols for evaluation of spinal implants. J. Bone. Joint Surg. Am. 88 (**Suppl 2**), 103–109.

Gortchacow, M., Wettstein, M., Pioletti, D.P., Müller-Gerbl, M., Terrier, A., 2012. Simultaneous and multisite measure of micromotion, subsidence and gap to evaluate femoral stem stability. J. Biomech. 45 (7), 1232–1238.

Hammel, J., Hall, K., Lees, D., Leifer, L., Van der Loos, M., Perkash, I., et al., 1989. Clinical evaluation of a desktop robotic assistant. J. Rehabil. Res. Dev. 26 (3), 1–16.

Hillman, M., 2004. Rehabilitation Robotics from Past to Present – A historical perspective. In: Bien, Z.Z., Stefanov, D. (Eds.), Advances in Rehabilitation Robotics. Springer, pp. 25–44.

International Organization for Standardization (ISO), 1989. ISO 8549-1 Prosthetics and Orthotics - Vocabulary, Part 1: General Terms for External Limb Prostheses and External Orthoses. ISO, Geneva.

International Organization for Standardization (ISO), 1989. ISO 8549-1 Prosthetics and Orthotics - Vocabulary, Part 3: Terms Relating to External Orthoses. ISO, Geneva.

Jasty, M., Bragdon, C., Burke, D., O'Connor, D., Lowenstein, J., Harris, W.H., 1997. In vivo skeletal responses to porous-surfaced implants subjected to small induced motions. J. Bone. Joint Surg. Am. 79 (5), 707–714.

Kienapfel, H., Sprey, C., Wilke, A., Griss, P., 1999. Implant fixation by bone ingrowth. J. Arthroplasty. 14 (3), 355–368.

Kurtz, S.M., Muratoglu, O.K., Evans, M., Edidin, A.A., 1999. Advances in the processing, sterilization, and crosslinking of ultra-high molecular weight polyethylene for total joint arthroplasty. Biomaterials. 20 (18), 1659–1688.

Kwee, H., Duimel, J., Smits, J., Tuinhof de Moed, A., van Woerden, J., 1989. The MANUS Wheelchair-Borne Manipulator: System Review and First Results. Proc. IARP Workshop on Domestic and Medical and Healthcare Robotics, Newcastle.

Lum, P., Reinkensmeyer, D., Mahoney, R., Rymer, W.Z., Burgar, C., 2002. Robotic devices for movement therapy after stroke: Current status and challenges to clinical acceptance. Top. Stroke Rehabil. 8 (4), 40–53, 2002.

Morrey, B.F., 2003. Joint Replacement Arthroplasty. Churchill-Livingstone, Philadelphia.

National Joint Registry, 2011. National Joint Registry for England and Wales 8th Annual Report 2011. Hemel Hempstead, U.K., The NJR Centre.

Nordin, M., Frankel, V.H., 2001. Basic Biomechanics of the Musckuloskeletal System. Lippincott, Williams & Wilkins, Philadelphia.

Ratner, B.D., Hoffman, A.S., Schoen, F.J., Lemons, J.E., 1996. Biomaterials Science: An Introduction to Materials in Medicine. Academic Press, San Diego.

Ries, M.D., 2011. Highly crosslinked ultrahigh molecular weight polyethylene in total hip arthroplasty: no further concerns-opposes. Semin. Arthroplasty. 22 (2), 82–84.

Robins, B., Dautenhahn, K., Boekhorst, R., Billard, A., 2004. "Effects of repeated exposure to a humanoid robot on children with autism". Published. In: Keates, S., Clarkson, J., Langdon, P., Robinson, P. (Eds.), Designing a More Inclusive World. Springer Verlag, London, pp. 225–236.

Rodriguez, J.A., 2006. Acetabular Fixation Options: Notes from the Other Side. J. Arthroplasty. 21 (Suppl. 4), 93–96.

Solomon, L., 1992. Hip replacement: Prosthetic fixation. Curr. Orthop. 6 (3), 153–156.

Suzuki, K., Mito, G., Kawamotot, H., Hasegawa, Y., Sankai, Y., 2007. Intention-based walking support for paraplegia patients with robot suit HAL. Adv. Robot. 21 (12), 1441–1469.

Swedish Hip Arthroplasty Register, 2011. Swedish Hip Arthroplasty Register Annual Report 2010. Gothenburg, Sweden, Swedish Hip Arthroplasty Register.

Swedish Knee Arthroplasty Register, 2011. Swedish Hip Arthroplasty Register Annual Report 2011. Lund, Sweden, Swedish Knee Arthroplasty Register.

Topping, M., 2001. Handy 1, A robotic aid to independence for severely disabled people. In: Mokhtari, M. (Ed.), Integration of Assistive Technology in the Information Age. IOS, Netherlands, pp. 142–147.

Van der Loos, M., 1995. VA/Stanford Rehabilitation Robotics Research and Development Program: Lessons learned in the application of robotics technology to the field of rehabilitation. IEEE Trans. Rehabil. Eng. 3 (1), 46–66.

Van der Loos, M., Michalowski, S., Leifer, L., 1986. Design of an Omnidirectional Mobile Robot as a Manipulation Aid for the Severely Disabled. Foulds R. (Ed.), Interactive Robotic Aids. World Rehabilitation Fund Monograph #37, New York.

Wang, M.L., Sharkey, P.F., Tuan, R.S., 2004. Particle bioreactivity and wear-mediated osteolysis. J. Arthroplasty. 19 (8), 1028–1038.

World Health Organization (WHO), 2001. International Classification of Functioning, Disability and Health <www.who.int/classifications/icf/en/> (last accessed October 2012).

Zivic, F., Babic, M., Grujovic, N., Mitrovic, S., Favaro, G., Caunii, M., 2012. Effect of vacuum-treatment on deformation properties of PMMA bone cement. J. Mech. Behav. Biomed. Mater. 5 (1), 129–138.

Zoss, A.B., Kazerooni, H., Chu, A., 2006. Biomechanical design of the berkeley lower extremity exoskeleton. IEEE/ASME Trans. Mechatronics. 11 (2), 128–138.

Zysset, P.K., Guo, X.E., Edward Hoffler, C., Moore, K.E., Goldstein, S.A., 1999. Elastic modulus and hardness of cortical and trabecular bone lamellae measured by nanoindentation in the human femur. J. Biomech. 32 (10), 1005–1012.

Further Reading

Prosthetics

Levine, D., Richards, J., Whittle, M.W., 2011. Whittle's Gait Analysis. fifth ed. Butterworth-Heinemann.

Lusardi, M., Nielsen, C., 2012. Orthotics and Prosthetics in Rehabilitation. third ed. Elsevier.

Ratner, B.D., Hoffman, A.S., Schoen, F.J., Lemons, J.E., 2008. Biomaterials Science. third ed. Elsevier.

Özkaya, N., Nordin, M., Goldsheyder, D., Leger, D.L., 2012. Fundamentals of Biomechanics — Equilibrium, Motion and Deformation. third ed. Springer.

Smith, D.G., Michael, J.W., Bowker, J.H., 2004. Atlas of Amputations and Limb Deficiencies — Surgical, Prosthetic and Rehabilitation Principles American Academy of Orthopaedic Surgeons.

Tortora, G.J., Derrickson, B.H., 2011. thirteenth ed. Wiley.

Orthotics

Morris, Dias, (Eds.), 2007. *Paediatric Orthotics* Clinics in Developmental Medicine No. 175. MacKeith Press.

Mosby, C.V., 1985. *Atlas of Orthotics, Biomechanical Principles and Applications* American Academy of Orthopaedic Surgeons, St. Louis.

Rose, G.K., 1986. Orthotics - Principles & Practice. William Heinemann, London.

Robotics

Hillman, M., 2004. Rehabilitation robots from past to present - A historical perspective. In: Bien, Z., Stefanov, D. (Eds.), Advances in Rehabilitation Robotics. Lecture Notes in Control and Information Sciences #306. Springer, New York.

Resources

Aids for Daily Living

As described in this chapter's "Aids for Daily Living" section, there is a vast array of assistive technology available commercially. A simple Internet search will yield plentiful results. In the United Kingdom, the Disabled Living Foundation (*www.dlf.org.uk*) is a charity offering independent advice on these types of products and can be an extremely useful resource.

Index